I0034804

Understanding and Designing Structures without a Computer

emerald PUBLISHING · ice Publishing

Understanding and Designing Structures without a Computer

Spatial structural systems, dynamics and foundations

Leonidas Stavridis

Konstantinos Georgiadis

Published by Emerald Publishing Limited, Floor 5,
Northspring, 21–23 Wellington Street, Leeds LS1 4DL.

ICE Publishing is an imprint of Emerald Publishing Limited

Other ICE Publishing titles:
Structural Design of Buildings: Holistic Design
Edited by Feng Fu and David Richardson. ISBN 978-1-8354-9561-2
*Conceptual Structural Design: Bridging the gap between
architects and engineers, Third edition*
Olga Popovic Larsen. ISBN 978-0-7277-6598-7
Empirical Design in Structural Engineering
Thomas Boothby. ISBN 978-0-7277-6633-5

A catalogue record for this book is available from the British Library

ISBN 978-1-83662-945-0

© Leonidas Stavridis and Konstantinos Georgiadis 2026 publishing
under exclusive licence by Emerald Publishing

Permission to use the ICE Publishing logo and ICE name is granted
under licence to Emerald from the Institution of Civil Engineers.
The Institution of Civil Engineers has not approved or endorsed
any of the content herein.

All rights, including translation, reserved. Except as permitted
by the Copyright, Designs and Patents Act 1988, no part of this
publication may be reproduced, stored in a retrieval system or
transmitted in any form or by any means, electronic, mechanical,
photocopying or otherwise, without the prior written permission
of the publisher, Emerald Publishing Limited, Floor 5, Northspring,
21–23 Wellington Street, Leeds LS1 4DL.

This book is published on the understanding that the author is
solely responsible for the statements made and opinions expressed
in it and that its publication does not necessarily imply that such
statements and/or opinions are or reflect the views or opinions of
the publisher. While every effort has been made to ensure that the
statements made and the opinions expressed in this publication
provide a safe and accurate guide, no liability or responsibility can
be accepted in this respect by the author or publisher.

While every reasonable effort has been undertaken by the
author and the publisher to acknowledge copyright on material
reproduced, if there has been an oversight please contact the
publisher and we will endeavour to correct this upon a reprint.

Cover photo: iStock/omersukrugoksu.

Commissioning Editor: Michael Fenton
Content Development Editor: Ryan Molyneux
Books Production Lead: Benn Linfield

Typeset by KnowledgeWorks Global Limited
Index created by David Gaskell

Contents

Foreword

Methods of structural analysis have experienced an explosive growth during the last 40 years. But it was the advent of powerful personal computers, along with the evolution of numerical tools (based mainly on the finite element method) and the parallel development of numerous reliable, comprehensive, commercially available computer software, that have enabled engineers to tackle very complex structural systems. As a consequence, in today's design offices, analysis of even some rather simple systems is performed (especially by the younger generation of engineers) with the use of such computer codes. Classical as well as modern methods of structural analysis (based on the principles of virtual work, compatibility of deformations, matrix analysis) are rather rarely invoked in everyday practice. Yet, these theoretical tools often constitute the major (if not the only) part of the curriculum in civil engineering schools.

Several problems may arise from this state of affairs. First, the danger of the 'black-box syndrome': when a sophisticated code is used without the analyst having the ability to check whether the results are indeed reasonable and to spot any errors in the physical meaning of their implicit assumptions and how these assumptions are materialised in the model. Second, there is little if any training to help the young engineer develop a deeper understanding of how structural systems behave, let alone to sharpen their physical intuition; such understanding and intuition are necessary especially in the conception and preliminary design stages. Indeed, conceptual clarity and physical insight are rarely mentioned as key objectives of structural analysis courses.

This two-volume book by Professor Leonidas Stavridis and Dr Konstantinos Georgiadis offers a much-needed addition to classical computational structural analysis. A physical approach is developed in which a structural system is decomposed into elements whose behaviour to the applied loads is easily computed 'from the basics'. Starting in the first chapters with fundamental concepts and applications, the step-by-step exposition becomes progressively more advanced. Structural analysis blends naturally with mechanics of materials – the latter include reinforced and prestressed concrete, steel and composites. The in-depth analysis of standard structural systems (such as simply and multi-supported beams, frames, arches, cabled beams) is followed by the exposition of some more advanced topics such as buckling, slabs and shells, thin-walled and box girders, grids and curved beams, laterally loaded multi-storey frames and shear walls.

It is amazing how the analysis of such complex systems is made so simple, clearly understandable even to a non-specialist civil engineer, as the present writer. This is accomplished to a large extent thanks to the numerous illustrative figures (sketches) that go far beyond the usual 'formalistic' figures of most available textbooks: they are imaginative, vivid, self-explanatory. What a difference they make when trying to comprehend difficult topics! For instance, the chapter on 'Shells' contains 56 elaborate figures, most of which comprise several sketches while a few of them are a whole page long. The three-dimensional nature of cylindrical, spherical, paraboloid and conical shells is elucidated with the help of ingeniously selected isometric views and numerous cross-sections so that the reader feels that this is a rather simple subject.

As an engineer with a special interest in soil–foundation–structure interaction, I was particularly happy with the comprehensive treatment of foundations. Viewed mainly from a structural engineer's viewpoint, the pertinent chapter deals not only with some classical deformation–settlement and stress–distribution problems, but also with the interplay between foundation stiffness and structure distress.

I believe this book will prove invaluable to both students and practising engineers in helping them not only to absorb a huge volume of material, but also (more significantly) to cultivate 'engineering intuition' and develop insight into the physics of structural analysis. For students, in particular, all this will offer the motivation for further study and the desire to later apply in real-life projects both the material and the methodology developed in the book.

George Gazetas, Professor of Soil Mechanics and Foundation Engineering, National Technical University of Athens

Preface

Πεπαιδευμένον γαρ εστίν επί τοσούτω το ακριβές επιζητείν όσον η του πράγματος φύσις επιδέχεται.

Αριστοτέλης

Because it is the essence of education to seek as much accuracy as the nature of things allows.

Aristoteles

A technically educated person – whether an engineer, architect or builder – today understands 'structural design' in much the same way as their predecessors did 500 years ago: as a practical procedure that applies specialised knowledge to ensure a structure 'stands up' and 'does not fall down,' resisting whatever loads it encounters during its lifespan.

Yet, what has evolved across the centuries – transforming structural theory from empiricism into a rigorous scientific discipline – is the introduction of analytical methods. The capacity to assess structural behaviour systematically, and the advancement of computer-based methods and tools, have fundamentally shaped this discipline. Structural mechanics is now a highly demanding subject, spanning both the analytical evaluation of structural behaviour and its practical application in design. Though intimately related, these two realms retain distinct focuses.

Analytically, the central question is: Given a particular structural system and specified loads, what are its resulting forces and deformations? Answering this requires a strict scientific approach – one that can, at times, give the impression that the analysis itself is the end goal. Indeed, modern computing methods and software have made these calculations routine.

On the other hand, practical design emphasises the art of applying this understanding to create efficient load-carrying solutions that are economical, functional and visually pleasing. Given a particular set of service requirements and environmental conditions, what structural concept – using appropriate materials – will best satisfy the design criteria? This is where engineering insight and creativity matter most.

Although engineering curricula tend to focus on the scientific side, aspiring structural engineers often discover too late that true mastery requires more than rote reliance on computer analysis or prescriptive codes. Equally important is the ability to 'see' how forces flow through a structure – to perceive its behaviour as a coherent system. Without this intuitive understanding, engineers may find themselves ill-equipped to engage meaningfully with architects and builders on real-world projects.

This two-volume book aims to bridge that gap. It explores the behaviour of a wide range of structural systems – beams, frames, arches, cables, grillages, slabs, shells, thin-walled sections and multi-storey structures – placing particular emphasis on the underlying mechanisms that allow them to support loads. It discusses traditional materials like steel and concrete alongside composite and prestressed solutions, and introduces the principles of plastic analysis, second-order theory and structural stability in a simplified manner.

Special chapters address the design of statically determinate and indeterminate plane structures, dynamic response under seismic and human-induced actions, and the treatment of shallow and deep foundations – recognising that structural design is never complete without an understanding of soil–structure interaction.

The book adopts a progressive, concept-building approach: each chapter builds on earlier material, ensuring readers establish a strong intuitive and analytical base before moving on to more advanced topics. Some background knowledge of elementary mechanics is assumed.

As Vitruvius wrote more than two millennia ago, successful structural design must satisfy four core criteria: safety, functionality, economy and beauty. Technical safety requires that a structure's capacity exceed the demand; functionality requires limiting displacements and vibrations; economy requires choosing efficient structural forms and construction methods; and beauty requires sensitivity to proportion and elegance. Achieving these ideals depends as much on an engineer's creativity and judgement as on their technical prowess.

Ultimately, this book is for anyone – student, practising engineer or architect – who wishes to gain deeper structural insight. Our hope is that it will not only enrich readers' appreciation of structural behaviour, but also help them to design with greater understanding and confidence.

Finally, we wish to thank Emerald Publishing and its editorial team, led by Dr Michael Fenton, for their invaluable guidance and support in bringing this book to fruition.

Leonidas Stavridis

Konstantinos Georgiadis

About the authors

Professor Leonidas Stavridis obtained his Diploma in Civil Engineering and his PhD from the National Technical University of Athens (NTUA). Subsequently, he attended a postgraduate course in Bridge Engineering and Prestressed Structures at the Federal Institute of Technology of Zurich (ETH) where he obtained his diploma in 1989. In 2011, after more than 20 years of teaching structural analysis, bridge engineering and structural behaviour at both undergraduate and postgraduate level, he was elected Professor of Structural Engineering and Design at NTUA.

He is the author of *Structural Systems: Behaviour and Design* published in 2010 by ICE Publishing. In addition, his publications in peer-reviewed international journals cover a wide range of topics related to static and dynamic analyses of orthotropic slabs and shallow shells, prestressed cable structures, structural behaviour of multi-storey buildings, dynamic behaviour of curved thin-walled beams, soil–structure interaction problems including prestressed foundations, the treatment of external partial prestressing in slab bridges and the behaviour of suspension and stress ribbon.

Professor Stavridis has been an active freelance structural engineering consultant for more than 45 years and has been involved in various design projects, mostly in Greece, including multi-storey buildings, prestressed concrete bridges, space covering roofs and special foundations as well as structural restoration and strengthening of traditional monumental structures.

Dr Konstantinos Georgiadis is a Chartered Engineer (CEng MICE) with over 10 years of experience in bridge engineering and structural design. He graduated from the National Technical University of Athens (NTUA) in 2011 and earned his MSc in Steel Design and Business Management with distinction from Imperial College London in 2012.

He began his professional career in 2013, gaining experience in the UK railway sector with Bouygues. In 2014, he joined AECOM as a Bridge Engineer in the structures team. In 2016, he was awarded a prestigious EPSRC scholarship to pursue a PhD in Bridge Design at Imperial College London. During his doctoral studies, he served as a Teaching Assistant in Steel Design at Imperial and worked as a part-time external consultant for AECOM. After completing his PhD in 2020, he joined Arup in Hong Kong as a Chartered Senior Bridge Engineer and later transferred to their London office.

Dr Georgiadis has led and contributed to feasibility studies, preliminary and detailed designs, structural assessments and inspections of numerous bridges, including long-span structures such as cable-stayed and suspension bridges. His international project experience spans the UK,

Hong Kong, Macau, the Philippines and Portugal. He is well-versed in multiple design codes, including the Eurocodes, British Standards and AASHTO.

With a strong technical background, Dr Georgiadis specialises in advanced linear and nonlinear, static and dynamic analysis and is a proficient user of various finite element analysis software. He also brings substantial site experience, having supervised bridge construction, inspections and surveys. He has authored multiple technical and research papers, which have been published and presented in leading journals and international conferences and has served as member of the Scientific Committee of IABSE.

Introduction

This book aims to cultivate a deep understanding of the structural behaviour and design principles of civil engineering structures. Its focus is on fundamental concepts concerning the design and analysis of various structural types, including different types of bridges, long-span roof structures and multi-storey buildings, among others. As the selection of the appropriate construction material plays an important role in the design, this book focuses on the main construction materials – that is, steel and concrete – as well as their combinations in the form of prestressed concrete and composite structures.

It covers the conceptual design of different structural forms, fostering a solid mindset to develop a sound preliminary design. Readers will gain an understanding of load-transferring paths and the ability to estimate the internal forces in structures – both qualitatively and quantitatively – without the need to rely on computer software or design codes. Of course, these tools will eventually be used when producing the detailed design and final construction drawings, which remain the ultimate goal of any structural project.

This second volume expands the discussion to include structures in space (i.e. three-dimensional structures) under vertically (e.g. gravity) or laterally applied loads (e.g. wind or seismic) in the form of large roofs, horizontal spanning systems or multi-storey buildings. In particular, the behaviour of the latter structures under dynamic seismic forces is explored through lumped-mass representations, highlighting the influence of material ductility on response and design.

This volume starts with the investigation of grillages, introducing the concept of torsion and its implications. This naturally leads to the examination of concrete slabs – including flat slabs and ribbed configurations – and the role of prestressing in enhancing their performance. Next, thin-shell structures are discussed, with an emphasis on their primary membrane action as well as the bending effects that may lead to localised stresses. Hypar shells receive particular attention due to their efficiency in construction, which is achieved through straight-line generatrices.

The behaviour of thin-walled beams and box-girder bridges is also covered, examining torsional effects, warping and the resulting axial stresses in open- and closed-sections. Multi-storey structures, with a focus on lateral load resistance and the interplay between horizontal and vertical elements – frames, shear walls and cores – are also presented.

The penultimate chapter focuses on the dynamic behaviour and seismic response of multi-storey buildings, with a particular emphasis on the role of material ductility in design loads. Vibrations induced by human activity or machinery are also discussed, as they often influence the serviceability of beams and plates.

The final chapter addresses mainly shallow foundations, presenting their design principles as well as basic soil–structure interaction concepts. Pile foundations are also briefly discussed.

Throughout this book, the underlying philosophy is that engineering judgement and structural insight is more important than computer analyses. Critical checks – informed by an understanding of the basic principles outlined here – always remain essential. Ultimately, every design process must lead to one final outcome: a well-conceived set of structural drawings that clearly communicate the configuration and construction method for building the project.

emerald
PUBLISHING

ice
Publishing

Leonidas Stavridis and Konstantinos Georgiadis
ISBN 978-1-83662-945-0
https://doi.org/10.1108/978-1-83662-942-920251001
Emerald Publishing Limited: All rights reserved

Chapter 1
Grillages

1.1. Introduction to grillages and torsion

1.1.1 Equilibrium conditions

This section introduces the fundamental concepts related to the stress and deformation behaviour of grid structures, so-called *grillages*. In contrast with typical plane structures, which are loaded within their plane, grillages are plane structures that are loaded perpendicularly to their own plane, as shown in Figure 1.1, thus creating a spatial structural system.

For analytical purposes, a grillage is assumed to lie within a horizontal plane defined by two arbitrary axes, OX and OY, as part of a three-dimensional coordinate system $OXYZ$. For any cut-out part of a grillage, the following conditions should be satisfied for equilibrium:

$$\Sigma \overrightarrow{M_x} = 0, \qquad \Sigma \overrightarrow{M_y} = 0, \qquad \Sigma \overrightarrow{P_z} = 0$$

These expressions indicate that the sum of the moment vectors of all external forces about the OX and OY axes must be zero. In addition, the sum of all vertical (out-of-plane) forces acting on the considered portion of the grid must also equal zero.

1.1.2 Internal forces and deformations

To accurately describe the internal forces and deformations at a specific point along a grillage member, it is useful to consider a vertical plane to the member at the point of interest. The intersection of this plane with the member exposes two adjacent cross-sections (free edges). These edges exhibit identical deformation vectors, reflecting the continuity of the structure. However, the vectors of internal forces at these sections are equal in magnitude but opposite in direction, ensuring equilibrium within the member.

The internal forces at any point along a grillage member are most effectively described by referring to a local orthogonal coordinate system (x, y, z), as illustrated in Figure 1.2. This coordinate system is centred at the centroid of the member's cross-section, with the x-axis aligned with the member's longitudinal axis. The y- and z-axes lie in the plane of the cross-section: the y-axis oriented horizontally and the z-axis vertically. It is assumed that the cross-section is symmetric about the horizontal y-axis.

The internal forces and the corresponding deformations are expressed using vector notation. The moment and rotation vectors shown in Figure 1.2 adhere to the *screw rule*: the direction of each vector indicates the sense in which a right-hand-threaded screw would rotate to move in that direction. When a grillage member is subjected to loading within its vertical plane $(x–z)$, the following internal forces are developed (see Figure 1.2):

1

Figure 1.1 Equilibrium conditions for a loading perpendicular to a plane structure (grillage)

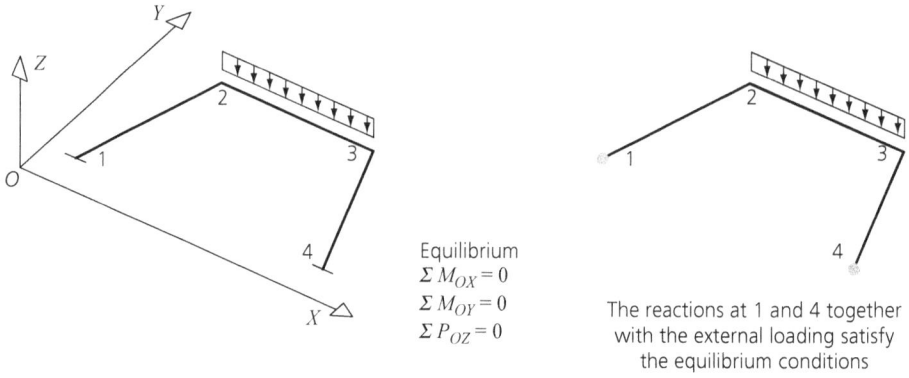

Equilibrium
$$\Sigma M_{OX} = 0$$
$$\Sigma M_{OY} = 0$$
$$\Sigma P_{OZ} = 0$$

The reactions at 1 and 4 together with the external loading satisfy the equilibrium conditions

Figure 1.2 Internal forces and deformations developed in a grid structure

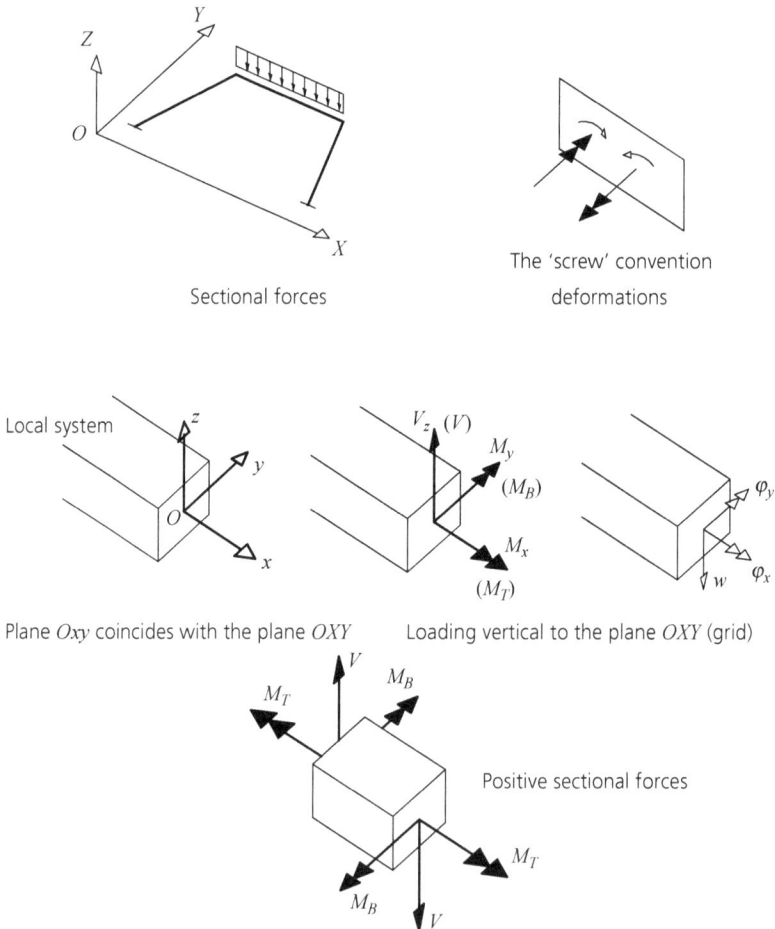

Sectional forces

The 'screw' convention
deformations

Local system

Plane Oxy coincides with the plane OXY

Loading vertical to the plane OXY (grid)

Positive sectional forces

- shear force, V_z (i.e. V)
- bending moment, M_y (i.e. M_B)
- torsional moment, M_x (i.e. M_T).

The bending moment, M_B, in combination with the shear force, V, induces flexure in the vertical plane (x–z) of the member. The torsional moment, M_T, on the other hand, causes twisting about the member's longitudinal axis. At each point along the member, the resulting deformations include a vertical deflection, w, along the z-axis and a rotation represented by a horizontal vector. This rotation has two key components

- φ_y, which is the projection of the rotation vector onto the y-axis, associated with flexural (bending) rotation
- φ_x, which is the projection onto the x-axis, representing the torsional (twisting) angle.

The signs of the internal forces are defined as follows (see Figure 1.2).

- *Shear force, V*, is positive when it, along with the corresponding force at an adjacent cross-section, produces a clockwise moment couple.
- *Bending moment, M_B*, is positive when it induces tension in the bottom fibres of the cross-section.
- *Torsional moment, M_T*, is positive when the rotation induced by its physical action on the cross-section results in a screw motion giving 'tension' for the cross-section itself.

These internal forces can be determined from the equilibrium conditions described previously, provided that all externally applied forces are known (see Figure 1.3).

For the determination of any deformation, the principle of virtual work is always applicable, which, in order to allow for the contribution of torsional moments in carrying the vertical loads, is now written in the following form:

$$\sum P^{virt} \cdot \delta^{resp,real} = \int M_B^{virt} \cdot \frac{M_B^{real}}{EI}\, ds + \int M_T^{virt} \cdot \frac{M_T^{real}}{GI_T}\, ds$$

Figure 1.3 Determination of the internal forces in a grid structure

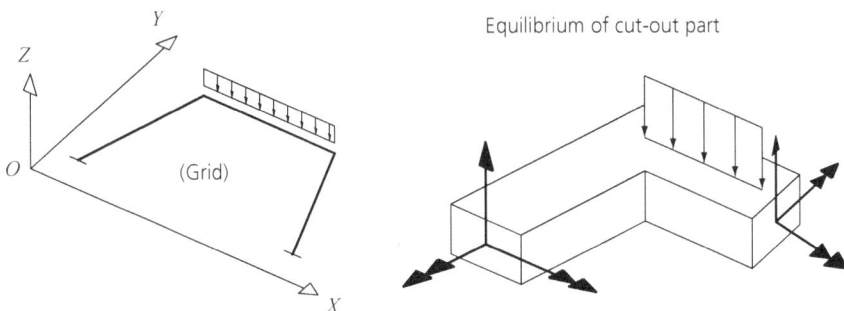

Equilibrium of cut-out part

Y

Z

O (Grid)

X

For the determination of sectional forces, the number of unknown magnitudes must not exceed 3

where by the term $\int (M_T^{real} / GI_T) ds$ expresses the twisting angle $\Delta\varphi_{real}$, according to Section 1.1.4 in analogy with the discussion in Stavridis and Georgiadis (2025, Section 2.3.3).

1.1.3   Types of supports
There are two main ways of supporting a grillage structure (see Figure 1.4).

- *Simple support*: which simply prohibits vertical movement. The deformations φ_x and φ_y are freely developed. The only reaction developed is a vertical force, R.
- *Fixed support*: which restrains all possible deformations. The developed reactions are a vertical force, R, and the two moments M_x and M_y.

If the whole grillage is supported in such a way that only three reaction forces are developed, then the reactions can be determined from the three equilibrium conditions, as mentioned in Section 1.1.1. If more than three reaction forces exist, the equilibrium conditions are insufficient for the determination of these forces. The structure is statically indeterminate and the reaction forces can be determined following the methods discussed in Stavridis and Georgiadis (2025, Chapter 3) (i.e. the method of forces or the method of deformations).

1.1.4   Torsion
It is already clear that the internal force that differentiates grids from typical plane structures is torsion. For a member of length, L, the relation of the developed torsional moment, M_T, at its free end to the corresponding twisting angle, φ_T, of the end section (see Figure 1.5) is:

$$\varphi_T = M_T \cdot \frac{L}{GI_T}$$

where I_T is the *torsional moment of inertia* (which is a geometric property of a cross-section that measures its resistance to torsion (twisting) about its longitudinal axis); G is the shear modulus of the member's material, $G = \dfrac{E}{2(1+v)}$ and v is Poisson's ratio (see Stavridis and Georgiadis, 2025, Section 2.3.2).

For a rectangular cross-section having dimensions b/t ($b>t$), the torsional moment of inertia, I_T, is calculated according to the expression $I_T = b \cdot t^3/\eta$, where the coefficient η varies from 3.0 to 7.0 for oblong to square sections, respectively.

Figure 1.4 Two ways of supporting a grid structure

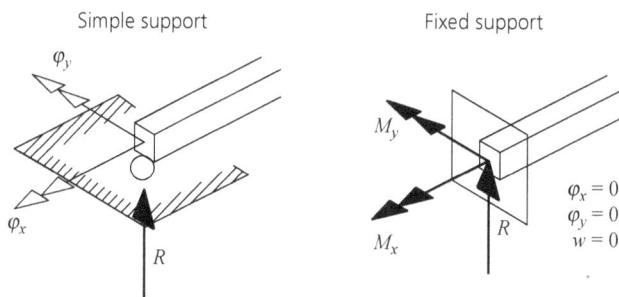

Simple support   Fixed support

Figure 1.5 Torsional response of a beam with an orthogonal solid section

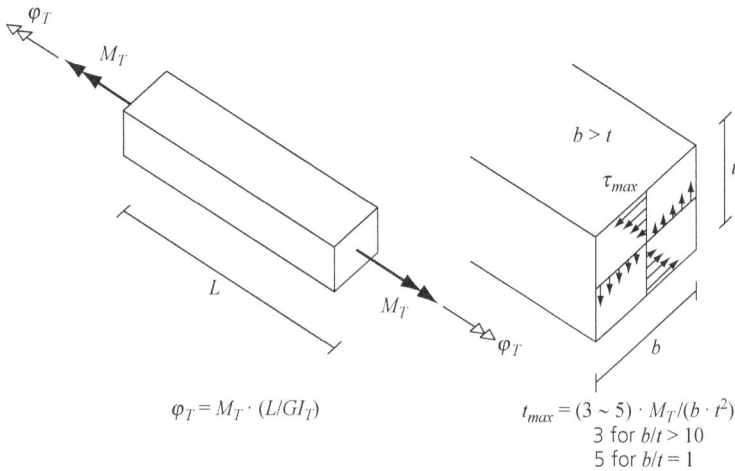

$$\varphi_T = M_T \cdot (L/GI_T)$$

$$t_{max} = (3 \sim 5) \cdot M_T/(b \cdot t^2)$$
3 for $b/t > 10$
5 for $b/t = 1$

From the above equation of φ_T it can be concluded that the magnitude (GI_T/L) expresses the *torsional stiffness* of the member, which is the torsional moment required in order to produce a unit twisting angle or, in other words, the torsional resistance offered by the member if subjected to a unit twisting angle $\varphi_T = 1$.

A torsional moment, M_T, applied to a rectangular cross-section causes a flow of shear stresses on the full cross-section, which diminish towards the centre of the cross-section. The resultant moment vector with respect to the centre of the cross-section from these stresses is obviously identical to M_T (see Figure 1.5). The maximum shear stress, τ_{max}, appears at the middle of the longest side and is proportional to the stress magnitude $M_T/(b \cdot t^2)$ multiplied by a factor 3.0 for oblong sections and a factor 5.0 for square sections. It must be noted that τ_{max} is inversely proportional to the square of the thickness of the cross-section (see Figure 1.5).

Apart from rectangular cross-sections, another common type of cross-section that is used in practice when torsion is present is the so-called *hollow box*, as shown in Figure 1.6. Each wall of the cross-section has a certain thickness, t, and takes up the torsional moment, M_T, through a peripheral constant shear stress, τ, which, according to the so-called *Bredt formula* is equal to:

$$\tau = \frac{M_T}{2 \cdot F_k \cdot t}$$

where F_k is the area enclosed by the middle line of the walls – that is, slightly less than the area of the full (solid) cross-section (see Figure 1.6).

Each wall of the hollow box cross-section is subjected to a constant shear flow $v = \tau \cdot t = M_T/(2 \cdot F_k)$ acting per unit length (kN/m) of the wall, which thus causes a shear force $V = v \cdot l$ over the length, l, of each wall, where l is either h or b (see Figure 1.6). It can be seen from the equation of the shear flow, v, that the shear flow, v, corresponding to a certain torsional moment, M_t, remains constant,

Figure 1.6 Torsional response of a beam with hollow box section

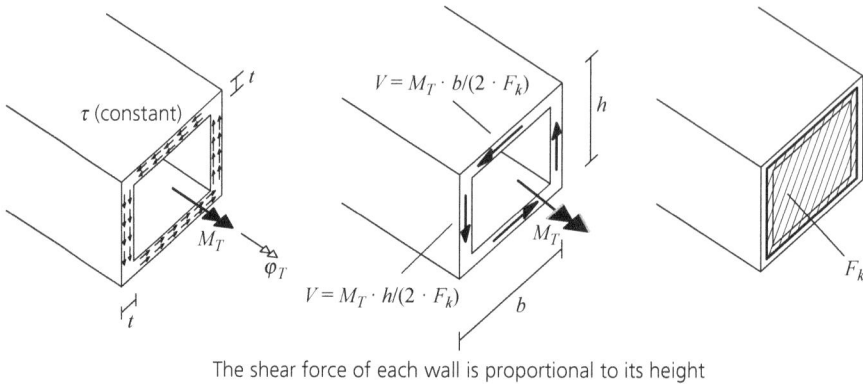

The shear force of each wall is proportional to its height

even if the walls' thicknesses are different. It should be pointed out that this shear flow provides the whole section with the torsional moment $(2 \cdot v \cdot F_k)$, which is statically equivalent to the applied moment, M_T.

Of course, the previous relation between the twisting angle, φ_T, and the torsional moment, M_T, is also valid for hollow box cross-sections, the torsional moment of inertia, I_T, being now equal to $I_T = 4 \cdot t \cdot F_k^2 / L_\Omega$, where L_Ω is the perimeter of the middle line of the section. It is pointed out that the torsional moment of inertia I_T – and, consequently, the torsional stiffness (GI_T/L) – is, for the same cross-sectional area, much greater in a hollow than in a solid section.

1.1.5 Design for torsion

As previously explained, in a rectangular cross-section subjected to torsion, the highest shear stresses are concentrated near the periphery of the section. This observation justifies modelling a solid section as a thin-walled hollow section, with the assumption that the entire torque is carried by a narrow region around its outer boundary of the section, as discussed in Section 1.1.4. This modelling approach is conservative as it results in higher stresses, thereby remaining on the safe side for design purposes. Consequently, the analysis of thin-walled hollow sections has broader applicability and practical significance in torsional design.

The shear flow developed around the thin wall that resists the applied torque gives rise to principal stresses, which are the actual physical stresses that are developed in the member (see Figure 1.7). The tensile principal stresses in steel sections can be resisted by the tensile capacity of the steel, but in concrete sections an appropriate reinforcement layout must be provided. This is typically achieved through the use of closed rectangular stirrups in combination with longitudinal reinforcement, which together provide the required oblique tensile force paths to resist the induced stresses (see Figure 1.7).

Allowing a service stress, σ_s, for the reinforcement steel results in a required section, a_s, per unit length for the stirrups equal to $a_s = v_l/\sigma_s$ (cm²/m), and in a total cross-sectional area, A_l, for the longitudinal bars (uniformly distributed along the perimeter) equal to $A_l = (v_l/\sigma_s) \cdot L_\Omega$.

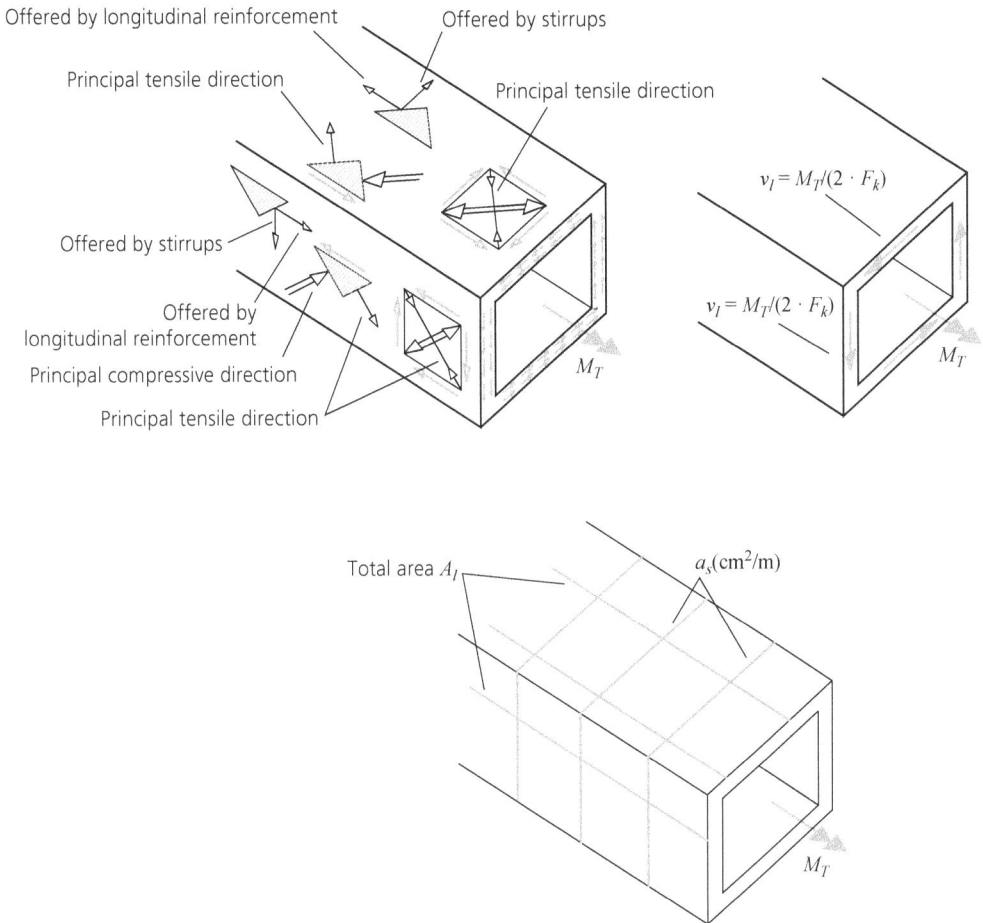

Figure 1.7 Thin-walled model for torsional design of reinforced concrete members

Finally, it should be mentioned that, apart from grillages, torsion can also appear in typical plane structures, as presented in Stavridis and Georgiadis (2025). In any case, the torsional requirements (e.g. torsional reinforcement) must be considered in addition to the design requirement from other actions such as bending and shear.

1.1.6   Static equilibrium and torsional deformation

If a beam of length, L, with fixed torsional support is subjected to a uniformly distributed torsional load, m_T, it develops a torsional moment diagram, M_T. Expressing the torsional equilibrium of an element of length, ds, of the beam, the following relation in plausible differential form is obtained (see Figure 1.8):

$$\frac{\mathrm{d}M_T}{\mathrm{d}s} = -m_T$$

The analogy of the above relationship with the relationship between a distributed load and the developed shear force, as discussed in Stavridis and Georgiadis (2025, Section 2.2.3), suggests

Figure 1.8 Equilibrium of an element of length, ds, of a fixed-ended beam under uniformly distributed torsional load, m_T

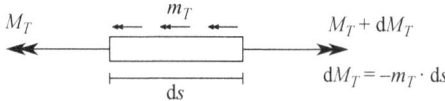

that the torsional moment diagram of the above beam is identical to the shear force diagram of an equivalent simply supported beam under a vertical uniformly distributed load equal to m_T (see Figure 1.9).

The same analogy is also valid for a concentrated torsional load applied at some point of the fixed-ended beam (see Figure 1.10).

Moreover, as the twisting angle, φ, actually changes along the beam in correlation with the corresponding torsional moment, M_T, it can be written in accordance with Section 1.1.3:

$$\frac{d\varphi}{ds} = \frac{M_T}{G \cdot I_T}$$

Regarding now the determination of the twisting angle diagram, a similar procedure to that given in Stavridis and Georgiadis (2025, Section 2.3.6) may be followed. By differentiating the above expression of $d\varphi/ds$, the following is obtained:

Figure 1.9 Analogy of a torsional beam under a uniformly distributed torsional load, m_T, with a simply supported beam with a vertical uniformly distributed load, $p(x)$

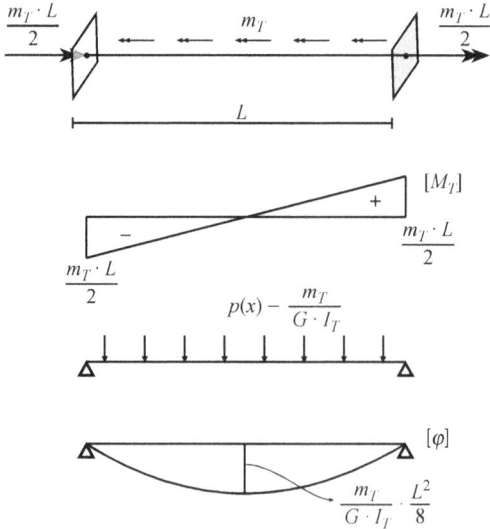

Figure 1.10 Analogy of a torsional beam under a concentrated torsional load, T, with a simply supported beam with a concentrated load, P

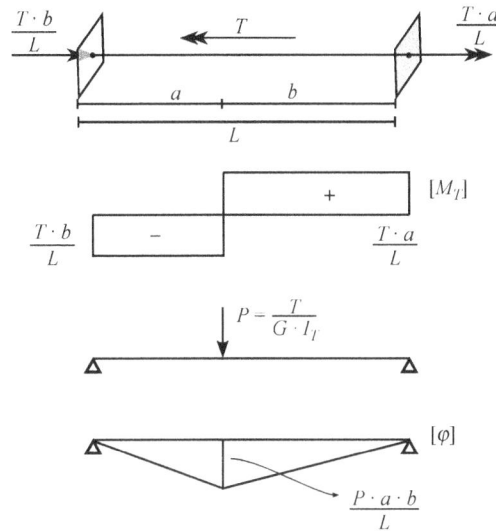

$$\frac{\mathrm{d}^2\varphi}{\mathrm{d}s^2} = \frac{\mathrm{d}M_T}{\mathrm{d}s} \cdot \frac{1}{G \cdot I_T} = \frac{m_T}{G \cdot I_T}$$

This relationship, in analogy with the one between a distributed load and the bending moment $\mathrm{d}^2 M / \mathrm{d}x^2 = p(x)$ in a simply supported beam (see Stavridis and Georgiadis, 2025, Section 2.3.6), reveals the following important conclusion: if the expression ($m_T / G \cdot I_T$) is considered as a uniformly applied vertical load $p(x) = (m_T / G \cdot I_T)$ of the equivalent simply supported beam, then the corresponding bending moment diagram of this beam represents the twisting angle diagram of the torsionally fixed beam (see Figure 1.9).

The above conclusion applies also in an analogous way for a concentrated torsional load, T, applied to a fixed torsional beam, whereby the twisting angle diagram coincides with the bending moment diagram of the equivalent simply supported beam under a concentrated point load $P = (T / G \cdot I_T)$ (see Figure 1.10).

1.2. General application and structural behaviour of grillages
1.2.1 General application of grillages
As described in Section 1.1.1, a grillage is a plane formation of beams connected to each other with joints aiming to receive loads acting perpendicular to the grillage plane and to transfer them in two or more directions (see Figure 1.1).

However, in general, grillages commonly have an orthogonal layout consisting of longitudinal and transverse members (see Figure 1.11). Although grillages are not very common structures on their own, they have been extensively used in bridge engineering to simulate the behaviour of continuous bridge decks (e.g. concrete slab decks, composite decks etc.). The idea is to divide a continuous structure (e.g. a bridge deck) into longitudinal and transverse beam elements with appropriate sectional properties to simulate the behaviour of the actual structure.

Despite the fact that the grillage nodes are almost always formed as rigid (monolithic), it may sometimes be useful to consider the longitudinal and transverse beams as simply supported to each other, thus transferring only an internal vertical force. In the case of monolithic joints, the beams of a grillage develop three types of internal forces according to Section 1.1.2 – that is, bending moments, torsional moments and shear forces. The grillage joints develop a vertical displacement and a rotation, represented by a horizontal vector (see Figure 1.2). In the case in which the longitudinal and transverse beams are simply supported to each other, they develop only bending moments and shear forces as the bending deflection of one beam does not introduce torsion to the perpendicular beam. It should be remembered that loads acting on any cut-out part of the grillage must satisfy three equilibrium conditions – that is, the equilibrium of vertical forces – as well as the equilibrium of the vector projections of the moments with respect to two arbitrary horizontal axes of the grid plane (see Section 1.1.2).

It is obvious that, in general, grillages constitute statically indeterminate structures with a large degree of redundancy and their analysis almost always requires the application of suitable computer software. Nevertheless, it is useful to examine some basic characteristics of their structural behaviour in order to gain a better understanding of their design.

1.2.2 The main characteristics of the structural behaviour of grillages
In order to examine the main structural function of a grillage – that is, the transfer of vertical loads in two horizontal directions – the simple model shown in Figure 1.12 is used. In this model, a

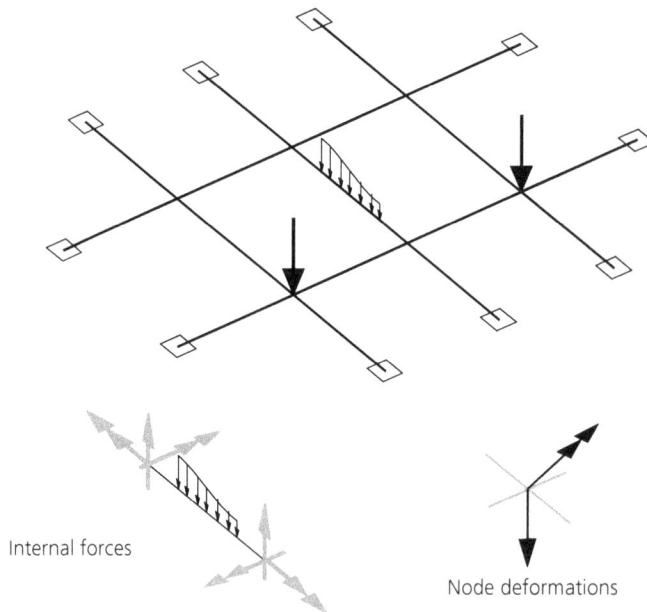

Figure 1.11 General layout and internal forces and displacements in a grillage

Internal forces

Node deformations

Figure 1.12 Load distribution depending on member stiffness

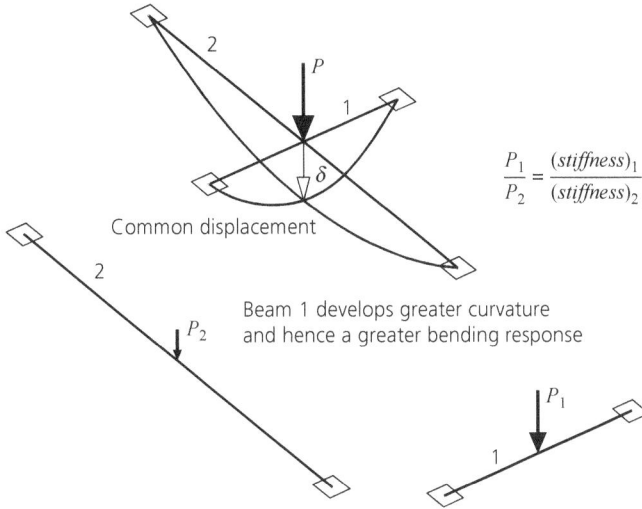

$$\frac{P_1}{P_2} = \frac{(stiffness)_1}{(stiffness)_2}$$

Common displacement

Beam 1 develops greater curvature
and hence a greater bending response

vertical point load, P, is transferred in two directions by way of two cross-beams. As the grillage is symmetric in both directions, it develops only one (vertical) displacement at the node. This displacement is common for both beams. However, the corresponding curvature $(1/r)$ is different. For the same displacement, the shorter beam develops a greater curvature than the longer one, which is given by the relationship $M = EI/r$. It also develops a higher bending moment, provided that the stiffness (EI) of both beams is the same.

As shown in Stavridis and Georgiadis (2025, Section 2.3.8), the bending stiffness or *rigidity* of a simply supported beam regarding its midpoint displacement is the force required to cause a unit displacement equal to $48 \cdot EI/L^3$. The factor 48 becomes 192 for a beam with fixed ends. It thus appears that doubling the length of the beam causes a decrease to the above stiffness by eight times, while it is proportional to (EI).

Analysing this simple system (using the method of forces or the method of deformations), the important conclusion is that if P_1 and P_2 are the forces carried by beams (1) and (2) at their midpoints, respectively, then it is

$$\frac{P_1}{P_2} = \frac{(stiffness)_1}{(stiffness)_2} \quad \text{(see Stavridis and Georgiadis, 2025, Sections 3.1.8 and 3.2.8)}$$

It can be easily concluded that if beam (2) is double the length of beam (1), then the load carried by this beam is 1/9 of the applied load, P, while the maximum moment of short beam (1) is four times greater than that of beam (2). If beam (2) becomes even longer, it is clear that its contribution to carrying the vertical load is negligible.

Figure 1.13 Influence of grid layout on load-carrying behaviour

The load is carried mainly by the short beam   Only the beams in the short direction
                                                                are practically stressed

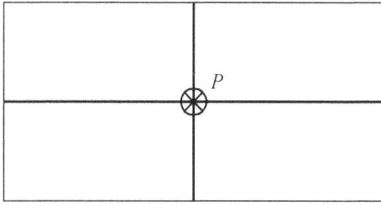

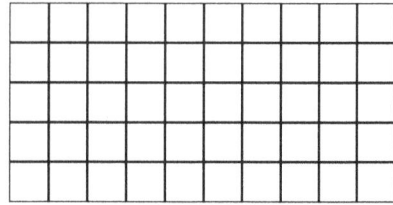

Plan view

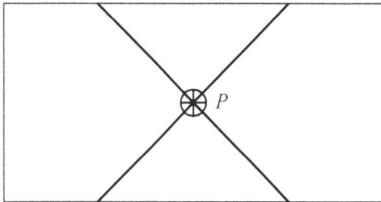

The load is carried equally by the two beams   Lower bending response of beams
                                                         Lower deformations

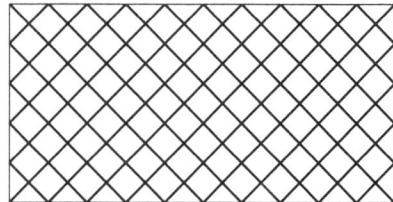

It is of course possible, even in the case of beams with different lengths, to achieve a balanced load distribution based on the above relation by designing the beams in such a way that $(stiffness)_1 = (stiffness)_2$. ∎

The conclusions resulting from the examination of the above simple model are also valid for regular grillages that cover areas with orthogonal outlines. If the beams have the same cross-sectional characteristics (EI), then those with shorter lengths are more stressed, while for an aspect ratio greater than 1.5, the beams in the longitudinal direction are clearly 'underworking' (see Figure 1.13).

The above relation between load and stiffness is obviously valid also in the case where the two cross-beams are not perpendicular to the supporting edges (see Figure 1.13). Based on this, it can be concluded that to cover oblong areas (regardless of the aspect ratio), the adoption of a skew layout of beams ensures practically the same stress for each group of beams since they have the same length. Thus, the balanced load transfer through the two groups of beams clearly implies smaller bending moments (i.e. smaller beam depths) and also smaller deflections compared with the orthogonal layout.

Moreover, the presence of beams of small lengths and hence of high stiffness near the corners provides fixed conditions for the corresponding diagonal beams, which is favourable for both the bending response and the deformation. ∎

The above model shows the contribution of cross-beams in transferring the load. An additional characteristic of the load-bearing action of grillages is that beams adjacent to the one where the load is applied contribute to undertaking and transferring this load. More specifically, if a point load is applied to the intermediate of three beams, denoted hereafter as main beams with parallel layout, as shown in Figure 1.14, the two edge main beams will obviously remain unloaded if they are not connected to a transverse beam. On the other hand, the presence of a transverse beam transfers the load also to the edge beams, thus relieving the intermediate beam (see Figure 1.14). In order to examine the influence of the transverse beam, this may be considered as a continuous beam of two spans resting on three flexible supports with spring stiffness $k_s = 48 \cdot (EI)_1 / L_1^3$, according to Stavridis and Georgiadis (2025, Section 2.3.7).

Under the applied point load, the intermediate main beam deflects downwards by Δ, imposing the same displacement to the central support of the transverse beam. At the edge supports of the transverse beam, upward reactions are developed, which are applied with the opposite sign to the edge main beams (see Figure 1.14). From the analysis (see Stavridis and Georgiadis, 2025, Chapter 3), this results in the downward end displacement:

$$v = \frac{\Delta}{1 + \dfrac{k_s \cdot L^3}{3 \cdot EI}}$$

The developed stresses of each main beam depend exclusively on its midpoint displacement, whereby the effect of the stiffness (EI) of the transverse beam according to the above relation becomes obvious. A small value of (EI) causes a small v, whereas with an increase of (EI), the displacement, v, tends to approach the value Δ (i.e. a very rigid transverse beam). This means that the load is also transferred to the two edge main beams due to the stiffness of the transverse beams. The utilisation of a very stiff transverse beam practically ensures that the load is equally distributed between the three parallel main beams.

Figure 1.14 Influence of a cross-beam (transverse beam) on the transfer of a concentrated load

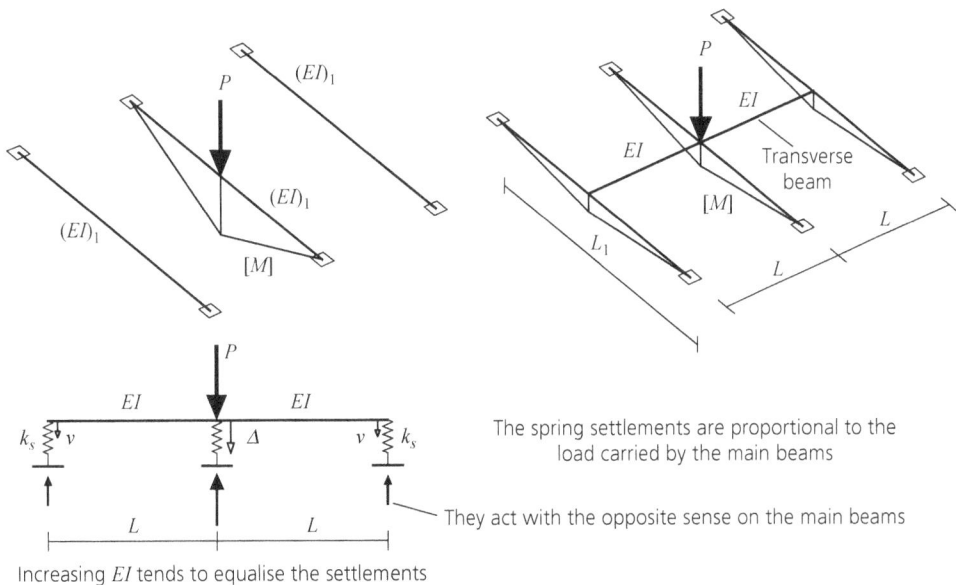

The spring settlements are proportional to the load carried by the main beams

They act with the opposite sense on the main beams

Increasing EI tends to equalise the settlements

Based on the above, it becomes clear that in oblong orthogonal areas where, as previously examined, only the beams of the shorter direction participate substantially in the distribution of loads, the beams in the longitudinal direction are necessary (despite the fact that they are ineffective for uniformly distributed permanent loads) in order to ensure by their rigidity the distribution of any concentrated load to the adjacent main beams. ∎

In the above considerations, it was assumed that the main and transverse beams of the grillage are simply resting on each other, transferring only vertical support forces. In practice, the connections between beams are not designed this way. Instead, they are usually rigid (or monolithic), which means the development of additional torsional moments in the beams. This fact does not alter the above conclusions, and contributes to a more favourable structural function of grillages with respect to bending stresses and deformations. The favourable contribution of torsion to the bearing capacity of grids is explained below.

In Figure 1.15, a symmetric model is considered, where the beams are monolithically connected to each other. The transverse beam is simply supported at its ends, while the two main beams are supported so as to be able to take up torsional moments. Two equal concentrated forces, P, are applied to the internal nodes of the model.

It is already clear from the above that the main beams offer an elastic support to the transverse beam at the corresponding nodes with a spring stiffness equal to $k_s = 48 \cdot (EI)_1 / L_1^3$. It is also clear that, besides the settlement of the support points, a certain rotation, φ, occurs at these points in the transverse beam, which would be freely developed without resistance if the transverse beam were simply supported, as in the previous example. However, in this case where the beams are connected monolithically, this rotation is not developed freely as it is restricted by the torsional resistance of the main beam. In order to impose the rotational angle, φ, to the main beam the application of a certain torsional moment is required. As this moment is provided by the transverse beam, according to action–reaction, the transverse beam will receive the same moment in the opposite sense. However, due to its direction, for the transverse beam this is a bending rather than a torsional moment, thus decreasing the bending stress state and deformation of the transverse beam (see Figure 1.15).

More specifically, the imposition of a rotation, φ, on the main beam requires the application of a torsional moment $(k_\varphi \cdot \varphi)$ where k_φ is the so-called *torsional stiffness* (see Section 1.1.3) and obviously this moment is also applied with the opposite sign to the bending moment to the member imposing this rotation – that is, to the transverse beam. Therefore, besides the translational springs, k_s, in its two internal nodes, the transverse beam also has the torsional springs, k_φ, that oppose the rotation of the nodes (see Figure 1.16). According to Section 1.1.6 (see Figure 1.10), it is:

$$k_\varphi = \frac{4 \cdot GI_T}{L_1}$$

For the particular layout and applied forces shown in Figure 1.16, by using the method of deformations (see Stavridis and Georgiadis, 2025, Chapter 3), it is found that the vertical deflection, δ, that is developed at the nodes, can be calculated from the relation:

$$\delta = \frac{PL^3}{3 \cdot EI} \cdot \frac{1}{1 - \dfrac{3}{4 + k_\varphi \cdot L / EI} + \dfrac{k_s \cdot L^3}{3 \cdot EI}}$$

Figure 1.15 Relieving effect of torsion in the bending and deformation states

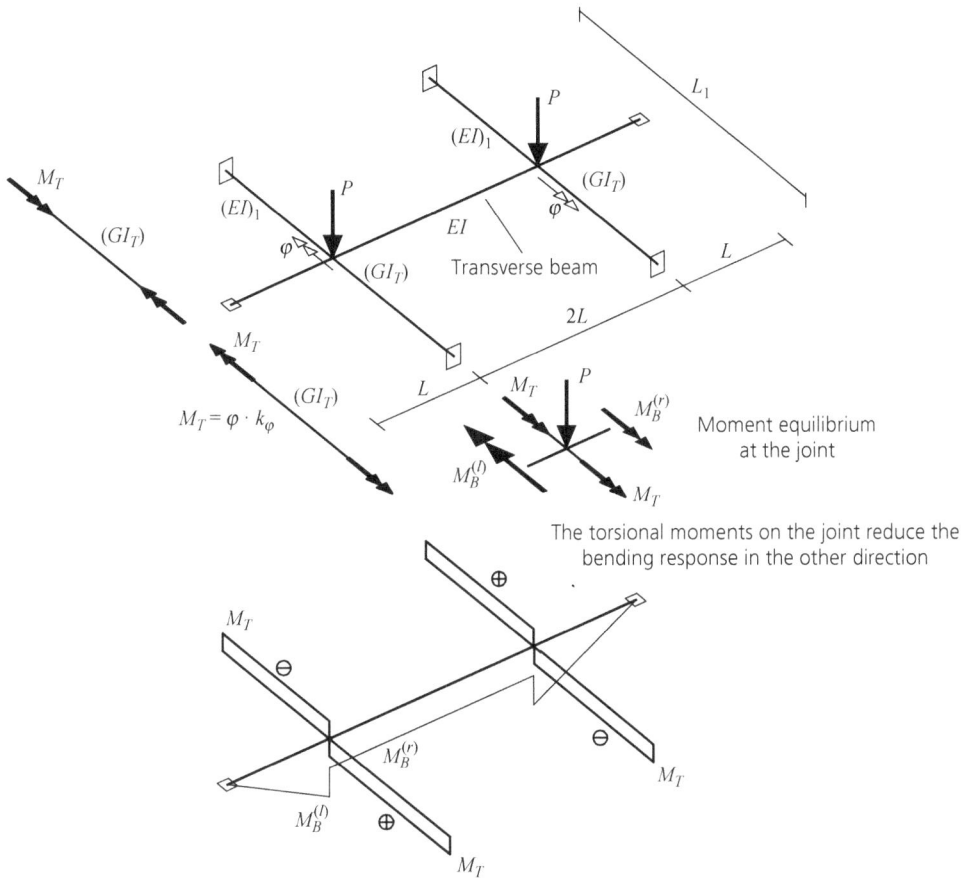

From this relationship, it can be concluded that the presence of torsional rigidity in the main beams that is expressed by the application of a rotational spring, k_φ, causes a reduction in the developing deflection, δ, of the transverse beam at the intersections with the main beams, in contrast to the case where this torsional rigidity does not exist ($k_\varphi = 0$). In other words, the developing torsion of the main beams decreases the bending stress in both the transverse and the main beams, given that this reduction of δ also decreases the corresponding curvatures (see Figure 1.12). Thus, a grillage with monolithic nodes possesses higher stiffness. This is done, of course, at the 'expense' of the introduction of torsional moments, which means additional shearing stresses.

It is clear that these conclusions are also applicable to grillages consisting of more beams. The presence of monolithic nodes implies the development of torsional moments that reduce the bending response of beams and increase the stiffness against vertical loads.

The above 'contribution' offered by torsion depends directly on the torsional rigidity of the beams, which is based on the value of the torsional constant, I_T, of the beams. Thus, in grillages with beams of I-sections, the developing torsional moments are much smaller than, for example, in

Figure 1.16 Justification of the relieving effect of torsion

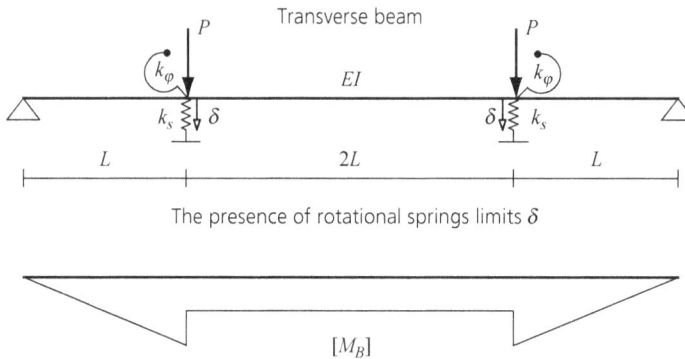

grillages of the same dimensions with hollow cross-sections of the same sectional area. In the latter case, the relieving influence of torsion on bending is stronger.

It becomes clear that the structural function of grillages actually constitutes an interaction between torsional and bending rigidity, and is generally based on the ratio, κ, of these values – that is $\kappa = (GI_T/EI)$. An increase of the ratio, κ, means an increased torsional contribution – and corresponding bending reduction – in carrying the applied loads. ∎

At this point, it should be pointed out that the development of torsion in grillages is not needed for equilibrium, as can be seen in all the models examined previously. This means that equilibrium in a grillage node, as shown in Figure 1.17, can be ensured – besides the requirement for vertical equilibrium – along with the presence of bending moments. The presence of torsional moments acts simply as an additional factor for equilibrium to each direction and their development is due merely to the requirements for compatibility of deformations. This has a direct repercussion in the design approach for a grillage with respect to the ultimate state, regarding the validity of the static theorem of plastic analysis (see Stavridis and Georgiadis, 2025, Section 6.6.2).

In particular, regarding a grillage consisting of beams of reinforced concrete with orthogonal sections, the developing cracking reduces their torsional stiffness much more than their bending stiffness, and as a result the torsional moments are reduced, whereas the bending moments are increased.

As an extreme situation, the torsional rigidity may be fully ignored, yielding to the highest possible bending. In any case, according to the static theorem of plastic analysis and provided that the conditions for equilibrium are kept, the design involving the omission of torsional moments (and corresponding increase of the bending moments) is, of course, on the safe side and it may be followed at least for the aims of preliminary design, since the estimate of the effective torsional stiffness due to cracking is rather uncertain.

1.2.3 Layout and structural function of skew grillages

According to the previous paragraph, it can now be understood how beams with relatively large spans and relatively high torsional rigidity can reduce their bending, thanks to the grillage action, through the bending stiffness of the transverse beams connected to them.

Figure 1.17 Figure 1.17 Load-bearing mechanism of a grillage with or without the development of torsion

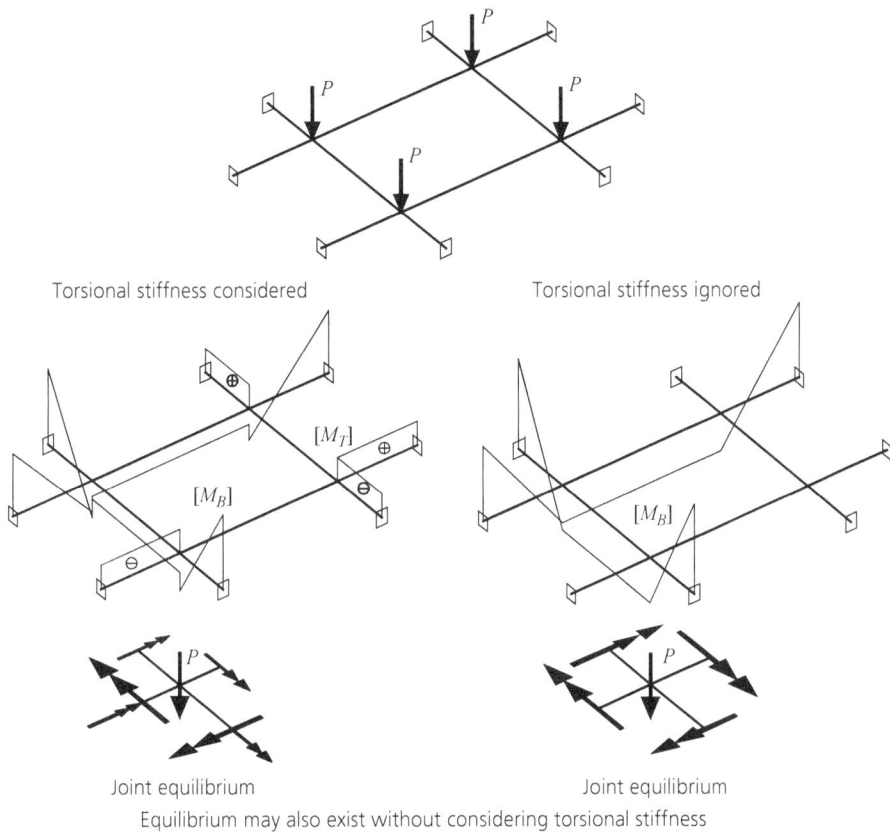

Torsional stiffness considered   Torsional stiffness ignored

Joint equilibrium   Joint equilibrium

Equilibrium may also exist without considering torsional stiffness

The grillage action is also useful in bridge design with the so-called *skew layout*, where the road alignment cuts the axis of an existing obstacle (river, under-passing road etc.) sideways and results in an increase of the length of the structure compared with what would exist if the two axes were crossing each other orthogonally. The beams used to support each end of the bridge are arranged parallel to the axis of the obstacle, which is sideways with respect to the main beam (see Figure 1.18). The support beams that are provided with high bending stiffness are simply supported on bearings that leave free (unhindered) the rotation about their connecting axis, so that no torsional stiffness is provided. To increase the efficiency of this system, the main beam must be able to carry significant torsional moments and, for this purpose, as mentioned in Section 1.1.4 (see also Chapter 4, Section 4.1), only hollow-type beam sections can meet this requirement.

The above structural system is shown in Figure 1.18. The main beam of length, L, is loaded with a uniformly distributed load, q. At each node, the vector of the transferred moment can only be perpendicular to the axis of the support beam since the deviation from the perpendicular direction would create a torsional loading that the beam is not able to carry. Considering this moment, X, as a statically redundant action, it is found from the requirement for compatibility of the developed rotations, by applying the method of forces (see Stavridis and Georgiadis, 2025, Chapter 3), that:

Figure 1.18 Skew bridge layout

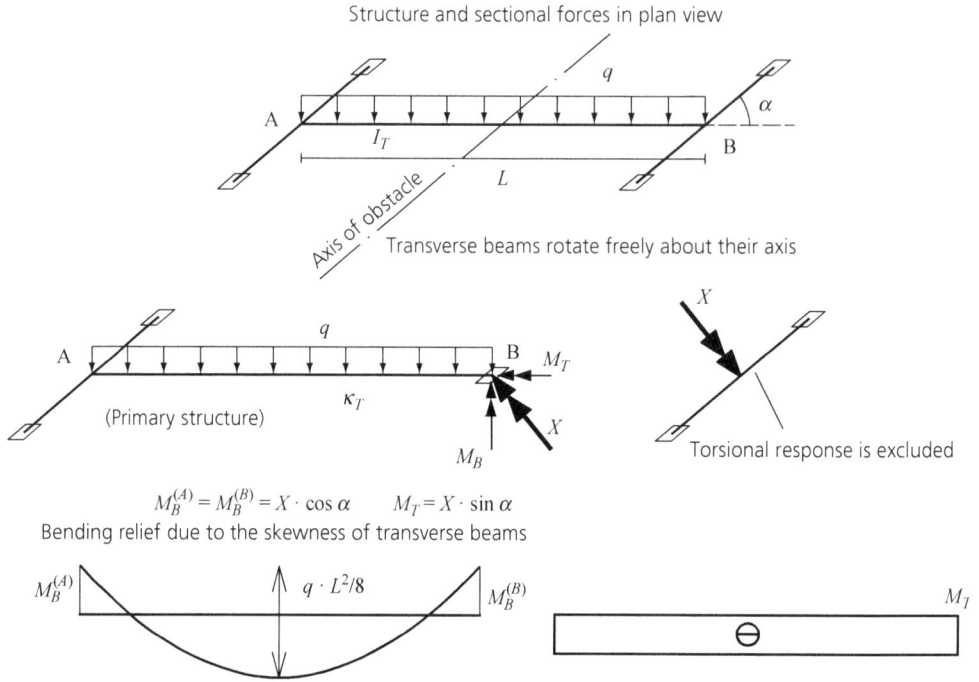

Structure and sectional forces in plan view

$$M_B^{(A)} = M_B^{(B)} = X \cdot \cos \alpha \qquad M_T = X \cdot \sin \alpha$$
Bending relief due to the skewness of transverse beams

$$X = \frac{q \cdot L^2}{12} \frac{1}{(1 + tan^2 \alpha / \kappa) \cdot \cos \alpha}$$

where

$$\kappa = \frac{GI_T}{EI}$$

It is clear that the moment, X, with its transverse component to the longitudinal axis of the main beam, provides the relieving bending moment, M_B, for the span, while with its component, M_T, on the axis provides the torsional moment of the beam. This action arises thanks to the skew angle. As this angle decreases, the bending relief in the main beam increases, while its torsional response also increases.

Regarding concrete structures, the use of prestressing, which is implemented to limit the bending response of the main beams, also contributes substantially to guarantee their full torsional rigidity, which is necessary for the appropriate development of the relieving action of torsional moments.

1.2.4 Worked example
What has been presented for grillages can now be applied to analyse a system consisting of a torsional steel beam which is connected with a parabolic arch through vertical struts which is

subjected to torsional loads (see Fig. 1.19). It is of interest to assess the torsional response of the beam and subsequently the developed twisting angle at midspan.

The span of the beam is $L = 40.0$ m, whereas the arch which is lying in a vertical plane has a rise $f = 5.0$ m. The vertical struts are spaced at $s = 5.0$ m and have a moment of inertia $I_m = 2492 \cdot 10^{-8}$ m^4. They are rigidly connected to the beam and hinged at their connection with the arch.

The arch having a moment of inertia about the weak axis equal to $I_A = 567 \cdot 10^{-6}$ m^4, is formed according to the pressure line (see Chapter 8 from Understanding and Designing Structures without a Computer: Plane Structural Systems) for a certain uniform permanent load, its geometry $y(x)$ following the curve $y(x) = 4 \cdot f \cdot x \cdot (L - x)/L^2$.

The beam, having a torsional moment of inertia equal to $I_T = 0.00136$ m^4, is subjected to a uniform torsional load m_D acting clockwise all along its length.

The torsional beam if considered free without any restrain from the struts, develops all along its length a twisting angle $\varphi(x)$, which according to Section 1.1.6 is identical to the fictitious bending moment diagram under the load m_D/GI_T. Thus, (being a parabola) it is given by the expression $\varphi(x) = (m_D/2GI_T) \cdot [x \cdot (L - x)]$. The angle $\varphi(x)$ is imposed on the relevant strut with a length of $y(x)$. As the strut is restrained by the arch, develops a restrain force R.

The arch deflects laterally and may be considered practically as a beam of length L , thus exhibiting at each point a transversal stiffness equivalent to a respective spring with stiffness equal to $k_s = 3 \cdot E \cdot I_A \cdot L /[x \cdot (L - x)]^2$.

The force R acting at the end of the vertical strut of length $y(x)$ must recover the created horizontal gap $\Delta = [\varphi(x) \cdot y(x)]$ with the arch. Firstly, the deflection for a cantilever with length $y(x)$ is $\delta_1 = R \cdot y^3/3 \cdot EI_m$. On the other hand the torsional moment $(R \cdot y)$ acting as a concentrated torsional load on the beam causes at its base on the beam a twisting angle $\Phi_1 = (R \cdot y) \cdot x \cdot (L - x) /(L \cdot GI_T)$ according to Section 1.1.6, and thus, acting on the strut's base too, rotates it as a rigid body and shifts horizontally its other end by the displacement $\Delta_1 = \Phi_1 \cdot y(x)$. In fact, the gap Δ must be

Figure 1.19 Arhc bridge with struts under torsional loads

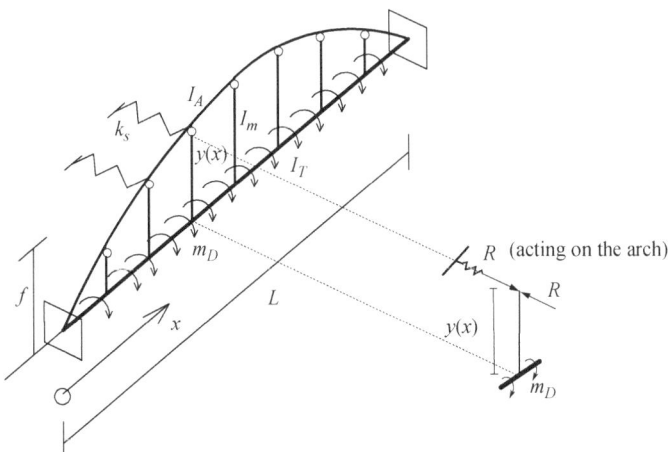

recovered by δ_l as well by Δ_l and also by the elongation of the spring (R/k_s), i.e. $\Delta = \delta_l + \Delta_l + (R/k_s)$, or more explicitly:

$$\varphi(x) \cdot y(x) = R \cdot y^3 / 3 \cdot EI_m + \{(R \cdot y) \cdot x \cdot (L - x) / (L \cdot GI_T)\} \cdot y(x) + R \cdot [x \cdot (L - x)]^2 / (3 \cdot E \cdot I_A)$$

The above equation allows to express the strut reaction R, by considering that $G = E/2.4$ as below:

$$R = \frac{1.2 \cdot 4 f / L^2}{\left[x(L-x)\right] \cdot I_T} \cdot \frac{m_D}{\dfrac{(\frac{4f}{L^2})^3}{3 I_m} + (4f / L^2)^2 \cdot \dfrac{2.4}{L \cdot I_T} + \dfrac{1}{3 I_A \cdot L \cdot [x(L-x)]}}$$

and thus, the relieving distributed torsional moment is:

$$m = R \cdot y(x) / s = \frac{1.2 \cdot m_D / s}{\dfrac{I_T}{I_m} \cdot \dfrac{4f / L^2}{3} + \dfrac{2.4}{L} + \dfrac{L^2 / 4f}{3 \cdot x(L-x)] I_A} \cdot I_T}$$

Substitution of the above values in the above expressions and considering the middle section $(x = 20.0$ m$)$ yields :

$$m = \frac{1.2 / 5.0}{54.57 \cdot \dfrac{4 \cdot 5.0 / 40.0^2}{3} + \dfrac{2.4}{40.0} + \dfrac{40.0^2 / (4 \cdot 5.0)}{3 \cdot 20.0 \cdot (40.0 - 20.0) \cdot 2.40}} \cdot m_D = 0.762 \cdot m_D$$

At $x = 10.0$ m the relieving torsional moment is:

$$m = \frac{1.2 / 5.0}{54.57 \cdot \dfrac{4 \cdot 5.0 / 40.0^2}{3} + \dfrac{2.4}{40.0} + \dfrac{40.0^2 / (4 \cdot 5.0)}{3 \cdot 10.0 \cdot (40.0 - 10.0) \cdot 2.40}} \cdot m_D = 0.740 \cdot m_D$$

Thus, the effective torsional load of the beam may be considered practically as constant along the beam and equal to $m_{eff} = m_D - m \approx 0.25\, m_D$

The maximum twisting angle at the middle is:

$$\varphi_{max} = (m_{eff} / GI_T) \cdot L^2 / 8 = \frac{0.25 \cdot m_D}{\left(\dfrac{2.1}{2.4}\right) \cdot 10^8 \cdot 0.00136} \cdot \frac{40.0^2}{8} = 0.00042\,\text{rad}$$

instead of $(m_D/GI_T) \cdot L^2 / 8 = 0.0016$ rad if the torsional beam would carry the torsional load m_D alone.

REFERENCE

Stavridis L and Georgiadis K (2025) *Understanding and Designing Structures without a Computer: Plane structural systems*. Emerald Publishing, Leeds, UK.

emerald PUBLISHING   ice Publishing

Leonidas Stavridis and Konstantinos Georgiadis
ISBN 978-1-83662-945-0
https://doi.org/10.1108/978-1-83662-942-920251002
Emerald Publishing Limited: All rights reserved

Chapter 2
Slabs

2.1. Introduction

Slabs or *plates* are flat structures that extend in two dimensions, having a relatively small, thickness. Slabs are designed to undertake loads applied perpendicular to their plane, but they also possess a particularly high stiffness in receiving loads within their plane (working as diaphragms). In this chapter, only slabs receiving loads perpendicular to their plane are examined. Such slabs are made almost exclusively using concrete. The supports of slabs – where naturally the applied loads are transferred – can be placed (*a*) along the whole length or a part of the edges or (*b*) in pre-selected points (see Figure 2.1).

The load-bearing behaviour of slabs can be considered to primarily arise from the load-bearing behaviour of grillages, as examined in Chapter 1 (see Figure 2.2). However, the fact that slabs constitute a continuous medium adds particular structural characteristics that are important for understanding both the load-bearing mechanism and their design.

The concept of simulating a slab as a dense grillage made of beams suggests that its load-bearing behaviour is governed primarily by bending in two directions, along with the associated torsional effects. This simulation enables an understanding of the load-bearing behaviour of a slab, in analogous terms as for beams.

In examining the load-bearing mechanisms, it becomes evident that a distributed load $p(x,y)$ applied to a unit square element is balanced by four *resistance* forces. These are the forces developed through bending in directions x and y, denoted as $p_{b,x}$ and $p_{b,y}$, respectively, and the forces developed through torsion about the same axes, denoted as $p_{t,x}$ and $p_{t,y}$, respectively. Thus, it is:

$$p = p_{b,x} + p_{b,y} + p_{t,x} + p_{t,y} \qquad \text{(see Figure 2.3)}$$

To better understand this behaviour, the orthogonal slab shown in Figure 2.2 and the corresponding grillage with its beams placed at unit distances are considered. The deformation in the region of a considered point A shows that the deflections increase along the length of each direction, whereas the slopes decrease towards the centre of the slab, with a gradual increase in curvatures in both x- and y-directions. In the cut-out square element, the vector bending moments m_x, m_y applied to the corresponding sides are considered, with the moments on the opposite sides determined according to the law of differential increase, as $m_x + (\partial m_x/\partial x) \cdot 1$ and $m_y + (\partial m_y/\partial y) \cdot 1$, respectively (see Figure 2.3). From the overall picture of deformation, it is clear that all the above moments causing tension at the bottom fibres of the slab increase toward the centre. This means that in the x- and y-directions, the element is under the resultant moments $(\partial m_x/\partial x) \cdot 1$ and $(\partial m_y/\partial y) \cdot 1$, respectively (see Figure 2.3).

Considering first the x-direction, the moment $(\partial m_x/\partial x) \cdot 1$ requires – for equilibrium – the development of shear forces $q_{b,x} = (\partial m_x/\partial x) \cdot 1/1$ that decrease toward the centre. These forces offer

Figure 2.1 A slab as a continuous plane structure loaded perpendicular to its plane with differently supported edges

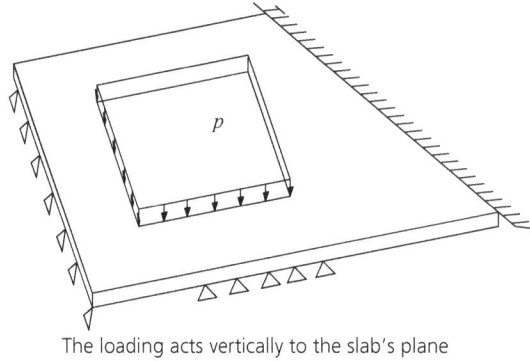

The loading acts vertically to the slab's plane

Figure 2.2 Deflected shape of an orthogonal slab under perpendicular loads

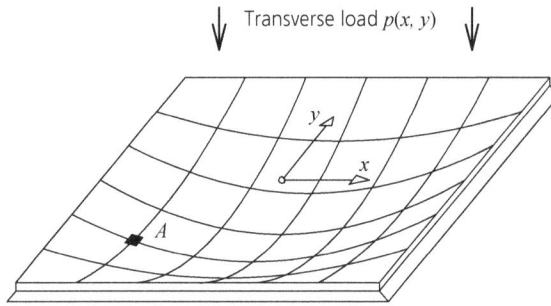

through their variation along x the corresponding *resistance force*, $p_{b,x}$, to the load $p(x,y)$ – that is, $p_{b,x} = -\partial^2 m_x/\partial x^2$ (see Stavridis and Georgiadis, 2025, Section 2.2.3)

On the other hand, if $w(x,y)$ expresses the deflection of the slab, then the change of slope $\partial w/\partial x$ at point A along y is $\partial^2 w/\partial x \partial y$. Given, however, that this slope change is of torsional nature along the y-direction, expressing the relative twisting angle of the unit length element, the development of a torsional moment m_{xy} along the y-direction of the slab is according to the relation:

$$m_{xy} = -\frac{\partial^2 w}{\partial x \partial y} \cdot GI_T \qquad \text{(see Section 1.1.4)}$$

Since the relative twisting angle $\partial^2 w/\partial x \partial y$ decreases along the y-axis, the torsional moment m_{xy} will decrease, so its change $(\partial m_{xy}/\partial y)$ requires for equilibrium the shear forces $(\partial m_{xy}/\partial y)/1$, which, decreasing along the x-direction, offer by their variation the corresponding resistance force, $p_{t,x}$, to the load, $p(x,y)$, with the same sign as $p_{b,x}$ and equal to $p_{t,x} = -\partial^2 m_{xy}/\partial x \partial y$.

In a similar manner, the consideration of the y-direction leads to the calculation of two other resistance forces (with the same sign as the previous ones) concerning the corresponding parallel sides of

Figure 2.3 Development of resistance forces due to the applied load perpendicular to the slab

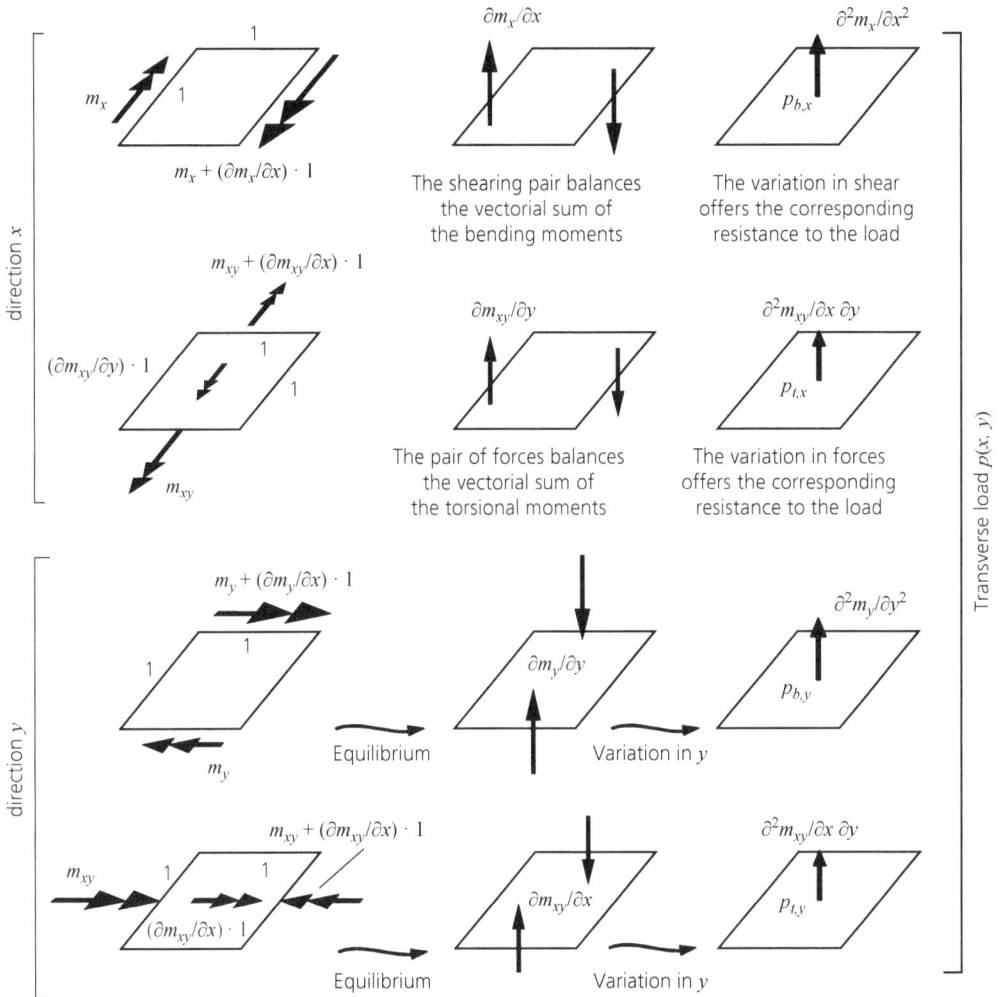

direction x

$\partial m_x/\partial x$

$\partial^2 m_x/\partial x^2$

m_x

$m_x + (\partial m_x/\partial x) \cdot 1$

The shearing pair balances the vectorial sum of the bending moments

$p_{b,x}$

The variation in shear offers the corresponding resistance to the load

$m_{xy} + (\partial m_{xy}/\partial x) \cdot 1$

$(\partial m_{xy}/\partial y) \cdot 1$

m_{xy}

$\partial m_{xy}/\partial y$

The pair of forces balances the vectorial sum of the torsional moments

$\partial^2 m_{xy}/\partial x\,\partial y$

$p_{t,x}$

The variation in forces offers the corresponding resistance to the load

direction y

$m_y + (\partial m_y/\partial x) \cdot 1$

m_y

Equilibrium

$\partial m_y/\partial y$

Variation in y

$\partial^2 m_y/\partial y^2$

$p_{b,y}$

$m_{xy} + (\partial m_{xy}/\partial x) \cdot 1$

m_{xy}

$(\partial m_{xy}/\partial x) \cdot 1$

Equilibrium

$\partial m_{xy}/\partial x$

Variation in y

$\partial^2 m_{xy}/\partial x\,\partial y$

$p_{t,y}$

Transverse load $p(x, y)$

the element, namely the *bending resistance force* $p_{b,y} = -\partial^2 m_y/\partial y^2$ and the *torsional resistance force* $p_{t,y} = -\partial^2 m_{xy}/\partial x\partial y$. The torsional moment, m_{xy}, is the same in both directions (Girkmann, 1963).

Thus, the above formulated equilibrium of load, p, with the four resistance forces can be written on the basis of their corresponding expressions (see Figure 2.3):

$$\frac{\partial^2 m_x}{\partial x^2} + 2\frac{\partial^2 m_{xy}}{\partial x \partial y} + \frac{\partial^2 m_y}{\partial y^2} = -p$$

In this way, the relieving role of twisting moments, m_{xy}, for the bending stress state of the slab becomes evident, either by offering the corresponding resistance forces, $p_{t,x}$ and $p_{t,y}$, or through the consideration of a bent strip of unit width – for example, along the direction x (or y) – receiving

23

Figure 2.4 Influence of the twisting moments on the bending stress state and deformation

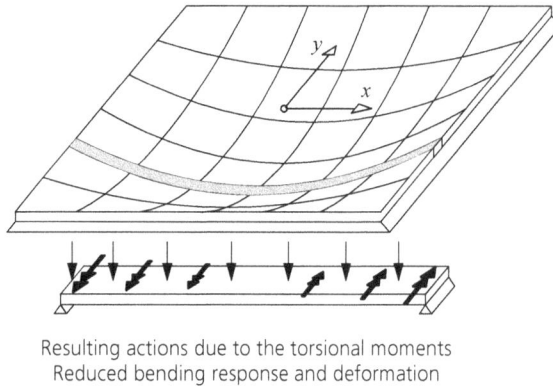

Resulting actions due to the torsional moments
Reduced bending response and deformation

along its length the action of the resultant of moments, m_{xy}, leading to a decrease of the corresponding bending (see Figure 2.4).

It must be emphasised that the moments m_x, m_y and m_{xy} do not represent bending moments (kNm) concerning a certain specific section as in the case of beams, but they refer to a unit length expressed in kNm/m – that is, in kN. The bending stress state is proportional to the corresponding curvature $(1/r)$ (see Stavridis and Georgiadis, 2025, Section 2.3.2) – that is:

$$m_x = -\frac{\partial^2 w}{\partial x^2} \cdot EI$$

and

$$m_y = -\frac{\partial^2 w}{\partial y^2} \cdot EI$$

while the expression for m_{xy} has been determined previously. In this respect, the moment of inertia is $I = d^3 \cdot 1/12$ and the torsional moment of inertia is $I_T = d^3 \cdot 1/6$ (see Section 1.1.4), where d is the depth of the slab and $G = E/2$. In slabs, only half the torsional inertia, I_T, is considered due to the fact that the shear stresses receiving the torsional moment, m_{xy}, are activated only through the thickness of the relevant section and not also through its width, as happens in a beam.

The above expressions for bending moments are not strictly correct, since they do not include the influence of Poisson's ratio related to the transverse deformation that generally accompanies an axial stress state. However, as concrete cracking reduces this effect, it can practically be ignored for preliminary design purposes.

The substitution of the expressions of moments m_x, m_y and m_{xy} into the above equilibrium equation leads to the classical equation for slabs with respect to deflection, w:

$$\frac{\partial^4 w}{\partial x^4} + 2\frac{\partial^4 w}{\partial x^2 \partial y^2} + \frac{\partial^4 w}{\partial y^4} = \frac{p}{K}$$

where K is the slab bending stiffness, practically equal to $Ed^3/12$.

This differential equation together with the effective support conditions, formulated also in terms of deflection, w, has in the past allowed the analytical determination of the stress state in slabs with orthogonal or circular boundaries. However, the increased computational power of computers in recent decades has allowed the implementation of numerical methods that were developed a long time ago. Hence, by using suitable structural analysis software, the reliable analysis of slabs is now possible, even with very complex geometry. Nevertheless, a physical interpretation of the load-bearing behaviour of slabs, as shown in this chapter, remains essential for their design. ∎

It should first be noted that the reactions along a straight boundary do not correspond solely to the shear force, q, associated with transverse bending, as in a grillage system. Instead, they also arise from the simultaneous development of twisting moments.

For example, the twisting moments, m_{xy}, that appear along the freely supported straight boundary of the orthogonal slab shown in Figure 2.5 can be represented from a physical point of view only through corresponding pairs of forces with a unit lever arm, being of course equal to m_{xy} (see Figure 2.5). The variation of these forces – for example, along the x-direction – involves the additional loading of this straight boundary of the slab by the distributed load $\partial m_{xy}/\partial x$ (upwards), with the value of this load increasing towards the ends of the boundary and resulting in a reaction equal to $(q + \partial m_{xy}/\partial x)$. Thus, it comes out that the sum of the reaction forces is higher than the total load applied to the slab. The explanation to this paradox is that, in the four corners of the slab, the above local 'fictitious' forces, m_{xy}, have the same direction and are added up in each corner to a resultant downward force, R, cancelling out the distributed load $\partial m_{xy}/\partial x$. Thus, the equilibrium between the external load, the distributed reactions $(q + \partial m_{xy}/\partial x)$ and the corner forces, R, is finally satisfied, resulting in $R = 2 \cdot m_{xy}$. Note that the force, R, is in the order of roughly 9% of the total uniform load applied to the slabs. To undertake this force, a suitable anchorage of each corner is required so that this downward force can be provided.

It must be pointed out that in the case in which the edge of the slab is fixed, the moments, m_{xy}, are not developed along it since the value of the corresponding zero slope is not altered in the other direction. Thus, the magnitude $\partial^2 w/\partial x \partial y$ is zero. Consequently, the reactions remain unaffected (see Figure 2.5). ∎

At this point it, should be mentioned that, similar to the loads in beams that are transferred to the supports through principal stresses (see Stavridis and Georgiadis, 2025, Section 4.1.1.6), the loads in slabs are also not taken up along the arbitrary directions x and y involving the development of bending and twisting moments m_x, m_y and m_{xy}, respectively, but through the so-called *principal moments*. These principal moments are the moments applied to each point of the slab in specific directions perpendicular to each other.

Thus, at each point of a slab there are only two orthogonal directions (principal directions), each one under the action of a moment vector parallel to the cut-out side, without any twisting action along it (see Figure 2.6). This fact has particular importance since these directions reveal the paths that are followed for an exclusively bending load-bearing behaviour. The two perpendicular directions at each point, as well as the principal moments m_1 and m_2 themselves, are determined through the values of m_x, m_y and m_{xy} that prevail in the point considered on the basis of the equilibrium of the two triangular cut-out elements (see Figure 2.6).

Figure 2.5 Development of reactions at the slab boundaries

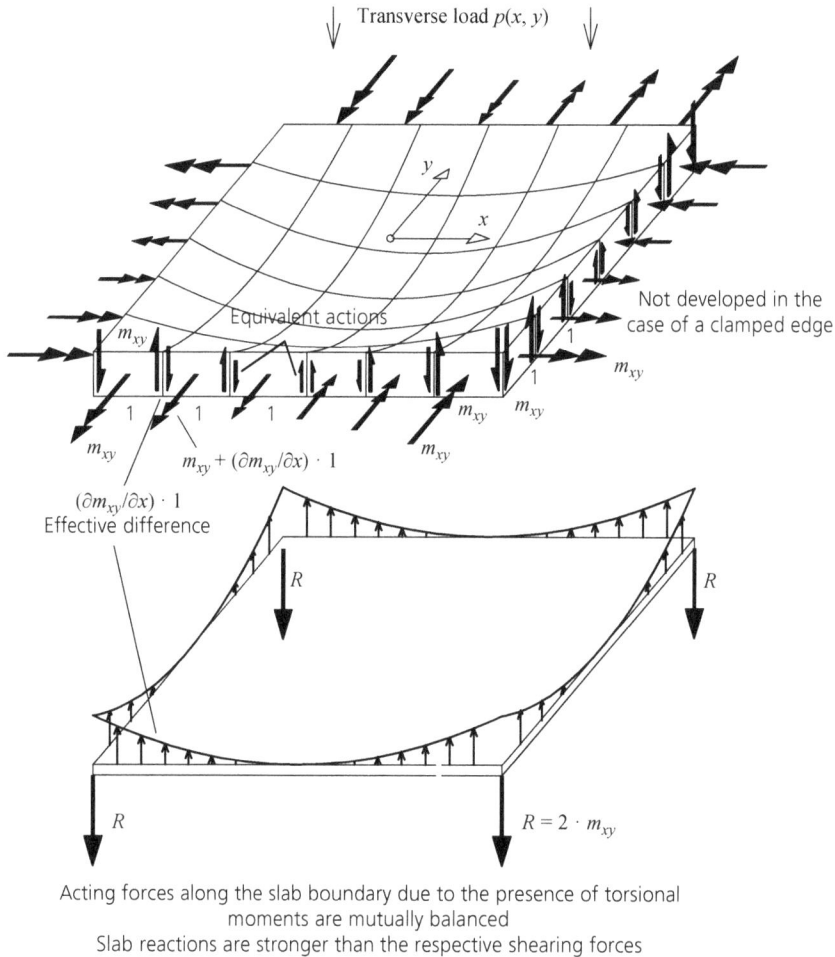

Acting forces along the slab boundary due to the presence of torsional
moments are mutually balanced
Slab reactions are stronger than the respective shearing forces

The fact that the developing cracks in a concrete slab correspond precisely to the principal moments considered confirms the above perception. Thus, it can be concluded that the essential characteristic of the structural behaviour of slabs results from their monolithic continuity, which, being obviously not valid for grillages, enables the development of the corresponding principal moments.

On the other hand, the fact that a slab transfers its loads only by bending in the directions of principal moments (see Figure 2.7) makes its simulation with a grillage model possible if the elements are aligned as closely as possible to the principal directions, transferring to each other only one vertical force – that is, only bending without any torsion. The utilisation of a grillage in an aesthetically very persuasive way can be seen in Figure 2.8. for a structure designed by the Italian architect and engineer Pier Luigi Nervi to cover a specific area using beams with negligible torsional stiffness (Nervi, 1956). ∎

Figure 2.6 The presence of principal moments

Principal moments

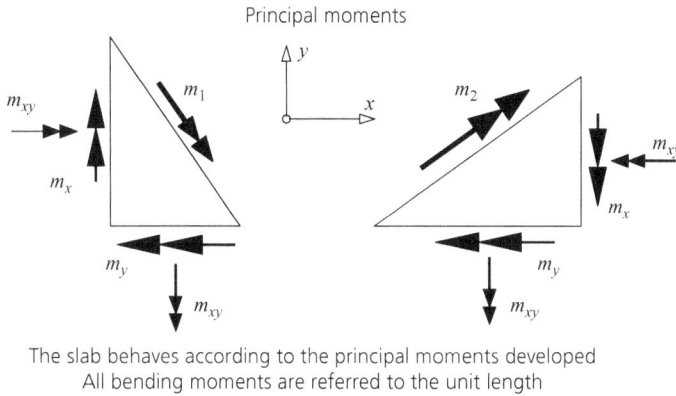

The slab behaves according to the principal moments developed
All bending moments are referred to the unit length

Figure 2.7 Trajectories of the principal moments under (a) free and (b) fixed support conditions

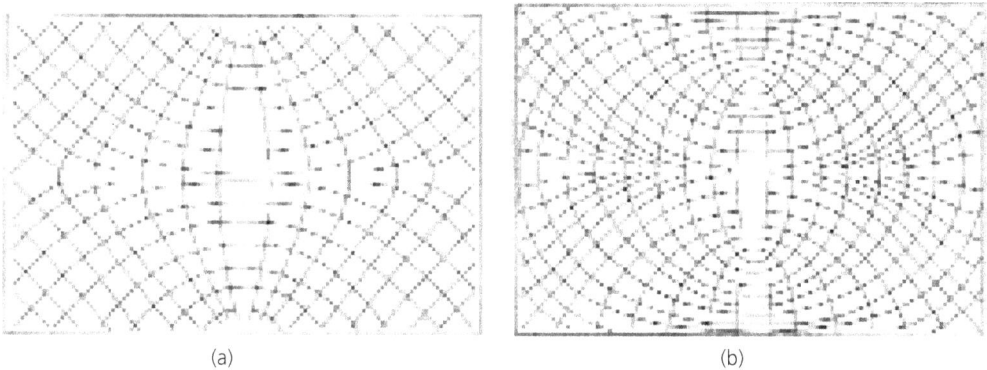

(a)   (b)

Considering the equilibrium of elements (see Figure 2.9), it becomes evident that the degree of divergence of the direction of principal bending moments from the directions x and y is measured by the magnitude of the twisting moment, m_{xy}. Thus, in the corner regions of a simply supported slab, where the greatest m_{xy} occurs and the bending moments m_x and m_y vanish, it is possible to have a clear picture of the principal moments m_1 and m_2 developing there (see Figure 2.9). Equilibrium of the two corner elements shown in Figure 2.9, with the equal twisting moments m_{xy} acting at the vertical sides, implies that at the 45° cuts, only the vector of principal bending moment m_1 and m_2 acts and, in particular, it is $m_1 = m_2 = m_{xy}$ (see Figure 2.9).

The sign of m_1 in the top left element in Figure 2.9 shows that the top face is in tension, with tensile stresses along the diagonal, and this is confirmed by the development of cracks in the oblique direction. Moreover, equilibrium of the top right element in Figure 2.9 shows that the vector of bending moment required for the equilibrium on the skew side is also of pure bending nature, representing the principal moment, m_2, causing cracks at the bottom face in the direction of vector m_2.

Figure 2.8 Configuration of a grillage following the trajectories of the principal moments

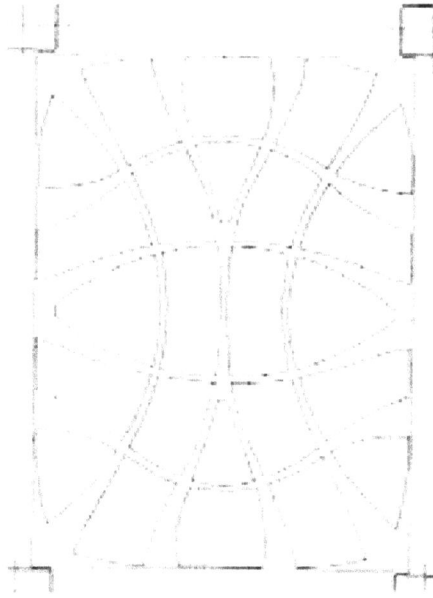

Figure 2.9 Equilibrium in the corner region of a slab

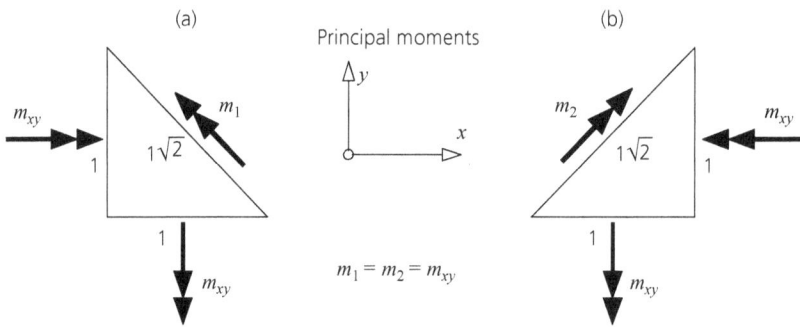

(a)

Principal moments

m_{xy}

m_1

$1\sqrt{2}$

1

1

m_{xy}

(b)

m_2

$1\sqrt{2}$

1

1

m_{xy}

m_{xy}

$m_1 = m_2 = m_{xy}$

All bending moments are referred to the unit length

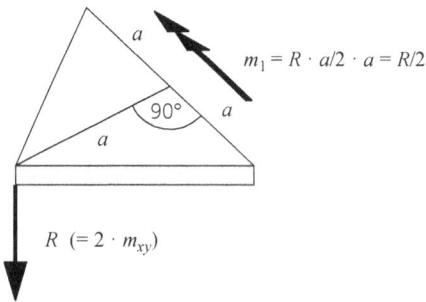

a

$90°$

a

a

$m_1 = R \cdot a/2 \cdot a = R/2$

$R \ (= 2 \cdot m_{xy})$

The principal moment, m_1, relates directly to the corner force, R, as this causes a moment ($R \cdot a$) at a distance, a, from the diagonal, applied over a length $2a$. Thus, it is $m_1 = (R \cdot a)/(2 \cdot a) = R/2$, a result that is justified by the above conclusions (see Figure 2.9).

2.2. Orthogonal slabs

Orthogonal slabs constitute the most common form of slabs in buildings as well as bridges. The presence of a concentrated load in an orthogonal slab causes curvature in both directions of the slab and, consequently, a proportional bending response. The paradox is that this bending response is practically independent from the dimensions of the slab (see Figure 2.10). The reason is that for a specific concentrated load, P, the total bending moment that is developed in each direction is proportional to the corresponding length but, if referred to the respective unit width of application according to Section 2.1, it comes out practically as a constant bending moment (see Figure 2.10).

2.2.1 Two-side supported slabs

2.2.1.1   Stress state

Two-side supported slabs are supported at two opposite sides with either simple or fixed supports, while the other two sides are either free or supported at such a large distance apart that the loads are practically transferred only through the shorter span distance, L.

These slabs, when loaded with a uniform load, p, act as beams and are substantially stressed only in the supporting direction, with maximum span moments $m = p \cdot L^2/8$ and $m = p \cdot L^2/24$ for the simply supported and fixed case, respectively.

However, under a concentrated load, P (which, of course, is not referred to a geometrical point, but is considered here as distributed on a small square area equal to $0.0025 \cdot L^2$), the fact that the locally developed deflection also causes curvature in the transverse direction implies development of moments in this direction too (see Figure 2.11).

Figure 2.10 Action of a concentrated load on a slab

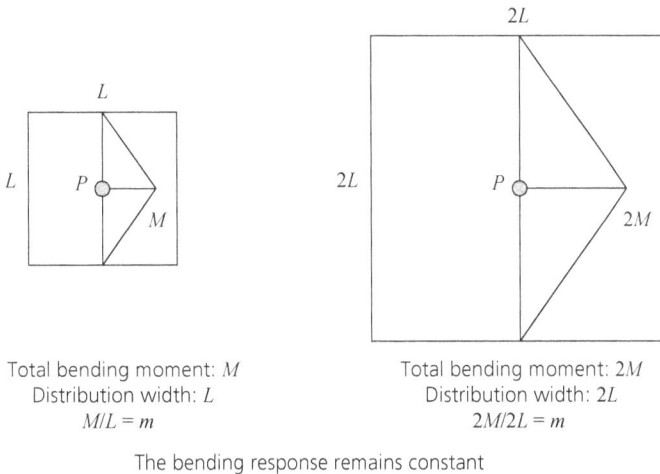

Total bending moment: M
Distribution width: L
$M/L = m$

Total bending moment: $2M$
Distribution width: $2L$
$2M/2L = m$

The bending response remains constant

Figure 2.11 Influence of a concentrated load on the stress state of a two-side supported slab

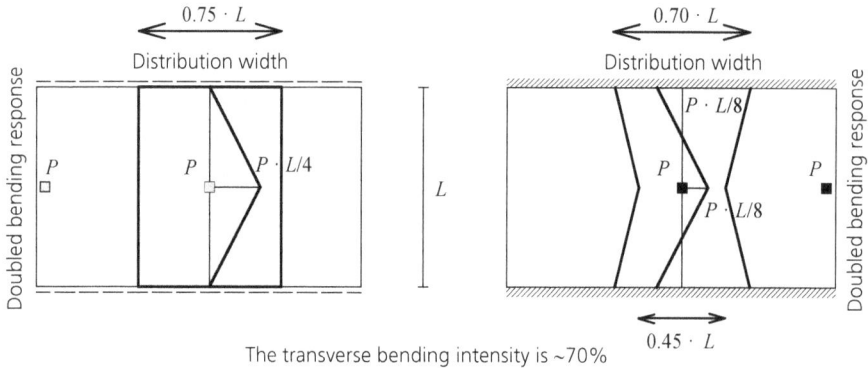

The transfer of the concentrated load to the supports can, in this case also, be considered similar to a beam through an effective zone of width, b_m, that is in the order of 75% of the span, L, for the simply supported slab, and 45% for the fixed supported one, assuming the load is applied at midspan. Thus, the statically developed total bending moment, M, which in the midspan is $P \cdot L/4$ for the simply supported slab and $P \cdot L/8$ for the fixed supported slab, is distributed over an effective width $0.75 \cdot L$ and $0.45 \cdot L$, respectively. Therefore, the bending moments that are developed can be estimated as $m = M/b_m = P/3$ and $m = M/b_m = P/3.60$, respectively. Special attention must be paid to the case in which a concentrated load is applied at the midpoint of the free edge. Then, the effective width is decreased by half and the bending moments are doubled accordingly (see Figure 2.11). It is pointed out that the transverse bending moments that are developed are about 70% of the longitudinal ones. Moreover, the total fixed support moment $M = P \cdot L/8$ in the fixed slab is distributed to the support over a greater width, b_m, in the order of 70% of the opening, L; therefore, the corresponding bending moment is estimated as $m = M/b_m = P/5.60$. Thus, it can be seen that, as mentioned initially, the bending moments due to concentrated loads are independent of the span length. This fact has a particular importance in bridge engineering, given the significant point loads applied from the wheels of vehicles.

Finally, it is pointed out that for a linear load, q (kN/m), extended over the span, L, of a simply supported slab (see Figure 2.12), the total developed bending moment $q \cdot L^2/8$ is distributed over a width $b_m = 1.35 \cdot L$, so that the bending moment, m, is roughly equal to $0.10 \cdot q \cdot L$, depending of course on the length, L.

2.2.1.2 Design

REINFORCED CONCRETE SLABS

Slabs are generally designed and reinforced mainly for bending rather than for shear. This is because the principal tensile stresses that are developed in the regions of high shear force – for the usual slab thicknesses – do not exceed the actual tensile strength of concrete. Regarding simply supported slabs, reinforcement is placed on the bottom region and is anchored at the ends. Thus, for an increasing load, with the developed cracking, the compact compression zone of concrete acts as an arch that takes up the loads without the need for stirrups, the shear action there being actually 'suppressed' (see Figure 2.13).

Figure 2.12 Influence of a linear load on the bending state of a two-side supported slab

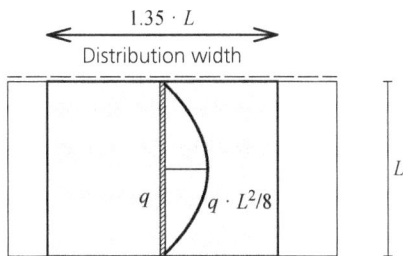

1.35 · L

Distribution width

q

$q \cdot L^2/8$

L

Figure 2.13 Action of a concrete slab in the cracked state

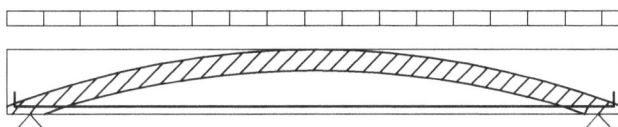

The design of bending reinforcement, a_s (cm^2/m), follows that examined in Stavridis and Georgiadis (2025, Sections 4.2.1.1 and 4.2.2.1), with the lever arm, z, of the internal forces considered now equal to $0.95 \cdot h$. Thus, for a design moment, m_d, it is $a_s = \gamma_R \cdot m_d/(f_y \cdot 0.95 \cdot h)$. ∎

The design of slabs must always include a check for deflections at serviceability. Regarding permanent loads, the concrete should not be considered as uncracked. Instead, a more detailed calculation, according to Stavridis and Georgiadis (2025, Section 4.2.1.1) should be performed. It must be noted that the influence of creep is always adverse. Creep increases the deformation, δ_L, due to the applied loads by $\varphi \cdot \delta_L$. Thus, the total deflection may be estimated as $\delta_L \cdot (1 + \varphi)$. This should be smaller than an acceptable ratio of the opening, L, usually $L/250$, while in the case in which brick walls are supported on the slab, this ratio must be smaller than $L/500$. These conditions are usually met for slab thicknesses of the order of $d \approx L/25$.

PRESTRESSED CONCRETE SLABS

Prestressing is mainly applied in slab bridges with spans over 15.0 m. The largest span that is economically acceptable for a slab bridge is roughly between 35.0 and 40.0 m. The maximum thickness of the slab should not exceed about 1.0 m.

For the design of prestressed concrete slab bridges, attention must be paid to the higher bending response compared with slabs for buildings, as well as to the deformation, developed at the free edges in comparison with the middle region of the slab, due to vehicle loading (see Section 2.2.1.1).

For an estimation of the required prestressing force, two basic approaches may be followed, namely full prestressing and partial prestressing (see Stavridis and Georgiadis, 2025, Section 4.3).

In full prestressing, the size of the prestressing force (i.e. the total section of cables) is determined, so that under permanent and live loads, the tensile stresses in the concrete are eliminated, or they are definitely smaller than the concrete resistance in tension (i.e. around 2 MPa). However, as explained in Stavridis and Georgiadis (2025, Section 4.3), this approach is uneconomical.

In partial prestressing, the prestressing force is limited to a value that allows only the permanent loads to be taken up, without developing tension. The live loads that are applied afterwards may lead to stresses higher than the tensile resistance of concrete and cause cracks (of limited crack width) that, however, disappear after their removal (see Stavridis and Georgiadis, 2025, Section 4.3.1.2). The approach of partial prestressing is clearly more economical.

Prestressing is usually distributed uniformly over the slab width – that is, the cables are placed and anchored at equal distances from each other (see Figure 2.14).

To fully undertake a permanent load, g (kN/m^2), over a span length, L, the required prestressing force, P_g, per metre results, according to Stavridis and Georgiadis (2025, Section 4.3.1.1), in:

$$P_g = \frac{g \cdot L^2}{8 \cdot (d/2 - c)}$$

where d is the slab thickness and c is the cable cover.

Under the permanent load, g, and the prestressing force, P_g, only a uniform compression across the slab thickness is developed without bending. As a consequence, this prestressing value usually offers, together with the necessary percentage of steel reinforcement (in the order of 1‰), a bending resistance, m_R, that is more than sufficient to satisfy the design criterion $\gamma \cdot (m_g + m_p) < m_R$, where $(\gamma \cdot m_g)$ and $(\gamma \cdot m_p)$ are the design moments for permanent and live loads, respectively.

However, by expressing the bending resistance in terms of P_0, defined as the applied prestressing force equal to 70% of the tendon's yield strength (in accordance with Stavridis and Georgiadis, 2025, Section 4.3.1.1), P_0 can be determined by satisfying the following equation:

$$m_R = (P_0/0.70) \cdot (d \cdot c \cdot 0.71 \cdot P_0/f_c) = \gamma \cdot (m_g + m_p)$$

where f_c is the concrete compressive strength.

Figure 2.14 Loads are taken up through distributed prestressing

Plan view

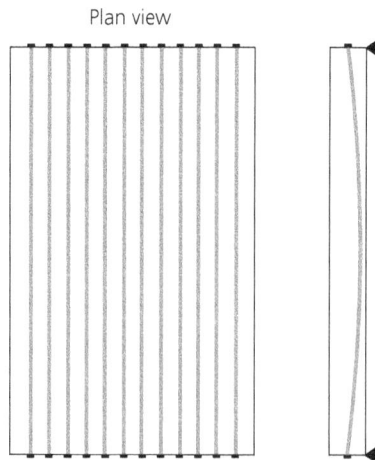

Additionally, of course, it should be ensured that under permanent load, g, no tensile stress are developed – that is, according to Stavridis and Georgiadis (2025, Section 4.3.1.1), the upwards shifted tendon force should fall within the borders of the central kern of the section, yielding the condition $m_g/(0.85 \cdot P_0) < = (2d/3 - c)$.

In the case of a bridge, it must be clear that the bending response state along the free edges due to traffic loads (m_p^r) is more critical than in the interior of the slab (m_p^m), as explained previously. Thus, the above checking of strength must be performed not only for the interior of the slab under the design bending moment $S_d^m = \gamma \cdot (m_g + m_p^m)$, but also for the zones at the edges with the moment $S_d^r = \gamma \cdot (m_g + m_p^r)$.

At this point, it must be pointed out that the force, P_g, is transferred as a compression force in the slab only if its support allows a movement along the force direction so that the corresponding shortening can take place. If this does not happen, then the slab cannot receive axial forces (since the prestressing force is absorbed by the immovable supports). Nevertheless, it is acted on by the full deviation forces and, thus, from a stress state point of view, the bending response, as previously considered, is still valid (see Stavridis and Georgiadis, 2025, Section 4.3.1.1). ∎

The saving in steel through partial prestressing may be even greater if the cables are concentrated in two bands at the outer edge zones of the slab over a width of the order of 20% of the total width, B, of the slab, as shown in Figure 2.15 (Menn, 1990; Stavridis, 2001).

In this case, due to prestressing the slab develops a bending moment, m_P, practically constant over the whole width of the slab, which is easily calculated on the basis of deviation forces as follows. If the width of each zone is $(\lambda \cdot B)$ and P is the required prestressing per unit width of the zone, then the applied deviation forces, u_P, on the zones are $u_P = 8 \cdot P \cdot f/L^2$ (kN/m²) where $f = (d/2 - c)$. Thus, the practically constant bending due to prestressing across the whole width of the slab is $m_P = (u_P \cdot L^2/8) [(2 \cdot \lambda \cdot B)/B] = P \cdot f \cdot 2\lambda$ (hogging). The above bending moment is composed from the so-called *statically determinate* part $m_{0P} = P \cdot f$ (hogging) that is limited only in the outer zones, affecting as already known only the bending resistance of these regions, as well as from the *statically redundant* part m_{SP} (see Stavridis and Georgiadis, 2025, Section 5.4.2). Hence, it is $m_P = m_{0P} + m_{SP}$ and, as $m_{0P} > m_P$, it can be written for the outer region as $m_{SP}^r = m_{0P}^r - m_P = P \cdot f \cdot (1 - 2\lambda)$ (sagging) and for the middle region as $m_{0P} = 0$, $m_{SP}^m = m_P$ (hogging).

Since this last part concerning m_{SP} is self-equilibrating ($m_{SP}^r \cdot 2 \cdot \lambda \cdot B = m_{SP}^m \cdot (B - 2 \cdot \lambda \cdot B)$), as explained in Stavridis and Georgiadis (2025, Section 5.4.2), it may be superposed with the bending moments due to the external loads in the strength check, according to the relations:

$$S_d^r + m_{SP}^r < m_R^r$$

and

$$S_d^m + m_{SP}^m < m_R^m$$

Figure 2.15 Taking up the load through prestressing of the edge zones

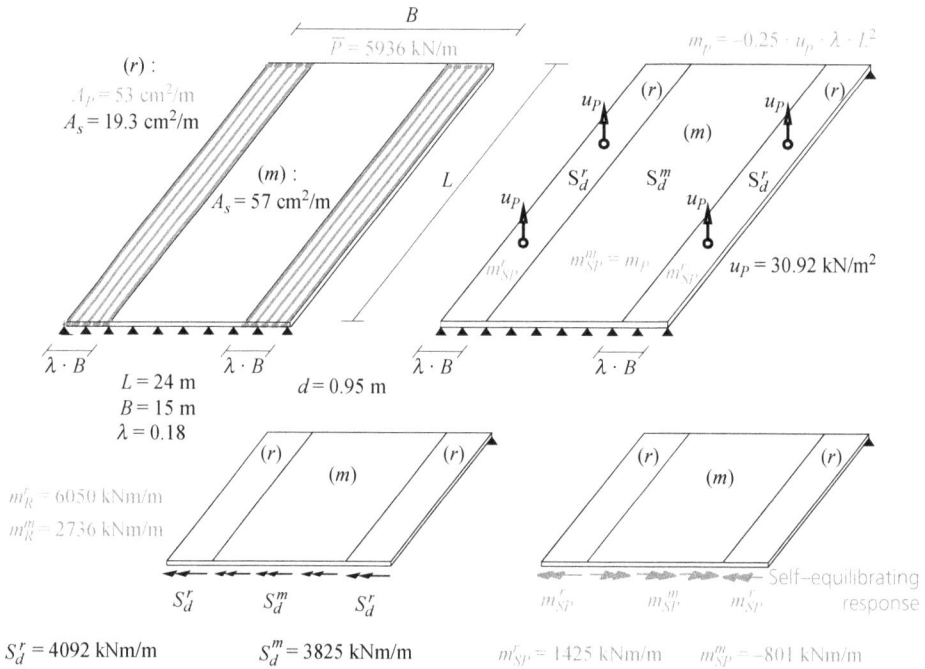

for the outer and interior regions, respectively, whereby the designed mild reinforcement should provide the appropriate value of the relevant bending resistance, m_R.

It may again be highlighted that the moment, m^r_{SP}, burdens the bending state, whereas the moment, m^m_{SP}, relieves it. It is also clear that the resistance moment, m^r_R, is formed by both the prestressing and the steel reinforcement, while the resistance moment, m^m_R, is formed only by the steel reinforcement, taking of course into account the favourable influence of the distributed compression force, n_p, due to prestressing, which is, in good approximation, equal to $n_p = 2 \cdot P \cdot \lambda$.

Figure 2.15 illustrates the numerical application of the above procedure for the design of an orthogonal prestressed concrete slab having a span of $L = 24.0$ m and a thickness of 1.0 m. Despite the obvious need for more steel reinforcement in the interior region than that required in the case of a uniform distribution of tendons over the whole width of slab, the layout examined leads generally to a more economical solution. Evidently, the most favourable combination of parameter, λ, and prestressing force, P, has to be found with trial and error.

2.2.2 Cantilever slabs

2.2.2.1 Stress state

The same introductory remarks made in Section 2.2 are also valid for cantilever slabs. A uniform load, p, causes bending that is proportional to the square of the cantilevered length, L – that is, $m = p \cdot L^2/2$. However, the concentrated load, P, applied at any point at a distance, L, from the fixed edge causes a total moment $M = P \cdot L$ at the fixed end that is distributed over a length roughly equal to $2 \cdot L$, meaning that the load is transferred to the support at an angle of about 90°. Thus, the bending moment is $m = 0.50 \cdot P$ – that is, independent of the length, L (see Figure 2.16).

It should be noted here that any stiffener or beam along the free edge somehow increases the distribution width (more favourable bending at the fixed end), while in the case of a cantilever with variable thickness, the distribution width is limited and the bending moment increases (see Figure 2.16).

The resulting stress state in the region of a cantilever corner is shown in Figure 2.17. For a uniform load, p, it is $m = 1.50 \cdot p \cdot L^2$, while for a concentrated load, P, on the external corner it is $m = 1.65 \cdot P$.

2.2.2.2 Design

Cantilever slabs are designed based on Section 2.2.1.2 for bending rather than shearing, with the corresponding reinforcement placed at the top side. As there is no reserve of bearing resistance in a cantilever (statically determinate structure), particular attention must be paid regarding its design, reinforcement and deformations, since even residential buildings constitute an area where higher live loads may be applied. Moreover, it is pointed out that in the case in which the cantilever is connected at its clamped end with relatively flexible slabs, this results in a compliance of its fixed end (elastically rotating support); a fact that increases the deformation of its free edge, in comparison with a completely fixed support (see Stavridis and Georgiadis, 2025, Section 2.3.7).

For the calculation of deflections, the same remarks as for the two-side supported slabs in Section 2.2.1.2 are valid, whereby particular attention must be paid to creep. For deformation due to permanent loads, the principle of virtual work may be applied, according to Stavridis and Georgiadis (2025, Section 4.2.1.1), by adopting the parabolic distribution of curvatures shown in Figure 2.18. In any case, as in the above section, the expression $\delta_L = (2 \cdot L)^2/(8 \cdot r) = L^2/(2 \cdot r)$ may be used as an acceptable estimate for the deformation δ_L.

The much stricter condition regarding the deformation at the free edge ($< L/250$) that should be respected can be met with a slab thickness $d > L/10$. It is finally noted that the deflection at the cantilever corner examined previously (see Figure 2.17) for a uniform load is twice that of the adjacent cantilevers.

Figure 2.16 Cantilever action under a concentrated load

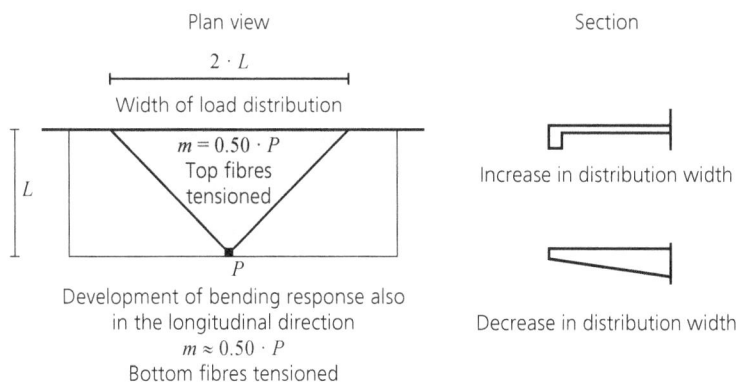

Figure 2.17 Bending state of a cantilever corner under a uniform load and a concentrated load

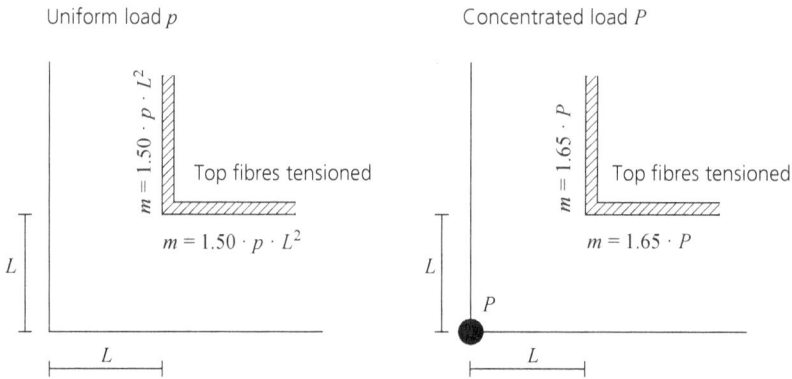

Figure 2.18 Cantilever deformation under a uniform load

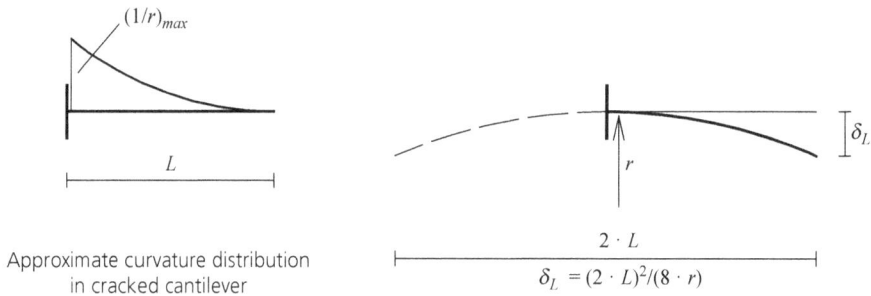

Approximate curvature distribution
in cracked cantilever

$\delta_L = (2 \cdot L)^2/(8 \cdot r)$

For a possible need for prestressing, the previous remarks in Section 2.2.1.2, as well as those in Stavridis and Georgiadis (2025, Section 4.6) are valid.

2.2.3 Four-side supported slabs

Four-side supported slabs constitute the most common type of slabs in buildings, supported perimetrically on beams or walls, and are usually ordered in orthogonal layout, being connected with each other in a continuous, monolithic way.

2.2.3.1 Stress state

Four-side supported slabs transfer their loads in both directions, differentiating again their load-bearing response depending on whether the load is uniformly distributed or concentrated.

In the case of a uniform load, the shorter direction receives the greatest bending response, according to that examined for grids (Chapter 1, see Figure 1.2). It should be clarified here, however, that, in respect to a uniform load, p, the concept of *stiffness*, k, along the x- or y-direction refers to an indicative strip with unit width with the correspondingly imparted total load ($p_x \cdot L_x$) or ($p_y \cdot L_y$) in each direction, respectively. This allows the determination of the deflection at its midpoint by dividing this total load by k, as discussed in Stavridis and Georgiadis (2025, Section 2.3.8). This stiffness is expressed as $k = \lambda \cdot (EI)/L^3$, where the factor λ is equal to 76.8 or 384, depending on whether the length, L, refers to a simply supported or to a fixed-end direction, respectively, and $I = d^3/12$. Thus, it may be considered that the load, p, is 'distributed' to the loads p_x and p_y (i.e. $p = p_x + p_y$) while the requirement of common deflection of the two mutually orthogonal strips induces that:

$$\frac{p_x \cdot L_x}{stiffness_x} = \frac{p_y \cdot L_y}{stiffness_y} \quad that\ is \quad \frac{p_x}{p_y} = \frac{k_x}{k_y} \cdot \left(\frac{L_y}{L_x} \right)$$

Thus, referring to uniform support conditions, in the case where $L_y = 2 \cdot L_x$, it is $p_x = 2^4 \cdot p_y = 16 \cdot p_y$.

As pointed out in Section 2.2.1.1, the bending stress state due to a concentrated load is independent of the dimensions of the slab and is proportional to the load.

Tables 2.1 and 2.2 referring to Figure 2.19 show the bending moments of (a) perimetrically simply supported slabs and (b) fixed slabs for both their midpoints and their fixed supports, respectively, under a uniform and a concentrated load, for three characteristic aspect ratios. The case of an extremely oblong ground plan has already been examined in the case of two-side supported slabs but is also included here for comparison.

Moreover, it may be confirmed that the twisting moment, m_{xy} (developed only at the free support), for the examined side ratios, is roughly equal to 4% of the total load of the slab.

With respect to a uniform load, it is observed that the moment, m_x, at the midpoint of a square slab is almost 30% of the corresponding two-side supported slab for the case of simple supports and 43% for the case of fixed supports. However, these percentages are increased if the beneficial influence of the twisting moments is ignored, which is equivalent to the omission of the corresponding term in the slab differential equation (Section 2.1). This influence in the case of ribbed slabs is deliberately not taken into account, as will be examined later. Thus, the corresponding table is presented below for the case of uniform load, where the twisting moments are ignored (Table 2.3).

Figure 2.19 Characteristic locations of bending moments in orthogonal slabs: (a) simple supports; (b) fixed supports

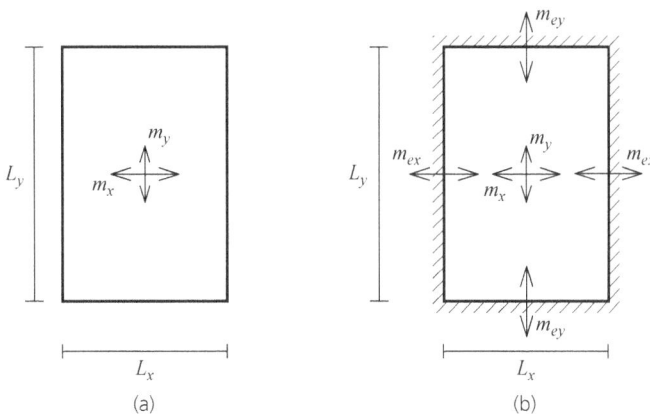

Table 2.1 Bending moments of orthogonal slabs under a uniform distributed load

L_y/L_x	Perimetrically simply supported slabs				Fixed slabs	
	$m_x/p \cdot L_x^2$	m_y/m_x	$m_x/p \cdot L_x^2$	m_y/m_x	$m_{ex}/p \cdot L_x^2$	$m_{ey}/p \cdot L_x^2$
1.0	0.037	1.0	0.018	1.0	−0.052	−0.052
1.50	0.073	0.38	0.034	0.30	−0.076	−0.057
∞	0.125	0	0.042	0	−0.083	−0.057

Table 2.2 Bending moments of orthogonal slabs under a concentrated load

L_y/L_x	Perimetrically simply supported slabs				Fixed slabs	
	m_x/P	m_y/m_x	m_x/P	m_y/m_x	m_{ex}/P	m_{ey}/P
1.0	0.274	1.0	0.233	1.0	−0.103	−0.103
1.50	0.32	0.83	0.257	0.88	−0.127	−0.034
1	0.327	0.76	0.26	0.85	−0.170	

Table 2.3 Bending moments, with the twisting moments ignored, under a uniform load

L_y/L_x	Perimetrically simply supported slabs				Fixed slabs	
	$m_x/p \cdot L_x^2$	m_y/m_x	$m_x/p \cdot L_x^2$	m_y/m_x	$m_{ex}/p \cdot L_x^2$	$m_{ey}/p \cdot L_x^2$
1.0	0.077	1.0	0.025	1.0	−0.056	−0.056
1.50	0.128	0.38	0.043	0.31	−0.085	−0.055
2.0	0.142	0.15	0.045	0.06	−0.088	−0.052

The developing bending moments are, of course, increased, but the ratio of moments in both directions remains essentially the same.

An estimate of the reactions on the sides of the perimeter can be practically obtained by evaluating the load acting on the regions created from the *bisectors* of the corners with supported sides and the mid-line of the two opposite larger sides (see Figure 2.20).

The bisectors in the case in which a fixed side is connected to a simply supported one are set to a 60° angle with the fixed side. ∎

In buildings, four-side supported slabs are usually set in a continuous orthogonal arrangement, being supported on beams or walls. Thus, it is considered that their support along the length of

Figure 2.20 Distribution of the reactions at the sides

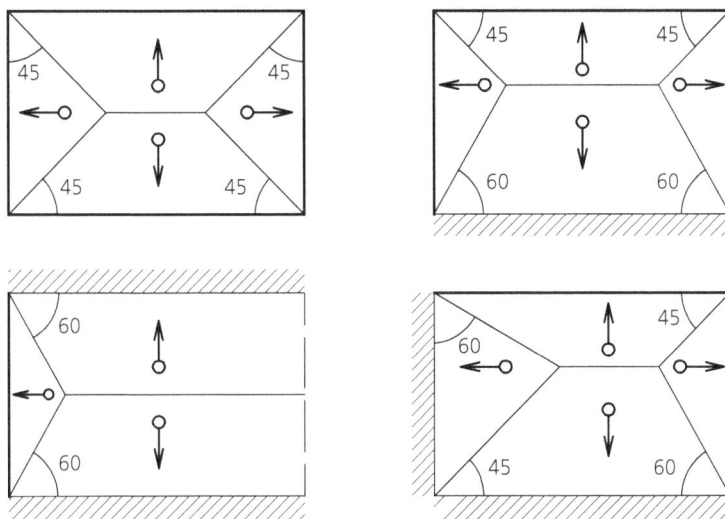

beams (or walls) is a simple one and transfers only a vertically distributed load, while the slabs themselves are mutually monolithically connected. Since slabs receive a permanent load, g, as well as a movable live load, p, the question of determining the most adverse bending moments arises both for the centre of the slabs and over their supports – that is, in the common contact lines (see Figure 2.21).

Figure 2.21 Evaluation of bending moments in continuous slabs

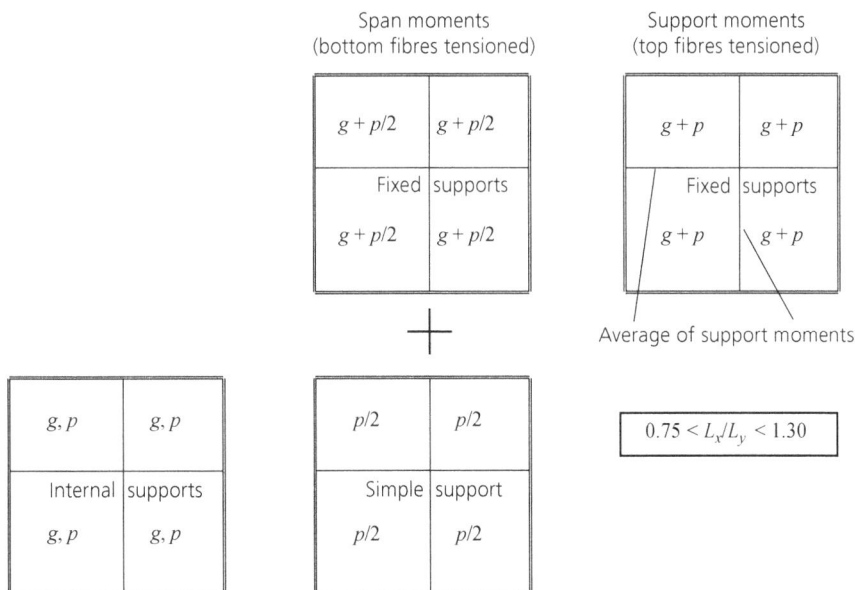

For this purpose, three distinct loading layout cases may be considered, namely: (1) loading with $(g + p/2)$, (2) loading with $p/2$ and (3) loading with $(g + p)$.

For each individual slab, the maximum bending moments in the centre can be estimated by super-imposing load case (1) with fixed conditions all over their perimeter and load case (2) assuming simple support conditions over the same support lines. The bending moments at each internal support (tension at the top side) can be estimated for load case (3) with fixed conditions all around and taking always the mean value from both sides (see Figure 2.21).

2.2.3.2   Design

The reinforcement is determined on the basis of the respective bending response. The fact that bending occurs in the centre of slab with respect to both directions implies that the reinforcement bars should be placed in two layers, one just over the other in the x- and y-direction, respectively, resulting obviously in a differentiated effective depth, h (or lever arm, z), for the two directions. As the bending moments develop only in the transverse direction along the length of the supports, they are obviously covered by one reinforcement layer only, normally placed on the top side. For an estimate of the reinforcement, the relation mentioned for two-side supported slabs (see Section 2.2.2.2) may be applied to each direction, respectively.

Regarding now the deformations, they can be determined through the considered distributed loads, p_x and p_y, on the basis of the previous discussion:

$$p_x = p\frac{L_x^{\,4}}{L_x^{\,4} + L_y^{\,4}} \qquad p_y = p\frac{L_y^{\,4}}{L_x^{\,4} + L_y^{\,4}}$$

The deformation, w, in the midpoint of a strip with unit width is:

$$w = \left(\frac{1}{\lambda}\right) \cdot p_{x,y} \cdot \frac{12 \cdot L_{x,y}^4}{E \cdot d^3} \quad (\lambda \text{ is determined according to Section 2.2.3.1})$$

However, the estimated w due to permanent loads must be increased by $w \cdot \varphi$ due to creep but the final result must not exceed $1/250$ of the respective span length. This condition is usually met for slab thicknesses $d > L/30$. ■

In the rather rare case in which the larger dimensions of a slab will require the application of prestressing, the permanent load, g, of the slab should be undertaken by the deviation forces of the cables.

The required deviation forces can be ensured by a uniform parallel arrangement of cables in one direction only. This direction is the shorter one, in order for the cables to offer the greatest deviation forces for the same prestressing force, given that $u = 8 \cdot P \cdot f/L^2$ (kN/m²) (see Figure 2.22). It should also be observed that, since the usually prevailing conditions do not allow the free movement of the prestressing front, uniform compressive stresses due to prestressing within the plane of the slab should rather be excluded, as explained previously (see Section 2.2.1.2).

2.2.4   Ribbed slabs

If in a slab only the compression zone of concrete is maintained and the remainder material is supplemented in the form of parallel ribs arranged in both directions, a statically more favourable

Figure 2.22 Layout for prestressing in a four-side supported slab

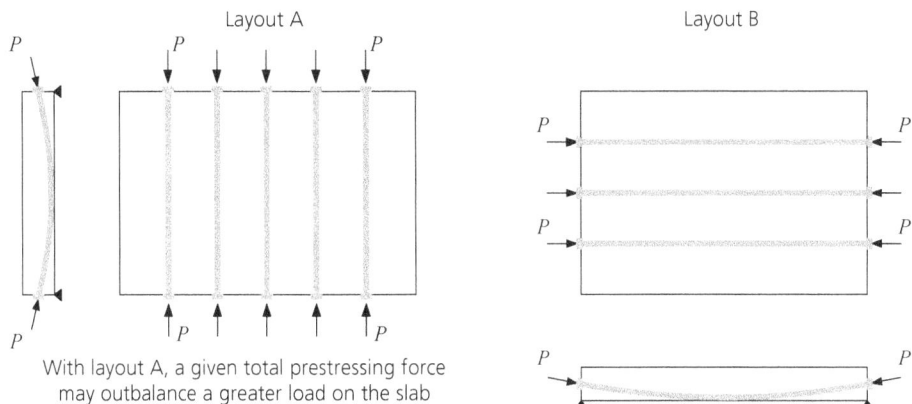

Layout A

Layout B

With layout A, a given total prestressing force may outbalance a greater load on the slab

system is obtained. This has, of course, a higher structural depth and hence smaller internal compressive and tensile forces, offering the possibility of even smaller thickness for the compressed slab and less quantity of tension reinforcement (see Figure 2.23). Moreover, the deformations of this structural system will obviously be smaller than the ones of a solid slab.

If the slab is two-side supported, the parallel ribs are ordered in a dense layout only in the load-bearing direction. However, up to three transverse ribs still have to be placed in the other direction (depending on the span) for reasons of wider distribution of eventual live loads, as explained in Chapter 1 for grids.

For an equal axial distance, e, between the ribs, under the self-weight, g, of a zone of unit length and width e, as well as a live load, p, the bending moment, M_r, that is developed on the considered rib is $M_r = (g + p \cdot e) \cdot L^2/8$ for a two-side supported slab. This moment is taken up, of course, by the beam-like rib with depth, d. The distance, e, should not exceed 70 cm (see Figure 2.23).

In a four-side supported slab, ribs are arranged, as already mentioned, along the x- and y-directions. Provided that the same distance, e (< 70 cm), exists in both directions, the slab may be considered to behave as having a constant thickness. One should only notice that, due to the relatively small thickness of ribs, which is limited practically to the degree of allowing the easy placement of reinforcement bars in only one layer, their torsional resistance is small, especially after cracking. For this reason, it is preferable, at least for preliminary design purposes, to ignore the beneficial influence of twisting moments and apply what has been previously examined in Section 2.2.3.1. It is, of course, clear that the bending moment, M_r, of the ribs results by multiplying the respective table values with the distance, e (see Figure 2.23). Moreover, the ribs corresponding to the longer direction will develop the least bending response, as indicated by the ratio m_y/m_x in Table 2.1. ∎

In a four-side supported slab, in order to achieve a preferable specific distribution of bending response, or even to facilitate construction, it is possible for the rigidities to vary in the two directions, thus creating an *orthotropic slab* (see Figure 2.24). This is usually the case if the ribs are ordered along one direction only, consisting either of concrete or of steel profiles, constituting thus

41

Figure 2.23 Use of a rib arrangement in an orthogonal slab

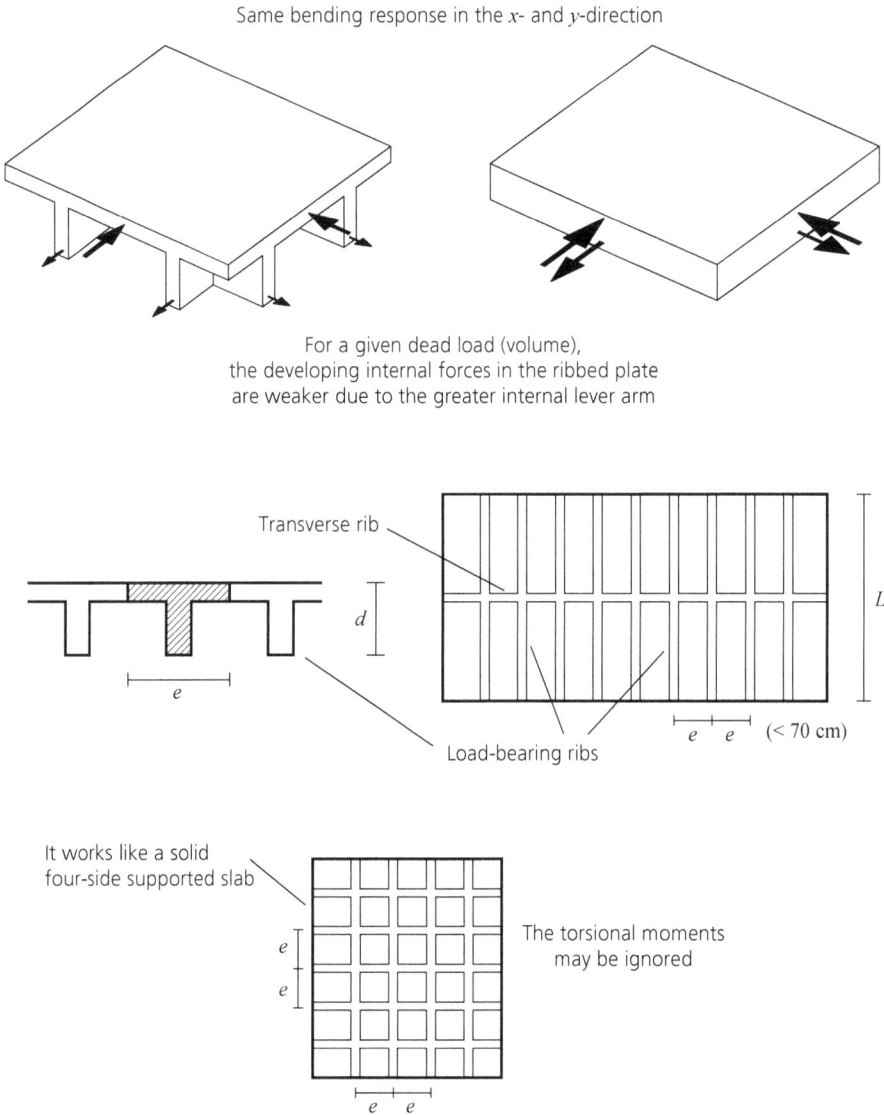

Same bending response in the x- and y-direction

For a given dead load (volume),
the developing internal forces in the ribbed plate
are weaker due to the greater internal lever arm

Transverse rib

d

e

Load-bearing ribs

L

e e ($<$ 70 cm)

It works like a solid
four-side supported slab

e

e

e e

The torsional moments
may be ignored

a composite structure (see Stavridis and Georgiadis, 2025, Section 4.5). Of course, the load distribution relation considered in Section 2.2.3.1 will be applied accordingly. For a total load, p, it is:

$$p_x = p \cdot \frac{(stiffness)_x}{(stiffness)_x + (stiffness)_y}$$

and

$$p_y = p \cdot \frac{(stiffness)_y}{(stiffness)_x + (stiffness)_y}$$

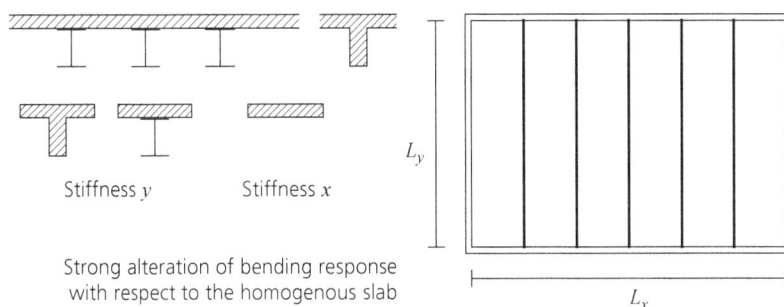

Figure 2.24 Stiffness differentiation in the two directions of a four-side supported slab

Table 2.4 gives some specific indicative results regarding four-side supported slabs, always ignoring the beneficial contribution of twisting moments (Stavridis, 1993). The slabs are considered perimetrically simply supported. Thus, it becomes clear how radically the distribution of bending moments can be influenced by the adopted stiffness ratio.

The decks of steel bridges may also be included in this category of so-called *orthotropic slabs*. These bridges are formed by two parallel main longitudinal beams with a distance between them equal to the width of the deck, connected with transverse beams at distances in the order of 5 m, thus forming a grillage of beams. The deck is formed by a steel slab with longitudinal stiffeners at distances in the order of 30 cm, thus creating a slab, the bending rigidity of which in the longitudinal direction is much higher than the one in the transverse direction (i.e. rigidities ratio in the order of 20). The ribs are usually selected having closed sections, so that with the existing torsional stiffness the bending response will be limited, as previously explained. Since the deck is designed to participate in receiving the intense compressive force of the top flange of the main beams, its contribution to avoid their lateral buckling may lead to an increase of its own bending rigidity (see Stavridis and Georgiadis, 2025, Section 7.6). ∎

In the case of oblong areas, a skew layout of orthogonal ribs may be applied, as examined in Chapter 1, Section 1.2.2 (see Figure 1.13), in order to achieve an equal distribution of forces in all ribs (see Figure 2.25).

In Table 2.5, the characteristic values of bending moment, M_r, of a rib with respect to the equal distances, e, are given for a 45° arrangement, both for (a) a simply supported slab and (b) a fixed

Table 2.4 Estimate of the bending moments in the two directions

| | stiffness$_y$ / stiffness$_x$ | | | | | |
| | 5 | | 10 | | 20 | |
L_y / L_x	$m_x / p \cdot L_x^2$	m_y / m_x	$m_x / p \cdot L_x^2$	m_y / m_x	$m_x / p \cdot L_x^2$	m_y / m_x
1.0	0.0225	5.76	0.0104	13.46	0.0038	37.7
1.50	0.0775	2.22	0.050	4.64	0.028	9.88
2.0	0.29	1.15	0.096	2.43	0.068	5.07

Figure 2.25 The use of a skew layout of ribs in oblong areas

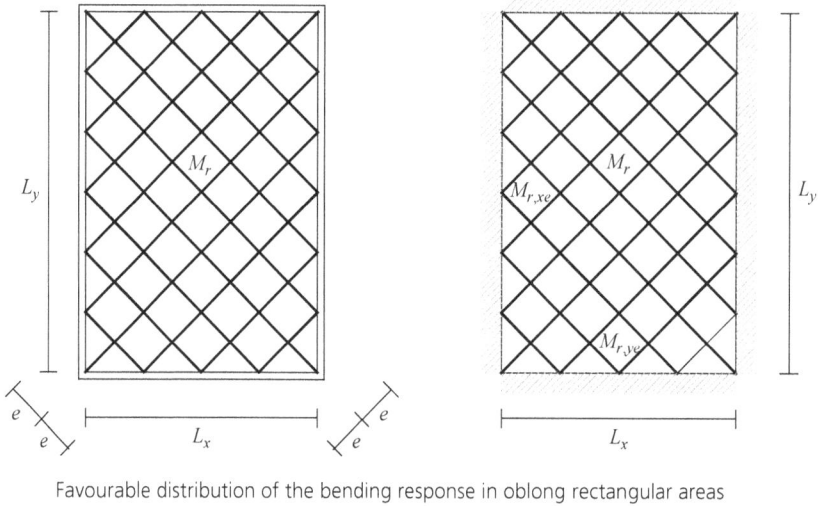

Favourable distribution of the bending response in oblong rectangular areas

Table 2.5 Estimated bending moments for a skew layout of the ribs

	Simply supported slab		Fixed slab	
L_y/L_x	$M_r/p \cdot e \cdot L_x^2$	$M_r/p \cdot e \cdot L_x^2$	$M_{r,xe}/p \cdot e \cdot L_x^2$	$M_{r,ye}/p \cdot e \cdot L_x^2$
1.0	0.035	0.021	−0.040	−0.040
1.50	0.052	0.029	−0.061	−0.046
2.00	0.069	0.034	−0.072	−0.044

slab (Stavridis, 1993). Again, for reasons already explained, the beneficial influence of the ribs' torsional rigidity has been ignored.

The quantity M_r/e corresponds to the moment, m, of the solid slab with constant thickness. Thus, it comes out, by comparison with the corresponding values in Table 2.1, that for the oblong areas, the bending moment of ribs in the skew layout constitutes only 40–50% of the respective moment, developed in the orthogonal arrangement.

2.3.   Circular slabs

The circular slabs used for covering circular areas are either simply supported or fixed at their perimeter. Under uniform load, they develop a symmetrical stress state with respect to their centre, which is basically expressed by the principal moments (m_r) in the radial direction.

As found for conditions of simple support, the highest span bending moment is developed at the centre of the slab and can be practically approximated by considering a fictitious square slab with sides 10% longer than the diameter (see Figure 2.26). For a slab with diameter, D, under a uniform load, p, it is $m_r = 0.047 \cdot p \cdot D^2$.

Figure 2.26 Assessment of the bending response in a circular slab

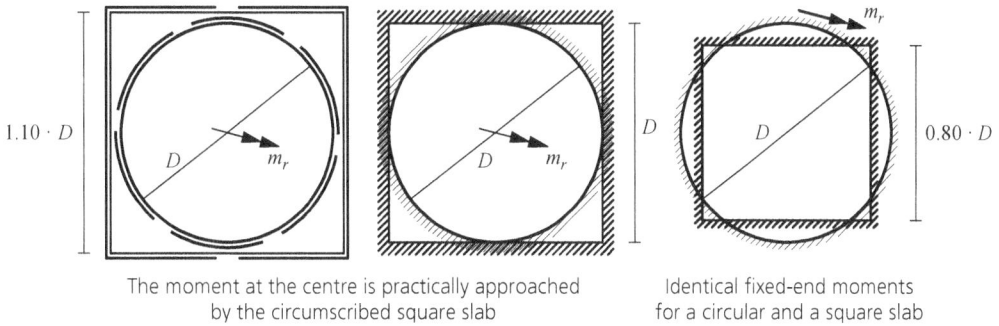

The moment at the centre is practically approached
by the circumscribed square slab

Identical fixed-end moments
for a circular and a square slab

For a fixed slab, the bending response at the centre of the slab is practically identical to that of a fixed square slab having sides equal to the diameter of the circle, since the cut-out corners of the square slab are essentially non-deformable, which is roughly valid also in the case of perimetrically simple support conditions (see Figure 2.26). The bending moment at the fixed support stems similarly to a fixed slab with sides equal to 80% of the diameter (see Figure 2.26). Thus, at the centre of the fixed slab it is $m_r = 0.018 \cdot p \cdot D^2$, while at the clamped support it is $m_r = -0.031 \cdot p \cdot D^2$.

2.4.   Skew slabs

Slabs with a parallelogram outline supported on two opposite sides are often used for covering skew transport crossings and are characterised as *skew slabs* (see Figure 2.27). The *skewness* becomes structurally more perceptible as the angle of the straight line that links the corners of the obtuse angles with the supported sides becomes greater (see Figure 2.27). The fact that the direction of this line is followed for the transferring of loads as being the shortest – that is, the stiffest – path to the supports leads to the conclusion that the directions of the principal moments are also skew with respect to the free edges of the slab, so that its bearing action is different in comparison with the orthogonal slab. It may practically be considered that the directions of the principal moments that cause tension at the bottom face of the slab are roughly found along the bisector of the obtuse angles. This fact results in a strongly unequal distribution of reactions as the skewness increases, with a particular 'accumulation' of forces at the obtuse corners, as well as a possible need for anchoring the regions at the acute corners due to the developing downwards reactions (uplift).

Moreover, an increasing skewness leads to a longer free edge with greater deformations, while, simultaneously, the conceivable continuance of the free edge zone to the immovable supports region at the obtuse corner creates there conditions of fixity. Thus, the principal moments that are perpendicular to the bisectors of the obtuse corners cause tension at the top face of the slab. The intense variation of bending moments along the free edge – that is, from the 'negative' moments at the supports to the 'positive' moments at the mid-length – signifies the development of intense shear forces ($Q = dM/ds$), thus re-confirming the 'accumulation' of reactions at the obtuse corners of the slab (see Figure 2.27).

A further characteristic of skew slabs is that the skew placement of principal moments directions with respect to the slab outline under a uniform load is practically not modified by the moments due to a concentrated load. This is of particular importance because it practically allows the direct

Figure 2.27 Load-bearing action of a skew slab

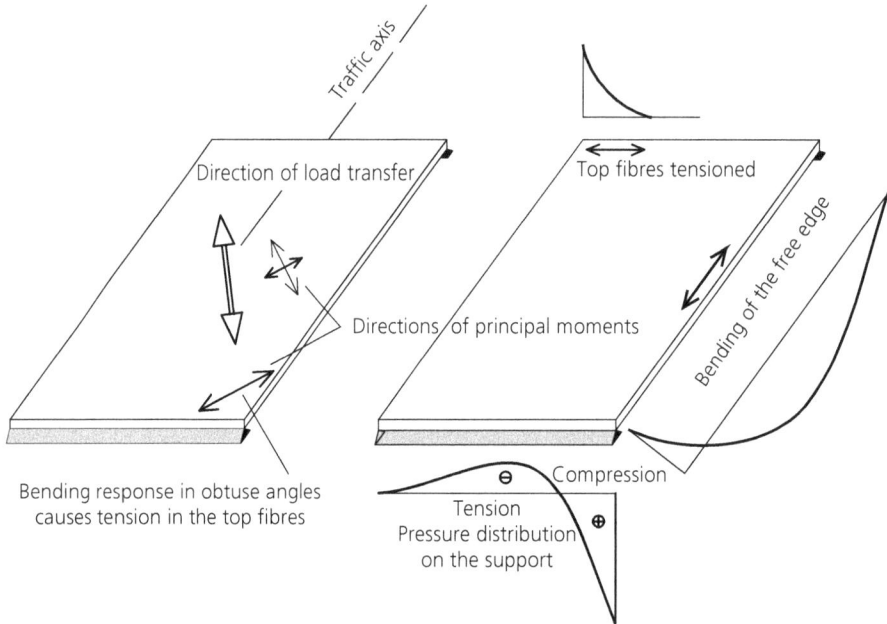

Traffic axis

Direction of load transfer

Top fibres tensioned

Directions of principal moments

Bending of the free edge

Bending response in obtuse angles
causes tension in the top fibres

Compression

Tension

Pressure distribution
on the support

superposition of the principal moments due to both uniform and concentrated loads. It should be noted, however, that for the final determination of bending moments in skew slabs, the employment of suitable computer software is required.

A direct consequence of the skew placement of principal moments with respect to the free edge is that, apart from bending according to the developed curvature, it is compelled to receive torsional moments too, as concluded from Figure 2.28, by following the equilibrium of the cut edge zone. The internal face of this zone in Figure 2.28 has been deliberately shaped as a 'saw' so that only the principal (bending) moments considered are applied to it. Since bending in the direction of the diagonal is stronger than in the transverse direction, it becomes clear that the resultant 'loading' moment vector at each point along the free edge of the slab always has the same sign and, consequently, creates a torsional loading that should be taken up accordingly. A lot of skew bridges that do not have a sufficient number of stirrups in their edge zones to undertake these torsional moments, according to Chapter 1, Section 1.1.5, will develop cracks due to the above torsional response. ∎

For spans above 15 m, the slab should be prestressed. Moreover, in this case as in orthogonal slabs, the sensitivity of the longitudinal edge zones regarding their response and deformations has to be taken into account, particularly for the applied vehicle loads. The slab thickness should again not exceed 1.1 m.

The cables are placed parallel to the free sides, usually uniformly distributed (see Figure 2.29). As examined for orthogonal slabs, the full uptake of the permanent load, g, by the prestressing force, P_g, per metre for a skew span equal to L according to the relation:

$$P_g = \frac{g \cdot L^2}{8 \cdot (d/2 - c)}$$

Figure 2.28 Torsional response in the free edge zone

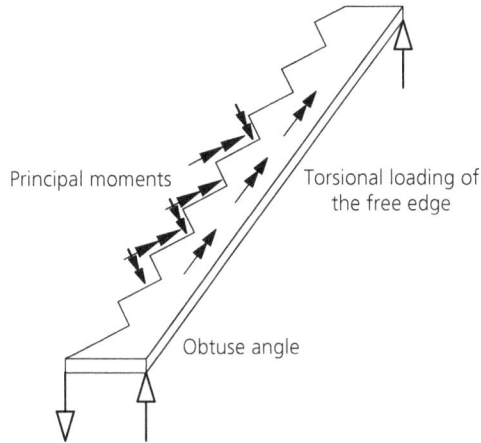

Principal moments

Torsional loading of the free edge

Obtuse angle

Figure 2.29 Uptake of loading by equally distributed prestressed cables

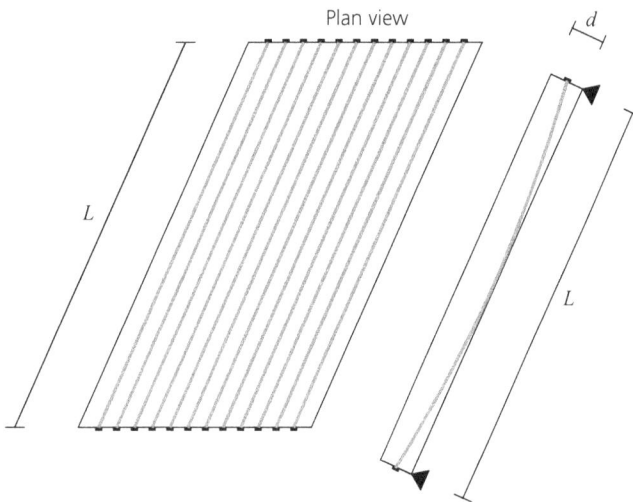

Plan view

d

L

L

together with the obligatory steel reinforcement, offers much greater bending resistance than required. For this reason, in order to follow a more economic design, only a percentage of the force, P_g, is applied, which, augmented with the skewness of the slab, may even be in the order of 50%.

It is pointed out that in a skew slab the full uptake of the load, g, by the cable deviation forces resulting from the prestressing force, P_g, ensures an equal distribution of the support reactions. The reason for this is that the deviation forces from the cables completely cancel out the load, g, of the slab, while the remaining equally distributed downward resultants of the prestressing force acting directly on the supports (being totally equal and opposite to the deviation forces) cause uniformly distributed reactions.

As in the case of orthogonal slabs, here there is also the possibility of arranging the prestressing cables in the edge zones in the parallel direction to the free sides (see Section 2.2.1.2). Thus, a stronger reduction of the required prestressing force can be achieved, and with analogously increased percentage of steel reinforcement, the corresponding resistance check can be satisfied (Stavridis, 2001). The process for determining the cross-sectional area of the cables (i.e. of the prestressing force, by assuming the cable is initially stressed with 70% of its yield stress) follows precisely the same course as for orthogonal slabs (see Section 2.2.1.2). In this way, a lower total cost may be achieved for the prestressed and steel reinforcement.

2.5.   Flat slabs supported on columns
2.5.1   Stress state
The load-bearing action of slabs in transferring their loads in at least two directions implies that, normally, at the ends of these directions, immovable continuous supports should exist in order to take up the loads. These continuous supports are usually offered by beams of sufficient stiffness that are supported on vertical elements (columns) transferring finally the loads to their bases (foundations). Alternatively, it may also be considered that the slabs transfer their loads to the horizontal girders of frames that are formed by the supporting beams and columns (see Figure 2.30).

Now, it can be intuitively seen that a slab can alternatively be supported directly on columns in a raster arrangement, without the use of beams, provided that the slab has the appropriate rigidity, thus permitting the utilisation of the full available height over all the covered area. As will be discussed next, such a slab, compared with that supported on beams, develops higher bending response and deformations, as well as a particular stress state, arising from the fact that the slab undergoes the risk of being punched by the columns. Of course, all these constitute disadvantages that can be overcome simply by increasing the thickness of the slab and by providing additional reinforcement, which leads to higher constructional cost, despite the simpler formwork needed. In any case, the cleaner aesthetic result offered by this layout has led to the relatively wide application of these slabs, particularly in cases with larger spans.

Figure 2.30 Transfer of slab loads through the beams to the columns

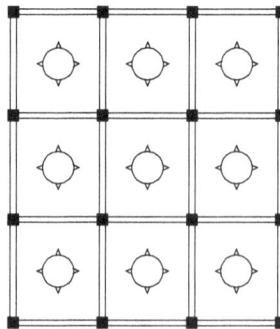

In the orthogonal raster of columns (see Figure 2.31), the natural tendency of the slab to transfer its loads along the stiffest – that is, the shortest – path to its supports clearly appears in the trajectories of the existing principal moments shown in Figure 2.32, where the regions between columns are formed in *quasi-beams* on which the remaining internal orthogonal panels are supported.

The load-bearing action of such a system may at first be considered, as shown in Figure 2.33, where each 'horizontal' row of columns constitutes, together with the imaginary strip of width, L_y, a one-storey frame with equal spans, L_x. (In the same way and quite independently, the action of the slab will be further taken into account in the y-direction too.) It is understood that the absence of twisting moments along the length of the 'free' edges of this strip, due to symmetry, allows it to be regarded as a free girder. Each span of this frame receives the total load $q \cdot L_x \cdot L_y$.

The bending response of such a strip is substantially depicted in the form of the moment diagram over the corresponding continuous beam with spans, L_x, and continuous uniform load $(q \cdot L_y)$ (see Figure 2.34), being characterised by the respective span moments, M_F (kNm), and support moments, M_S (kNm). These moments, obtained either from the frame or from the beam action, are globally referred to the whole width of the strip and are distributed between the included quasi-beam – that is, an appropriate column strip – and the remainder of the whole strip extending on both sides of the column axis. It is clear that the 'column strip', being more rigid than the remaining one, will receive the greater portion of the above moments (Franz and Schäfer, 1988).

According to German standard DIN 1045 (DIN, 2001), the width of the column strip may be considered equal to 40% of the width, L_y, symmetrically arranged with respect to the column's axes.

Thus, it is assumed that one half of the span moments, M_F (kNm), will be attributed to the above column strip, meaning a respective bending response of $m_F = 0.50 \cdot M_F/(0.40 \cdot L_y) = 1.25 \cdot (M_F/L_y)$ instead of the 'mean' bending moment (M_F/L_y) (kNm/m), whereas the other half of the span

Figure 2.31 Transfer of the slab load to the columns without the presence of beams

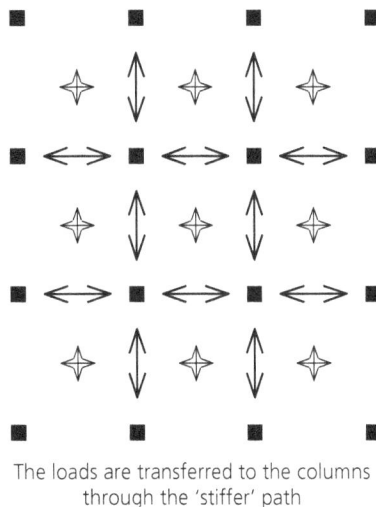

The loads are transferred to the columns
through the 'stiffer' path

Figure 2.32 Trajectories of the principal moments

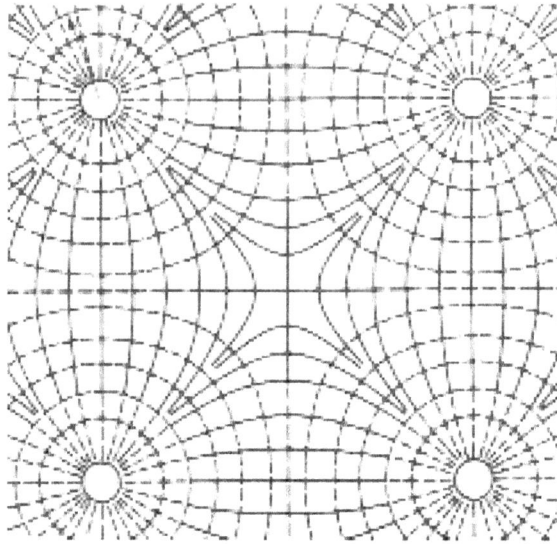

Figure 2.33 Establishment of the frame action between the slab and the columns

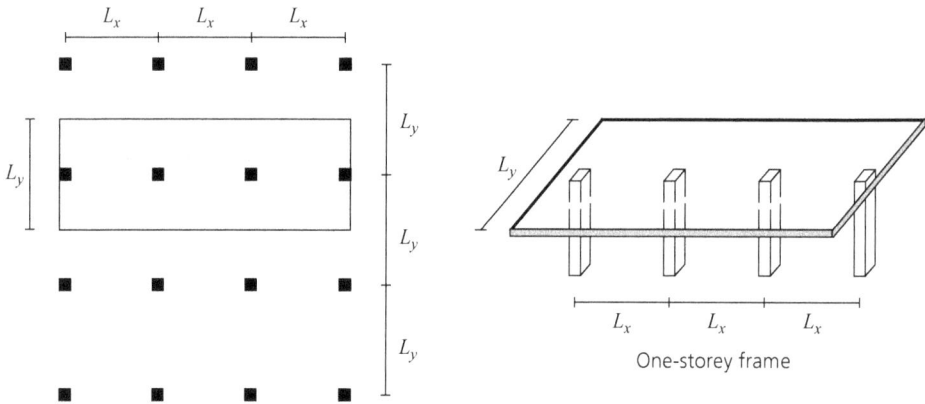

One-storey frame

Bending action of slab in the x–x direction: uptake of the total load
Respectively also in the y–y direction: uptake of the total load

moment, M_F, is attributed to the remaining width $(0.60 \cdot L_y)$ of the whole strip, meaning a corresponding bending response of $m = 0.50 \cdot M_F/(0.60 \cdot L_y) = 0.84 \cdot (M_F/L_y)$ (see Figure 2.34).

Regarding now on the other hand the support moments, M_S (kNm), the column strip considered above is stressed by 70% of this value, meaning a corresponding bending response equal to $m_S = 0.70 \cdot M_S/(0.40 \cdot L_y) = 1.75 \cdot (M_S/L_y)$ instead of the mean bending moment M_S/L_y (kNm/m),

Figure 2.34 Distribution of the bending moments of the strip

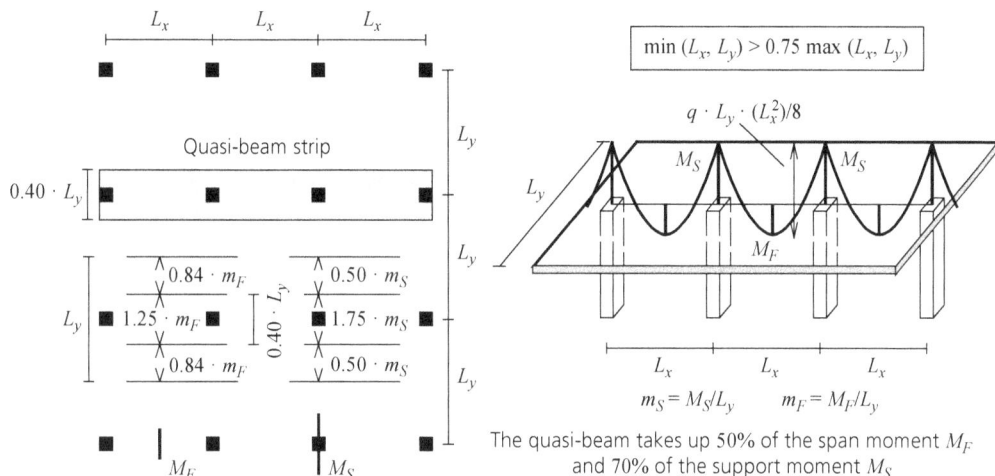

whereas the remaining part of the whole strip having a width of $0.60 \cdot L_y$ receives the remaining 30% of M_S, with a corresponding bending response equal to $m = 0.30 \cdot M_S/(0.60 \cdot L_y)$ (Figure 2.34).

Since, as mentioned above, the transfer of slab loads also happens in the transverse y-direction, the analogous frame with span lengths, L_y, and strip width, L_x, should also be considered (see Figure 2.35). This frame will again carry over each span the total load of the slab $q \cdot L_x \cdot L_y$, while the moment magnitudes, M_F and M_S, will be distributed in the same manner as in the other direction.

In this way, due to the absence of continuous supports, the entire slab load is necessarily taken into account in full, in both directions, in order to ensure equilibrium for each of the two directions. Thus, the resulting bending response of the slab is clearly less favourable than the one developed with a continuous beam support offered over the columns. It is, of course, clear that each column practically receives the total load ($q \cdot L_x \cdot L_y$) of the corresponding tributary area, according to Figure 2.35.

The above quantitative estimates concern the preliminary design and are only valid provided that the existing aspect ratio of each panel is not smaller than 0.75 (see Figure 2.34). It is obvious that for the complete determination of the state of stress of a flat slab, the use of appropriate computer software is required. ■

As stated at the beginning, a fundamental problem of flat slabs is the risk of punching. The column loads are applied to the slab practically as upward concentrated loads and, given their high value, they require particular attention for their safe introduction into the slab. The region around a column has higher shear forces than that with a continuous beam, as shown below in the example of a circular column (see Figure 2.36).

Figure 2.35 Establishment of the frame action in the other direction

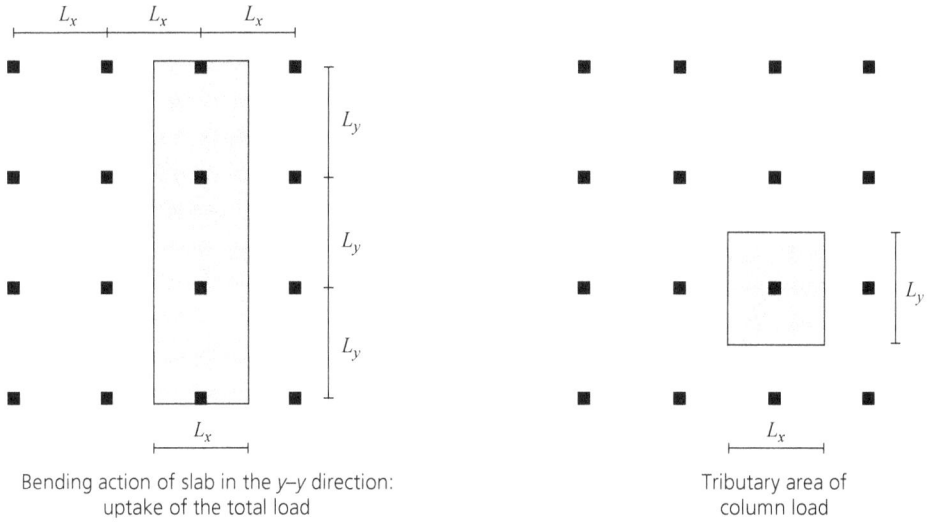

Bending action of slab in the y–y direction:
uptake of the total load

Tributary area of
column load

Figure 2.36 Variation in shear force in the region of a column

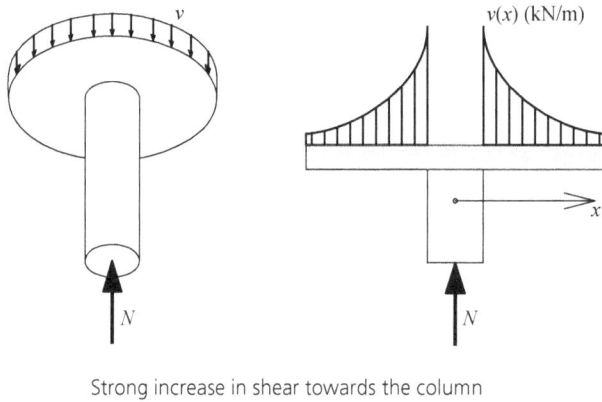

Strong increase in shear towards the column

In a flat slab over a column raster with span length, L, under a uniform load, q, the column load is $N = q \cdot L^2$ and the shear force, v (kN/m), of the slab at a radial distance, x, from the column centre is:

$$v = \frac{N - q \cdot \pi \cdot x^2}{2 \cdot \pi \cdot x} = q\left(\frac{L^2}{2\pi x} - \frac{x}{2}\right)$$

The presence of x in the denominator indicates the particularly intense increase of the shear force as the circular section shrinks to the outline of the column section.

2.5.2 Design for punching shear

The determination of reinforcement on the basis of the locally considered bending moments does not present any particularity in comparison with conventional slabs. However, as mentioned previously, designing for the bending moments only is not enough. The punching shear also needs to be taken into account, as examined below.

As explained previously, the development of large shear forces, and hence shear stresses in the vicinity of the column, induces skew tensile stresses that ultimately cause a perimetric detachment of the slab from the remaining column region over a surface in the form of a cone (see Figure 2.37). The failure does not occur at the perimeter of the column but at an angle roughly 30° from the column face. The reason is that in the direct region of the load introduction, the slab shear stresses are hazardless, due to the high vertical compressive stresses, while at some distance these relieving stresses practically vanish. The direct consequence of this fact is that a conical concrete mass is formed over the top of each column that transmits a certain part of the acting gravity load on the slab directly to the column through compressive stresses.

The punching check consists in assuring the equilibrium of the column together with the conical concrete mass at the top, which for additional safety is considered at an angle of 45°. Denoting by V_{eff} the total force transferred from the slab to the perimeter of the cone perimeter and expressing the vertical equilibrium together with the applied load on the column head, this can be written as:

$$V_{eff} + q \cdot (c + 2 \cdot h)^2 = N = q \cdot L^2$$

or, equivalently:

$$V_{eff} = N \cdot k \quad \text{where } k = \left[1 - \left(\frac{c + 2h}{L} \right)^2 \right] \qquad \text{(see Figure 2.37)}$$

where c is the dimension of the column and h is the slab thickness. The force, V_{eff}, is transferred to the column cone through the shear stress:

Figure 2.37 Region for checking effective shear against punching

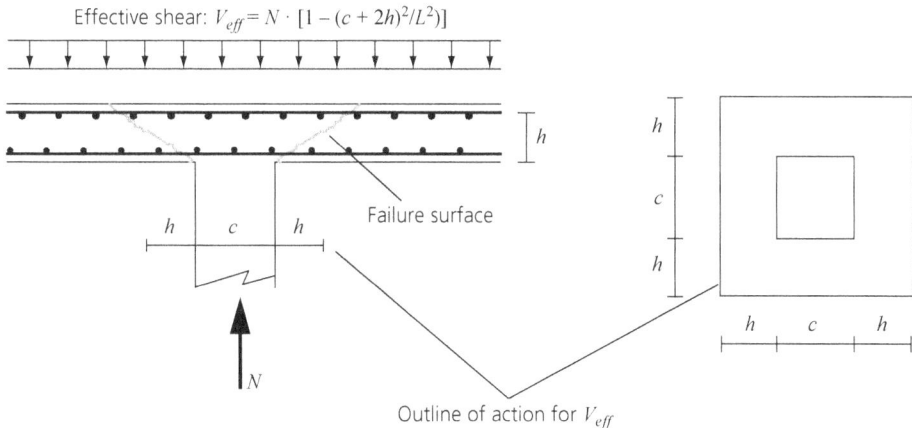

Effective shear: $V_{eff} = N \cdot [1 - (c + 2h)^2/L^2)]$

Failure surface

Outline of action for V_{eff}

$$\tau = \frac{V_{eff}}{4 \cdot h \cdot (c + 2h)}$$

The ultimate value, τ_u, of this stress is determined as a function of the ratio, μ, of the mean value, A_s (cm²/m), of reinforcement in the two top layers to the *effective* cross-section of the slab $100 \cdot h$ (cm²), the compression strength of concrete, f_c, and the yield stress of reinforcement, f_y, on the basis of the equation (Herzog, 1996):

$$\tau_u = f_c \frac{1.6 \cdot \mu \cdot (f_y / f_c)}{1 + 16 \cdot \mu \cdot (f_y / f_c)}$$

Thus, the maximum value ($max V_{eff}$) for the force, V_{eff}, that can be taken is $V_u = 4 \cdot \tau_u \cdot h \cdot (c + 2 \cdot h)$. For an assumed ultimate value $\gamma \cdot q$ of the service load, q, the above initial equilibrium condition can be written as $max V_{eff} = \gamma \cdot q \cdot L^2 \cdot k = \gamma \cdot N \cdot k$. If now $max V_{eff} < V_u$, then sufficient safety against punching is provided, whereas in the opposite case, vertical stirrups should be placed in the slab. The action of these stirrups, as shown in Figure 2.38, consists in offering a relieving *suspension* vertical force to the loaded slab around the column through the cracked conoidal surface under a 30° angle, being then by way of action–reaction 'safely' transferred as a compressive force through the remaining column head to the column and so contributing to its equilibrium as a free body. At the ultimate state, it may be considered with respect to column equilibrium:

$$V_u + \gamma \cdot q \cdot (c + 2 \cdot h)^2 + A_w \cdot f_y = \gamma \cdot q \cdot L^2$$

where A_w is the total cross-sectional area of the stirrups.

As now according to the above relations it is:

$$\gamma \cdot q \cdot L^2 - \gamma \cdot q \cdot (c + 2 \cdot h)^2 = \gamma \cdot V_{eff} = \gamma \cdot N \cdot k$$

the following expression of the required stirrup total cross-sectional area is:

$$A_w = \frac{\gamma \cdot N \cdot k - V_u}{f_y} \qquad \text{(see Figure 2.38)}$$

It should be pointed out that the punching failure is a brittle one, which happens without previous warning, and its repercussions are particularly dangerous (much worse for example than failure of an individual beam). For this reason, in the design of flat slabs, the exhaustion of the *calculative limits* should be avoided.

Worked example

A concrete flat slab is to be designed with raster dimensions 10.0 m by 10.0 m having a depth of 30 cm. The slab is supported on columns with dimensions 50 cm by 50 cm. The compressive strength of concrete is $f_c = 20$ MPa. The loads to be carried are shown below.

Slab self-weight: $0.30 \cdot 25.0 = 7.5$ kN/m²

Plaster ceiling: 1.0 kN/m²

Live load: 10 kN/m²

Total: 18.5 kN/m²

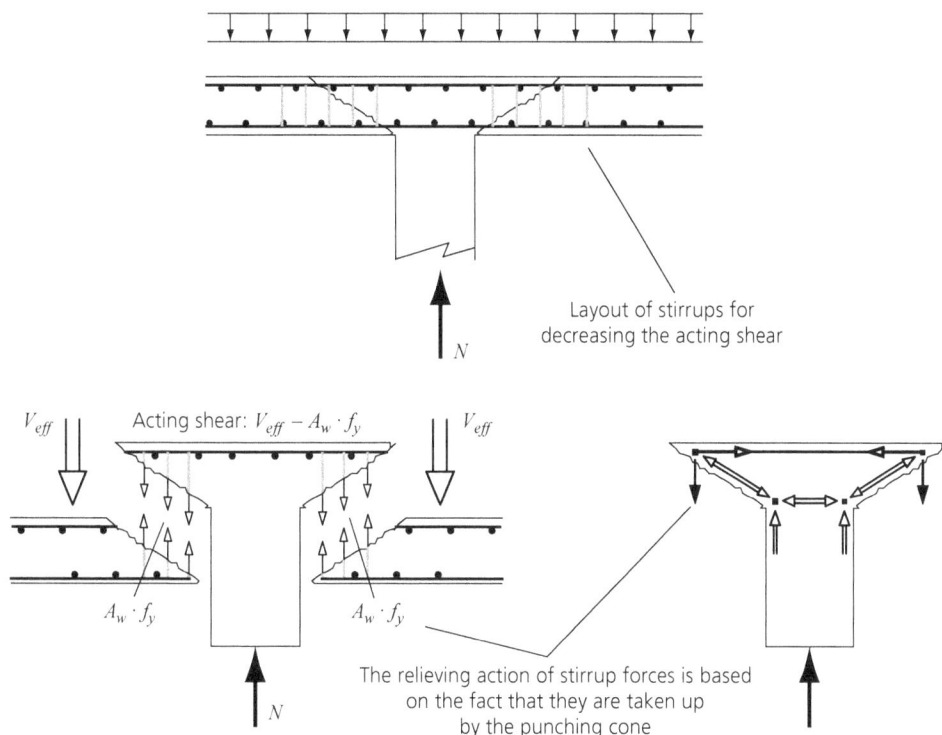

Figure 2.38 Contribution of stirrups to the uptake of the effective shear force

Layout of stirrups for
decreasing the acting shear

Acting shear: $V_{eff} - A_w \cdot f_y$

V_{eff} V_{eff}

$A_w \cdot f_y$ $A_w \cdot f_y$

N

The relieving action of stirrup forces is based
on the fact that they are taken up
by the punching cone

BENDING RESPONSE

Strip load: $18.5 \cdot 10.00 = 185.0$ kN/m

Span bending moment: $M_F = (185.0 \cdot 10.0^2/24) = 771$ kNm

Support bending moment: $M_S = (185.0 \cdot 10.0^2/12) = 1542$ kNm

SPAN REGION

Column strip (40%): $771 \cdot 0.50/(0.40 \cdot 10.0) = 96$ kNm/m

Remaining width (60%): $771 \cdot 0.50/(0.60 \cdot 10.0) = 64$ kNm/m

COLUMN REGION

Column strip (70%): $1542 \cdot 0.70/(0.40 \cdot 10.0) = 270$ kNm/m

Remaining width (30%): $1542 \cdot 0.30/(0.60 \cdot 10.0) = 87$ kNm/m

TOP REINFORCEMENT OVER THE SUPPORTS

x-direction: $A_s = 270/(0.90 \cdot 0.27 \cdot 200000) = 56$ cm^2/m

y-direction: $A_s = 56 \cdot 0.27/0.24 = 63$ cm^2/m

PUNCHING SHEAR

Axial force in the column: $N = 18.5 \cdot 10.0^2 = 1850$ kN

$$k = \left[1 - \left(\frac{c + 2h}{L}\right)^2\right] = 1 - \left((0.5 + 2 \cdot 0.30)/10.0\right)^2 = 0.99$$

$$V_{eff} = 1850 \cdot \left[1 - (0.50 + 2 \cdot 0.30)^2/10.0^2\right] = 1828 \text{ kN}$$

$$\tau = \frac{1828}{4 \cdot 0.30 \cdot (0.50 + 2 \cdot 0.30)} = 1385 \text{ kN/m}^2$$

ULTIMATE SHEAR STRESS CAPACITY

$$\tau_u = 20000 \cdot \frac{1.6 \cdot \dfrac{59.0}{100 \cdot 27} \cdot \dfrac{420000}{20000}}{1 + 16 \cdot \dfrac{59}{100 \cdot 27} \cdot \dfrac{420000}{2000}} = 1408 \text{ kN/m}^2$$

$$V_u = 4 \cdot 1408 \cdot 0.30 \cdot (0.5 + 2 \cdot 0.30) = 1858 \text{ kN} > V_{eff}$$

Because the values of V_{eff} and V_u are very close to each other, stirrups are required:

$$A_w = \frac{1.7 \cdot 18.5 \cdot 10.0^2 \cdot 0.99 - 1858}{420000} = 0.0030 \text{ m}^2 = 30 \text{ cm}^2$$

2.5.3 Use of prestressing

Prestressing is particularly suitable for dealing with the bending response and deformations of flat slabs, as well as for increasing their punching resistance. In order to counteract the bending influence of the load, q, on a flat slab over a column raster arrangement, the initial idea consists in taking it up through deviation forces $u_{x,F}$ and $u_{y,F}$ of prestressed cables in each panel (see Figure 2.39). It is reasonable to distribute this load equally in two directions – that is, $u_{x,F} = u_{y,F} = 0.5 \cdot q$. Therefore, considering the corresponding sag, f, of the cables in both directions as practically equal, the corresponding prestressing force per unit length is (see Stavridis and Georgiadis, 2025, Section 4.3.1.1):

$$P_{x,F} = \frac{0.5 \cdot q \cdot L_{x,F}^2}{8 \cdot f}$$

and

$$P_{y,F} = \frac{0.5 \cdot q \cdot L_{y,F}^2}{8 \cdot f}$$

The profile of each cable transversally crossing a number of column strips necessarily follows a path similar to that applied for a continuous beam (see Figure 2.39), which passes towards the ends of each panel through an inflection point to an opposite curvature over the column strips, in both x- and y-directions and continues in this way to the neighbouring panels (see Stavridis and Georgiadis, 2025, Section 5.4.1).

Figure 2.39 Uptake of loads through equally distributed prestressed cables

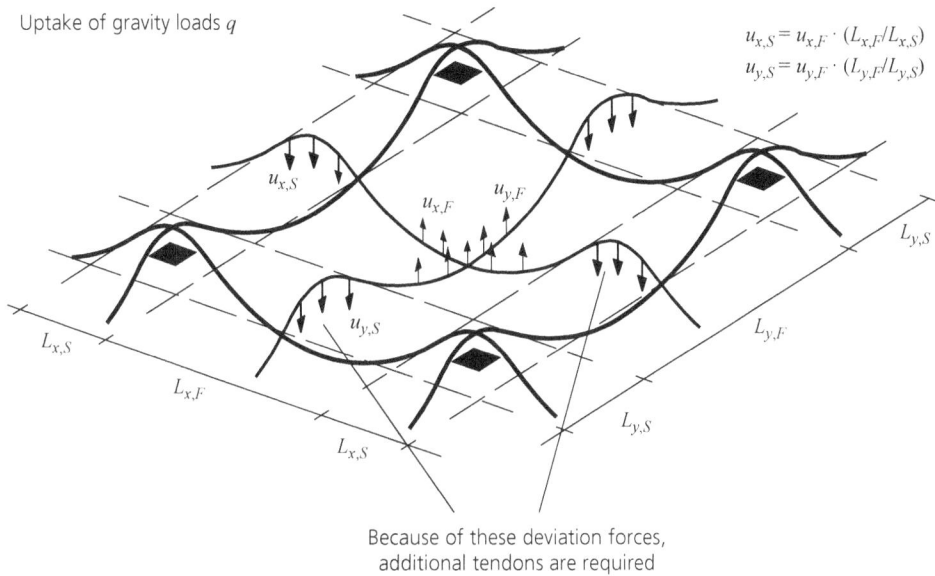

Uptake of gravity loads q

$$u_{x,S} = u_{x,F} \cdot (L_{x,F}/L_{x,S})$$
$$u_{y,S} = u_{y,F} \cdot (L_{y,F}/L_{y,S})$$

Because of these deviation forces,
additional tendons are required

Prestressing should take up the total load in each direction: uneconomic solution

As explained in Stavridis and Georgiadis (2025, Section 5.4.1), the deviation forces of each cable between the column axes constitute a self-equilibrating system and, thus, each column strip is under the downward applied deviation loads $u_{x,S} = u_{x,F} \cdot (L_{x,F}/L_{x,S})$ and $u_{y,S} = u_{y,F} \cdot (L_{y,F}/L_{y,S})$. These loads have to be taken up by prestressing, meaning that additional cables have to be placed in the support strip, as can be seen in see Figure 2.39. The distributed prestressing force of these cables per metre width in the corresponding column strip are:

$$P_{x,S} = \frac{u_{y,S} \cdot L_{x,F}^2}{8 \cdot f}$$

and

$$P_{y,S} = \frac{u_{x,S} \cdot L_{y,F}^2}{8 \cdot f}$$

Thus, the total prestressing force, for example in the x-direction, is:

$$P_{x,F} \cdot L_{y,F} + P_{x,S} \cdot L_{y,S} = \frac{(q \cdot L_{x,F} \cdot L_{y,F}) \cdot L_{x,}}{8 \cdot f}$$

clearly corresponding to the uptake of the total load, q, and not half of it, as was initially intended. Obviously, the same also happens in the y-direction and, consequently, the above cable arrangement does not represent an economic solution. ∎

A much more reasonable layout of tendons results from the placement of cables in the column strips only, making sure that the load, q, is fully taken up (see Figure 2.40).

Figure 2.40 Uptake of loads through prestressing of the column strips

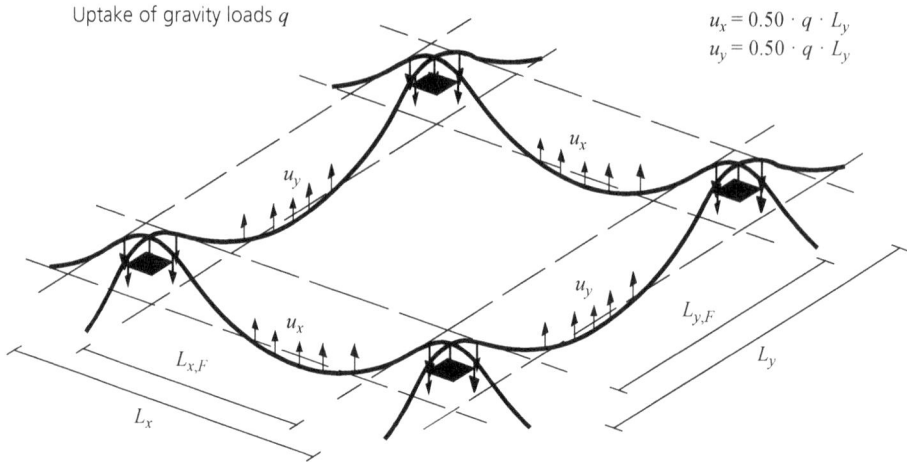

The total load is equally distributed to the deviation forces at the column strips in the x- and y-directions – that is, $u_x = 0.5 \cdot q \cdot L_y$ (kN/m) and $u_y = 0.5 \cdot q \cdot L_x$ (kN/m), respectively, leading to the total prestressing force:

$$P_x = \frac{u_x \cdot L_{x,F}^2}{8 \cdot f}$$

and

$$P_y = \frac{u_y \cdot L_{y,F}^2}{8 \cdot f}$$

respectively, for each strip.

Thus, the downwards acting deviation forces, being in equilibrium with those in the panels, load the columns directly and thus, on one hand, they do not create additional stresses that should be dealt with as happened in the previous case and, on the other, they relieve the punching stresses, as will be examined below. It is obvious that with this cable layout, the bending moments in the panels are met using steel reinforcement by considering them as four-side supported slabs.

Regarding the column strips, it is pointed out that the total bending moment, m_P, that is developed due to prestressing (i.e. due to the deviation forces) is split to a statically determinate part, m_0, and a statically redundant part, m_{SP} (see Section 2.2.1.2 and Stavridis and Georgiadis (2025, Section 5.4.2)). The bending moment, m_P, can be determined using an appropriate method for analysing a continuous beam subjected to relevant prestressing deviation forces, where $m_P = m_0 + m_{SP}$ and $m_0 = P \cdot e$. Both m_P and m_0 are hogging at the span and sagging over the column support. The m_{SP} is equal to $|m_P - m_0|$ and comes out always as sagging. It should be remembered (see Stavridis and Georgiadis, 2025, Section 5.4.2) that m_{SP} must be added to the 'loading side' of the strength relation, whereas m_0 is automatically taken into account in the determination of the bending resistance of the slab section, whereby the existing steel reinforcement must also be considered. ∎

Regarding punching shear, it is evident from the above layout that upward deviation forces reduce the effective shear demand. In contrast, downward deviation forces are directly transmitted to the columns as downward acting forces within the region of the punching cone (see Section 2.5.2) (see Figure 2.41).

The above upwards acting deviation forces are equivalent to the vertical components of the cable forces, appearing at their section points with the punching cone lateral surface, and are equal to $\sum P \cdot \sin \alpha$. The angle α is measured from the point where the cable intersects the failure surface and may be (unfavourably) considered equal to $45°$ (see Figure 2.41).

According to the previously formulated punching check in Section 2.5.2, considering the vertical equilibrium of the concrete punching cone at the ultimate state without the presence of stirrups, it must hold that:

$$V_u + \gamma \cdot q \cdot (c + 2 \cdot h)^2 + \sum P \cdot \sin \alpha = \gamma \cdot q \cdot L^2$$

and thus the final requirement is:

$$\sum P \cdot \sin \alpha > (\gamma \cdot N \cdot k - V_u)$$

which can easily be satisfied.

Worked example

A concrete flat slab is to be designed for prestressing with raster dimensions 10.0 m by 10.0 m having a depth of 30 cm. All data are the same as in the worked example in Section 2.5.2. The loads to be carried are shown below.

Slab self-weight: $0.30 \cdot 25.0 = 7.5$ kN/m^2

Plaster ceiling: 1.0 kN/m^2

Live load: 10 kN/m^2

Total: 18.5 kN/m^2

The strip prestressing is: $P_{x,F} = \dfrac{(0.5 \cdot 18.5 \cdot 10.0) \cdot (0.80 \cdot 10.0^2)}{8 \cdot 0.20} = 3700$ kN

Figure 2.41 Relieving action of prestressing cables against effective shear

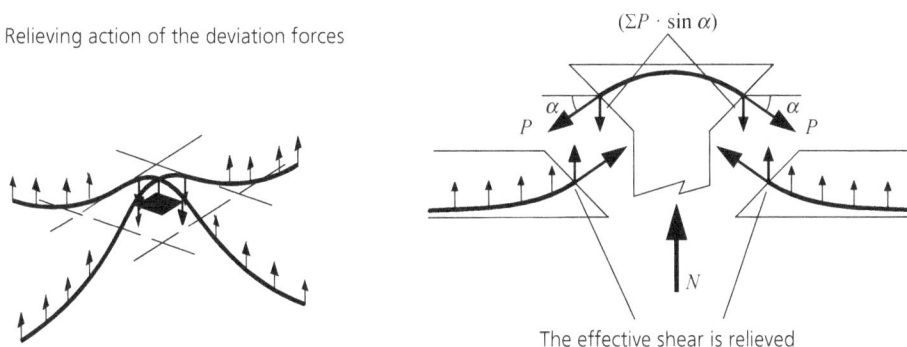

Relieving action of the deviation forces

$(\Sigma P \cdot \sin \alpha)$

The effective shear is relieved

PUNCHING SHEAR CHECK

Axial force in the column: $N = 18.5 \cdot 10.0^2 = 1850$ kN

$$k = 0.99$$

$$V_u = 4 \cdot 1408 \cdot 0.30 \cdot (0.5 + 2 \cdot 0.30) = 1858 \text{ kN} > V_{eff}$$

The inflection point is at $10.0/15 \approx 0.70$ m from the column axis

Sag of tendon over span: $f_F = 0.17$ (see Stavridis and Georgiadis, 2025, Section 5.4.1)

Tendon inclination at inflection point: $\alpha = 2 \cdot 0.17/(10.0/2 - 0.70) = 0.079$

Vertical component of prestressing force: $P_{x,F} \cdot \alpha = 3700 \cdot 0.079 = 292$ kN

Total relieving force: $\sum P \cdot \sin \alpha = 4 \cdot 292 = 268$ kN

Force to be taken up: $(\gamma \cdot N \cdot k - V_u) = 1.70 \cdot 1850 \cdot 10.99 - 1858 = 1255 \text{ kN} < \sum P \cdot \sin \alpha$

2.6.   Folded slabs

As pointed out at the beginning of this chapter, slabs are not only able to carry loads transversely to their plane through bending, but they can also offer a higher stiffness against loads applied within their plane. Exploitation of this attribute in covering large (mainly orthogonal) areas is possible by connecting oblong slabs along their long common sides, thus creating a folded form (see Figure 2.42). Therefore, the covering of larger spaces with a limited slab thickness can be achieved. However, this structural layout obviously needs much greater available height than a single compact slab.

A direct understanding of the behaviour of a folded slab is acquired if a paper sheet, supported on two opposite sides and deforming excessively because of its small thickness, is folded as shown in Figure 2.43. Its deformation then becomes negligible, even if a small additional load is added.

The load-bearing behaviour of a typical folded slab shown in Figure 2.42 is explained below. It is immediately clear that this structure is made up of Λ-shaped beams that, due to their high structural height and their much higher moment of inertia compared with a plane slab, constitute a considerably more rigid structure (see Figure 2.44).

The above remarks allow the conclusion that the folded edges may offer practically immovable supports to the oblong slabs that are formed by each plane surface. It becomes clear that the slabs considered act in bending as two-side supported slabs, as their length is much greater than their width. However, it should be pointed out that the longitudinal edges are 'immovable' only if frontal vertical diaphragms or appropriately formed frames with significant bending rigidity exist that are firmly connected to the ends of the slabs. The structural action of a folded slab is examined below and is illustrated in Figure 2.45.

If imaginary supports are considered along each fold, offering the same vertical upward reactions as those of a respective continuous beam representing the individual two-side supported slabs, it is obvious that the bending response of the 'altered' folded system will be identical to that of the continuous one and, thus, it may be directly determined.

Figure 2.42 Layout of a folded slab

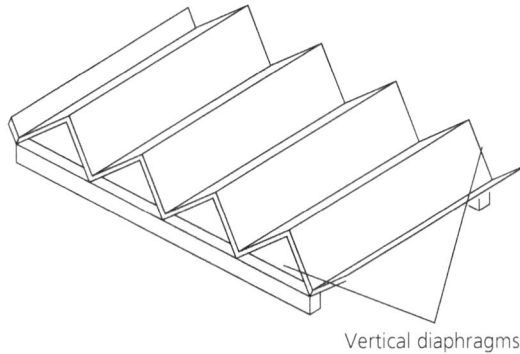

Vertical diaphragms

Figure 2.43 Drastic improvement in the load-bearing capacity of a sheet of paper due to folding

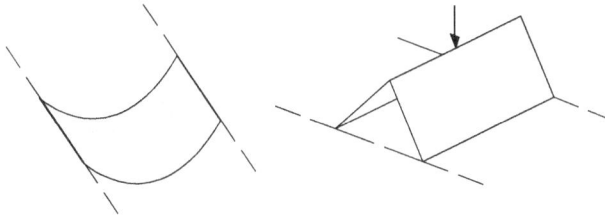

Of course, the fact that the edges of a folded slab are actually free from external forces implies that the above 'reaction' forces on each edge have to be applied again, in a second phase, as loads in the opposite direction, so that this state of stress may be superposed to the previous one (see Stavridis and Georgiadis, 2025, Section 3.2.1). In this second phase, each edge load may be analysed along the direction of the adjacent slabs and thus each slab is loaded within its plane by the resultant of the corresponding loads along its two longitudinal sides. These loads are taken up by the slab acting now as a deep beam, which transfers them to the aforementioned frontal walls through the corresponding shear forces of its end sections. It is clear that the above treatment of the downward

Figure 2.44 Drastic increase in the active moment of inertia due to folding

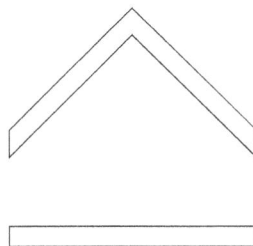

A much higher stiffness is achieved
by folding, while maintaining the same thickness

Figure 2.45 Load-bearing action of a folded slab

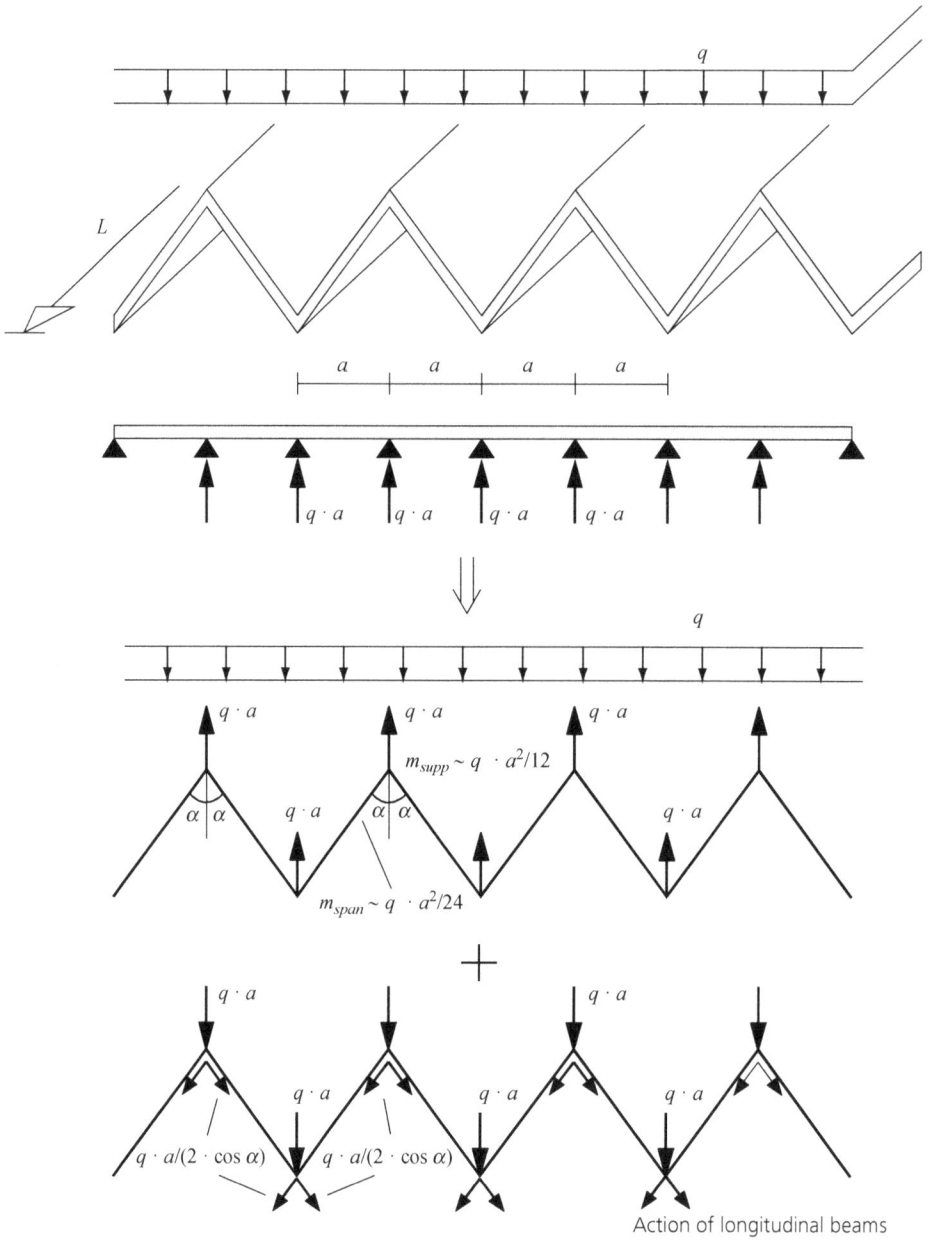

Action of longitudinal beams

edge loads refers to the taking up of these loads by the $\varLambda$-shaped folded beams constituting the whole folded slab.

Hence, a folded slab acts in a combined way – that is, as a combination of a system of rigid deep beams in longitudinal bending together with the formed oblong slabs continuously supported in transverse bending (see Figure 2.46).

Figure 2.46 Load-bearing action of a slab as a deep beam

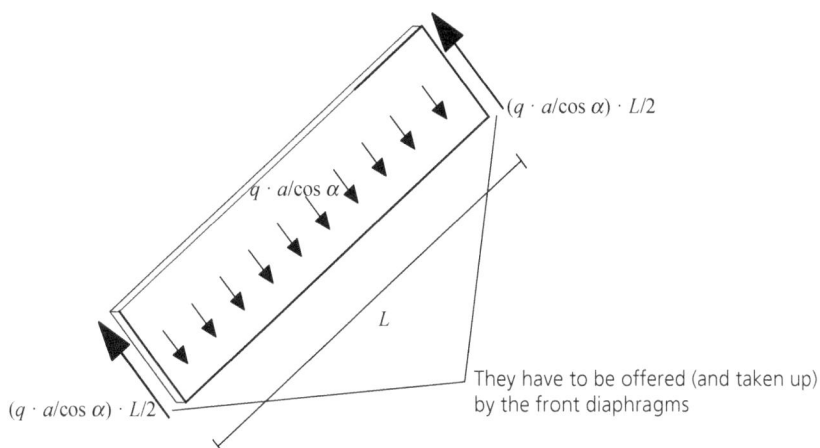

$(q \cdot a/\cos \alpha) \cdot L/2$

$q \cdot a/\cos \alpha$

L

They have to be offered (and taken up) by the front diaphragms

$(q \cdot a/\cos \alpha) \cdot L/2$

It must be pointed out that the above approach can be used only for preliminary design purposes. For a more accurate determination of the response, the use of suitable finite element software is necessary.

Since folded slabs take advantage of the rigidity of their slabs in their own plane, they allow – provided that the necessary space for the required folding height is available – a large variety of structural configurations so that surface loads can be taken up in the vertical as well as in the horizontal direction – for example, in retaining walls (see Figure 2.47).

Figure 2.47 The possibility of taking up earth pressure

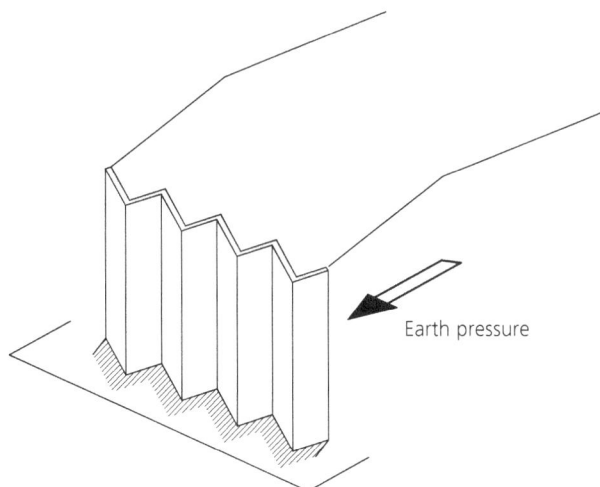

Earth pressure

REFERENCES

DIN (Deutsches Institut für Normung) (2001) DIN 1045 – Plain, reinforced and prestressed concrete structures. Deutsches Institut für Normung, Berlin, Germany.

Franz G and Schäfer K (1988) *Konstruktionslehre des Stahlbetons, Band II, B*. Springer, Berlin, Germany. (In German.)

Girkmann K (1963) Flächentragwerke, Wien, Springer Verlag. (in German.)

Herzog M (1996) Kurte baupraktische Festigkeitslehre, Düsseldorf, Werner Verlag. (in German.)

Menn C (1990) *Prestressed Concrete Bridges*. Birkhäuser, Basel, Switzerland.

Nervi PL (1956) *Structures*. McGraw-Hill, New York, NY, USA.

Stavridis LT (1993) Static and dynamic analysis of orthotropic rectangular slabs. *Stahlbau* **62**: 73–80. (In German.)

Stavridis LT (2001) Alternative layout for the prestressing of slab bridges. *Journal of Bridge Engineering ASCE* **6(5)**: 324–332.

Stavridis L and Georgiadis K (2025) *Understanding and Designing Structures without a Computer: Plane structural systems*. Emerald Publishing, Leeds, UK.

emerald PUBLISHING

ice

Leonidas Stavridis and Konstantinos Georgiadis
ISBN 978-1-83662-945-0
https://doi.org/10.1108/978-1-83662-942-920251003
Emerald Publishing Limited: All rights reserved

Chapter 3
Shells

3.1.   Introduction

Covering spaces of large dimensions that are more than 15 m long in both directions with a concrete slab is inefficient in practice and becomes impossible for larger areas. The reason for this inefficiency lies simply in the unavoidable bending that, apart from the inherent non-exploitation of the full stress capacity of the sections, induces by its nature large deflections. However, by introducing curvature into the structure and adopting a monolithic thin shell form shaped by an appropriately curved surface with the same basic outline, the structure can primarily resist gravity loads through in-plane forces within the surface, rather than through transverse internal forces that produce bending (see Figure 3.1). Such a type of structure is called a *shell*. A typical shell may have a span-to-thickness ratio in the order of 250.

The contour of the shell boundary does not necessarily lie on a plane. It may be represented by any properly shaped spatial curve whose projection on the horizontal plane corresponds to the boundaries of the floor space that needs to be covered. Of course, the shape of the shell is affected by its contour and this shape, in turn, has a major influence on its static behaviour. The morphological options for the problem of covering large spaces using shells are essentially unlimited and the aesthetic result always has a particular importance.

Figure 3.1   Hyperbolic paraboloid shell for covering a rectangular space

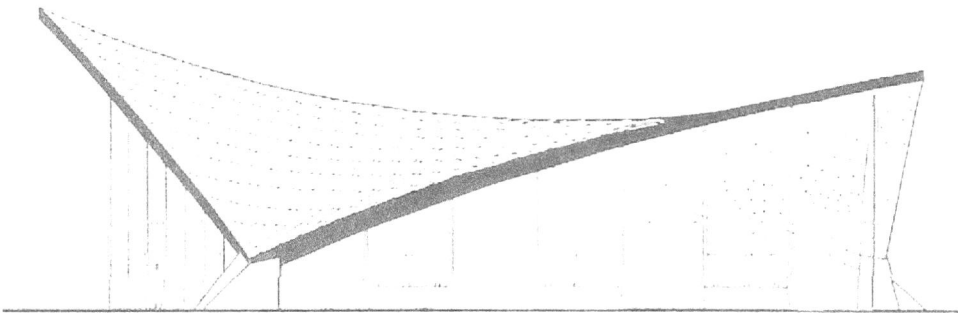

3.2.   The membrane action of shells and its importance for their design

A shell is always referred to by its middle surface, and for this reason it is necessary to define some fundamental geometric attributes that play an important role in its load-bearing behaviour.

It is clear that there is only one line passing from a point O to a surface that is perpendicular to it. Two planes are considered now to pass from that line, having arbitrary orientations and being perpendicular to each other (see Figure 3.2). These two planes intersect the tangent plane to the surface at point O along the straight axes x and y, and the surface of the shell along the curves $z = f_x(x)$ and $z = f_y(y)$, where z is measured in the normal direction, as shown in Figure 3.2.

Each of these two curves has at the common point O a specific radius of curvature and a curvature $1/R_x$ and $1/R_y$, respectively. If the radii of curvature R_x and R_y are found on the same side with respect to the surface, the surface is considered to be of positive (Gaussian) curvature, otherwise it is of negative (Gaussian) curvature. Whether the surface has positive or negative curvature is very important for the static behaviour of the shell.

At any point O of a surface, there is always a pair of mutually perpendicular planes that produce curvatures corresponding, simultaneously, to the maximum and minimum values that they can achieve as the pair of planes rotates about the normal line at point O. These curvatures are called *principal curvatures*.

Let a point O_1 be considered on the curve f_y at a small distance Δy from O and proceed with the intersection of the shell surface and a plane passing from point O_1 that is parallel to the plane defined by the curve f_x (see Figure 3.2). This intersection is another curve, and if its tangent at point O_1 forms an angle, α, with the direction O_1x, then the quantity $\alpha/\Delta y$ (i.e. the rate of change

Figure 3.2 Surface geometry in general

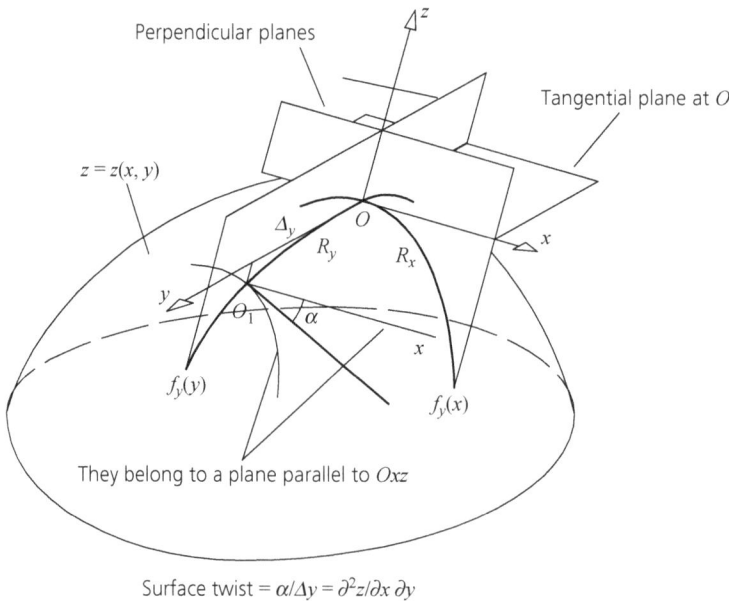

Surface twist $= \alpha/\Delta y = \partial^2 z/\partial x\, \partial y$

of the angle α) is called the *twist, t_{xy},* of the surface at point O with respect to the coordinate system Oxy and,

$$t_{xy} = \frac{\partial^2 z}{\partial x \cdot \partial y}$$

where $z = z(x, y)$ is the function describing the surface of the shell.

Although shells can exhibit unlimited morphologies, they can be placed into two categories with respect to their basic geometric configuration: shells generated by revolution and shells generated by translation (see Figure 3.3). The first category refers to shells whose midsurface is generated by revolving a planar line, either straight or curved, which is called the *generator,* about a constant axis. The shells of the second category are produced by sliding a planar line (straight or curved

Figure 3.3 Different ways of generating shells

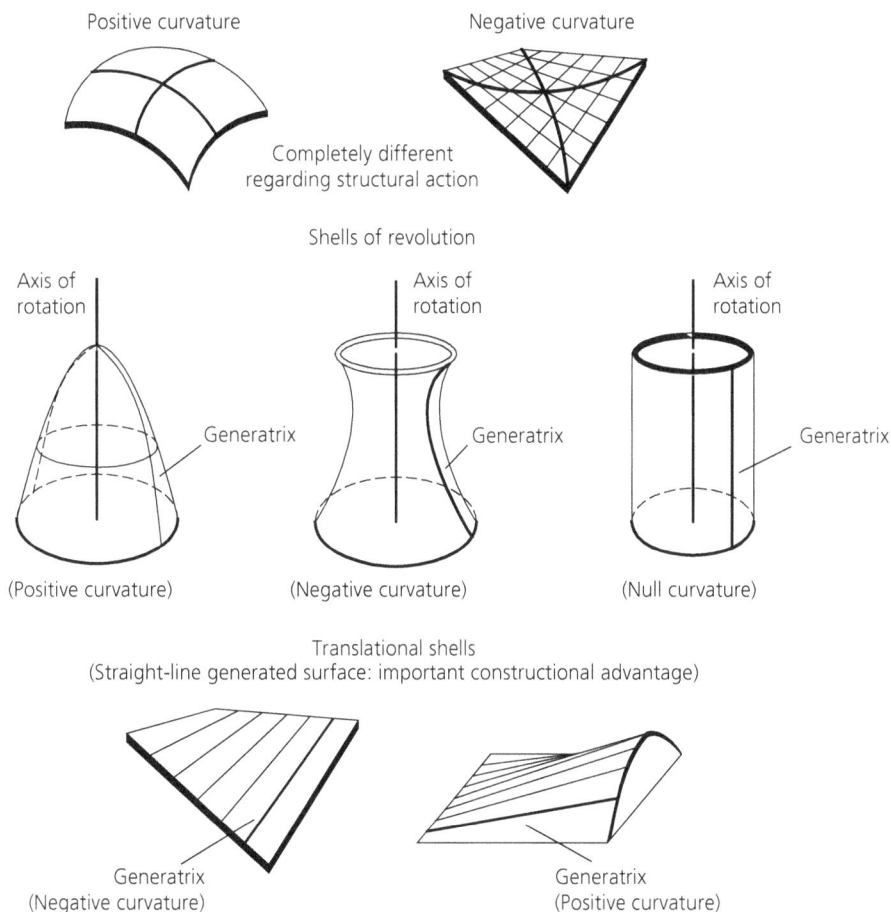

Positive curvature

Negative curvature

Completely different regarding structural action

Shells of revolution

Axis of rotation

Axis of rotation

Axis of rotation

Generatrix

Generatrix

Generatrix

(Positive curvature)

(Negative curvature)

(Null curvature)

Translational shells
(Straight-line generated surface: important constructional advantage)

Generatrix
(Negative curvature)

Generatrix
(Positive curvature)

generator) on another constant line, referred to as the *directrix*, maintaining its plane always parallel to itself. This distinction must be combined with that of positive or negative curvature. It becomes clear that the twist of the surface with respect to the directions of the principal curvatures is always zero, since in that case the angle, α, vanishes. ∎

It should be recalled here that in the case of an arch that carries the loads only through compressive internal forces, its shape is determined by considering the inverted shape of a fictitious cable that is subjected to exactly the same loads. Exactly the same procedure may be applied in the case of a shell.

More specifically, observation reveals that, if a membrane takes only tension and no bending at all and is firmly supported along the contour of the shell, application to the membrane of the vertical loads – which will be eventually received by the shell – will generate a specific shape for the membrane (see Figure 3.4). If this load is inverted and applied to a shell having the same specific shape as the deformed membrane, the shell will be stressed exactly like the membrane, but with the opposite sign, meaning solely compression. Considering its mirror image with respect to the horizontal plane leads to the solution of the problem (see Figure 3.5). This state of the shell is called the *membrane state of stress* and plays a dominant role in design since the described approach achieves uniform utilisation of the fibres of the section.

Adoption of the specific directions x and y is necessary in order to consider the membrane-bearing mechanism for a distributed load, q, acting along the z-direction – the normal at point O of the surface – through the cut-off quadrilateral element of the shell, as shown in Figure 3.6.

The state of stress of the membrane consists of an axial membrane force, N, which is normal to the element side and tangential to the shell surface, and a tangential shear force, S, acting along the side. Both forces refer to a unit side length.

Figure 3.4 Creation of a funicular membrane

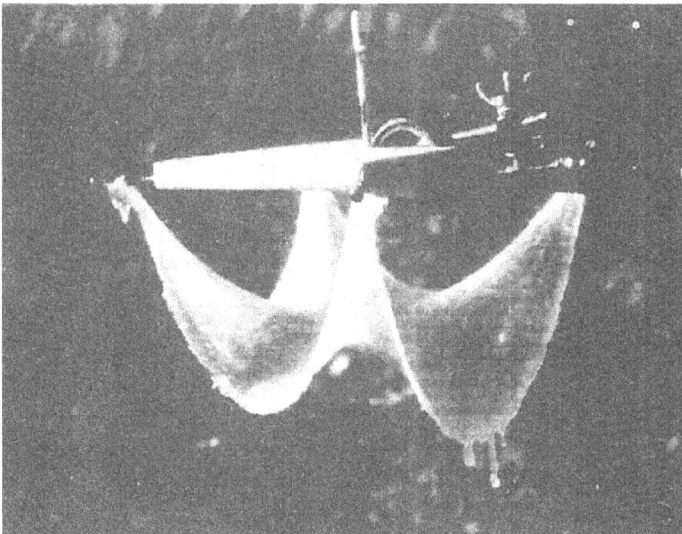

Figure 3.5 Shell formation by inverting the funicular membrane

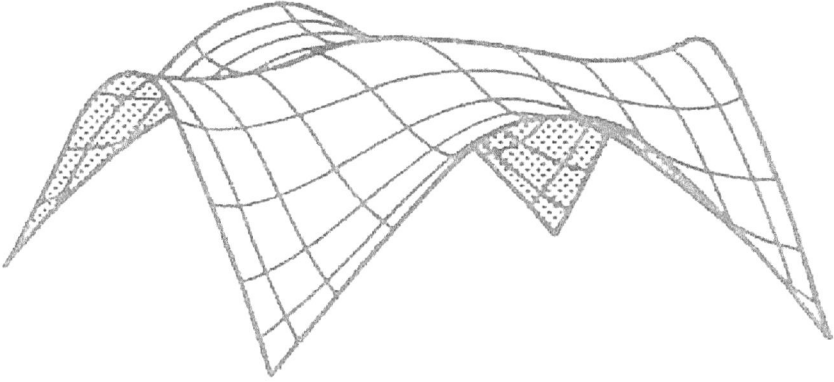

Figure 3.6 Membrane load-bearing mechanism

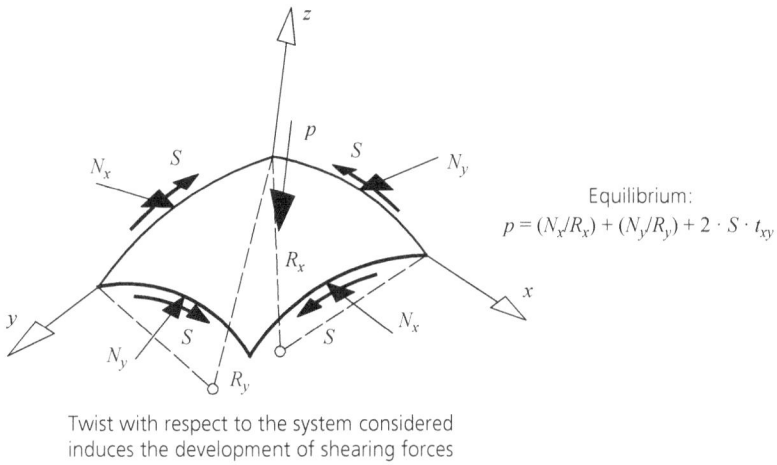

Twist with respect to the system considered
induces the development of shearing forces

Equilibrium of the quadrilateral element in the z-direction reveals that the load, p, is taken over by the membrane forces N_x, N_y and S, according to the following relation (see Figure 3.6):

$$p = \frac{N_x}{R_x} + \frac{N_y}{R_y} + 2 \cdot S \cdot t_{xy}$$

The first two terms on the right-hand side of this equation are in direct correspondence with the equilibrium equation of cables, as studied in Stavridis and Georgiadis (2025, Section 2.2.8) (see Figure 2.42). The third term refers to the vector resultant of the shear forces, S, along the z axis.

It is clear that the shear force, S, does not contribute to bearing the load when the twist of the surface at the point in question is zero and, consequently, there is no reason for its development ($S = 0$). Obviously, this happens when the directions x and y correspond to those of the principal curvatures.

For any distributed load, p, there is one and only one membrane state of stress (N_x, N_y and S), which is in equilibrium with the load, and its determination can always be considered as a statically determinate problem. It should be pointed out that the uniqueness of the membrane solution is attributed to the existence of curvature. If, for example, a slab is subjected to in-plane loads, the developed membrane state of stress cannot be determined simply through equilibrium considerations, because there are three unknown quantities, and there are only two available equilibrium equations for the plane case. This plane stress problem is addressed using the theory of elasticity. ∎

The construction of a shell usually requires an arrangement of beams along its boundaries, which should be designed to carry the opposite of the forces applied to the shell at the boundary (see Figure 3.7).

The shell, by developing the membrane action under the applied loads and the forces on its boundary required for equilibrium, exhibits some specific deformations. However, the same boundary forces of the shell act in the opposite direction on the beams or, more generally, on the supporting structural elements, causing other deformations there. The question that arises is whether these two different types of deformation are compatible. If this is not the case, then the membrane state of stress for the shell is not possible, and re-establishment of the displacement compatibility requires one of the following two procedures.

According to the first procedure, additional appropriate forces are applied to the shell boundary and exactly opposite forces to the supporting structure in order to make the final deformations equal.

Figure 3.7 The need for displacement compatibility at the boundary of the shell

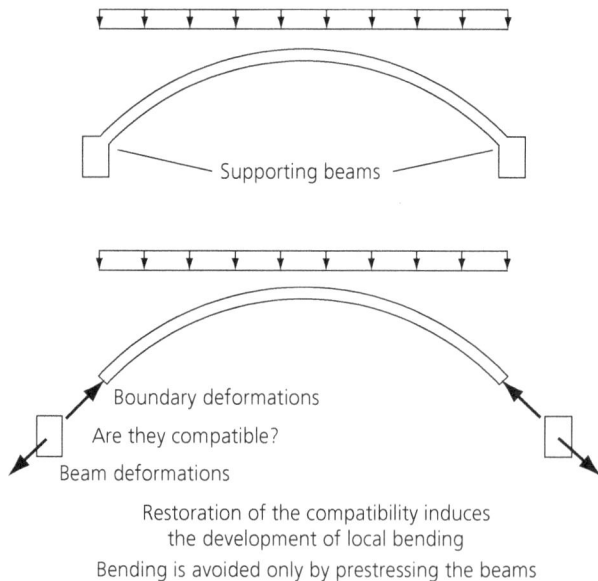

Application of such forces to the shell will cause bending, although bending is not acceptable for a shell whose natural state of stress is that of a membrane. Since displacement compatibility is restored through the introduction of boundary forces, the problem becomes statically indeterminate. As a result, bending arises naturally and depends on the shell's thickness. However, in many cases, this bending is confined to a small region near the boundary. It must be stressed that the shell's thickness has no influence on its membrane action.

In the second procedure, prestressing is introduced to the supporting structure so that its displacements are the same as those developed on the shell. This solution, whenever applicable, is statically preferable. It is also reasonable to apply prestress to the shell instead of the supporting structure in order to achieve compatibility for the displacement between the shell boundary and the supports.

The design of a shell – that is, the selection of its shape and dimensions – can and must be based on the membrane action in order to utilise its material to the maximum. Nevertheless, in cases of marked incompatibility between the deformations of the shell boundary and the supporting system, the state of stress due to bending should also be considered in the selection of dimensions, as explained previously, although for shells with positive curvature this is confined within a small area close to the boundary. However, a detailed computation of the stresses requires appropriate computer software.

Obviously, given that shells covering large areas are subjected to high compressive forces, particular effort should be made to avoid failure due to buckling, which may possibly become the dominant factor in the determination of the thickness. Since the critical compressive stress and the critical pressure applied to the shell are, in all cases, proportional to the modulus of elasticity, E, of the material (i.e. of the concrete) and given that there is always uncertainty about the implementation of the analytical results due to constructional defects, creep of concrete and so on, it is reasonable to adopt a reduced value for E in order to ensure the desired safety factor.

The basic characteristics of the load-bearing behaviour of the most common types of shells are presented in the following sections.

3.3. Cylindrical shells

Cylindrical shells of revolution are used either for storing liquids or granular materials (reservoirs or silos). Their longitudinal axis may either coincide with the vertical direction, in which case they rest on a circular concrete base, or may be horizontal in order to cover orthogonal areas, using only a part of their ring section.

3.3.1 Cylindrical shells under constant internal pressure

The perception of the load-bearing behaviour of a cylindrical shell – which is supported at its lower circular end and is loaded internally by constant pressure, p_r, along its height – is of particular importance, as will also be seen later for other shells of revolution (see Figure 3.8).

The internal pressure, p_r, considered initially as a self-equilibrating system acting on the free shell, produces the pure membrane tension, N_θ, in the ring, while the straight line generators, due to their infinite radius of curvature, do not develop any axial forces. All the foregoing may be directly concluded from the membrane equilibrium equation given in Section 3.2, which yields $N_\theta = p_r \cdot R$, where R is the radius of the cylinder.

Figure 3.8 Load-bearing behaviour of a free cylindrical shell under internal pressure

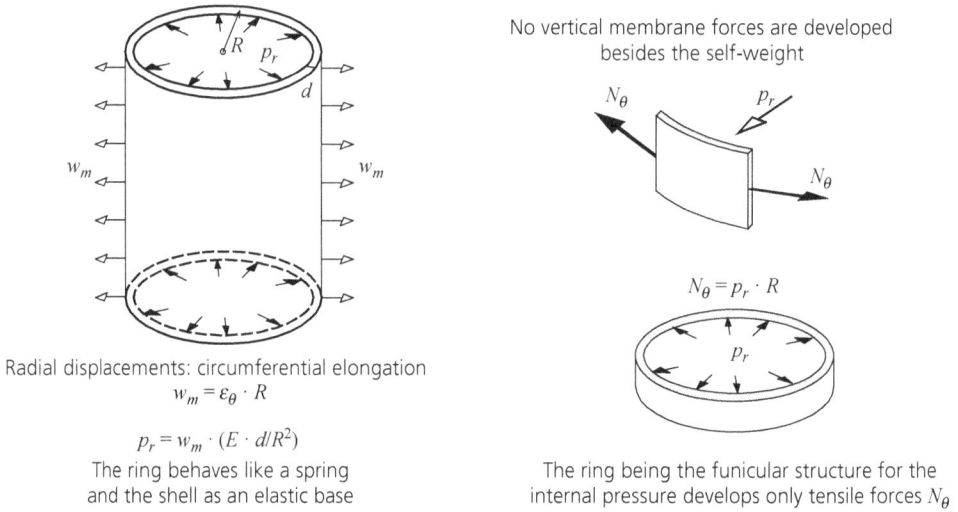

No vertical membrane forces are developed besides the self-weight

$N_\theta = p_r \cdot R$

Radial displacements: circumferential elongation
$$w_m = \varepsilon_\theta \cdot R$$

$p_r = w_m \cdot (E \cdot d/R^2)$
The ring behaves like a spring and the shell as an elastic base

The ring being the funicular structure for the internal pressure develops only tensile forces N_θ

This tensile state of stress results in an angular elongation strain, ε_θ, which depends on its thickness since $\varepsilon_\theta = N_\theta/E \cdot d$ and, consequently, the resulting outward displacement, w_m, according to the geometric relation $[2\pi \cdot (R+w_m) - 2\pi \cdot R = \varepsilon_\theta \cdot 2\pi R]$, is equal to $w_m = \varepsilon_\theta \cdot R$. These last relations result in:

$$p_r = w_m \cdot \frac{E \cdot d}{R^2}$$

This relation reveals that the pressure, p_r, is taken up by the shell through membrane action, just as in the case of an elastic base having stiffness modulus $k = (E \cdot d/R^2)$ (see Chapter 8, Section 8.3.2.2). This stiffness is offered by the fictitious consecutive shell rings through the whole height of the shell and is proportional to their thickness, d, and inversely proportional to the square of the shell radius, R. Obviously, w_m represents the deformation of the fictitious spring base (see Figure 3.9). ∎

It is now clear that the fixed support around the circular bottom end of the shell is not compatible with the development of w_m. As a consequence, bending moments and radial shear forces should act along the boundary in order to make its radial displacement vanish and guarantee the zero slope of the generators at the clamped support (see Figure 3.9).

Otherwise, the shell may be considered to consist of vertical straight beams (having, of course, a bending stiffness) and to be supported elastically by the horizontal shell rings along the shell height (see Figure 3.9). These vertical beams may be considered as fixed or hinged at their base. In the case in which they are not supported there, the constant pressure, p_r, is carried through the membrane action of the rings, much like springs with the aforementioned stiffness, k. The fact that all along the vertical beams there is an imposed constant displacement,

$$w_m = (p_r \cdot R^2)/(E \cdot d)$$

leads to the conclusion that these beams remain stress-free. If, however, their lower end is supported, the developed transverse displacements will obviously change, resulting in the activation of their bending stiffness. Thus, the load, p_r, is considered to be carried by these beams as if these were resting on an elastic foundation (see Section 8.3.3.1) with the modulus of the subgrade reaction $k = (E \cdot d/R^2)$ (see Figure 3.9).

Figure 3.9 Load-bearing behaviour of a cylindrical shell fixed at its base

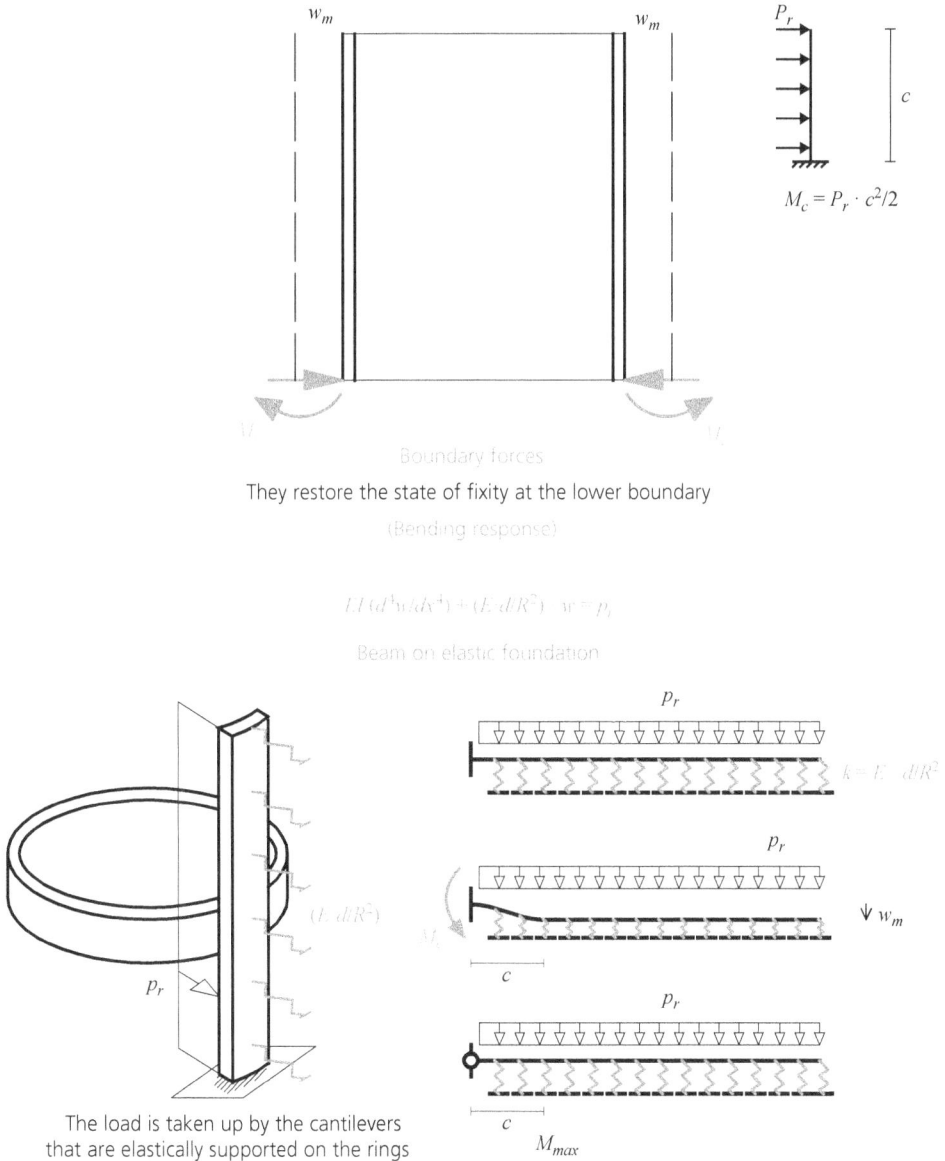

$$M_c = P_r \cdot c^2/2$$

Boundary forces
They restore the state of fixity at the lower boundary
(Bending response)

Beam on elastic foundation

The load is taken up by the cantilevers
that are elastically supported on the rings

M_{max}

The above consideration means that the vertical beams are also subjected, apart from the load, p_r, to the relieving transverse action $[(E \cdot d/R^2) \cdot w]$ of the shell ring and, consequently, obey the classical beam equation, so that it may be written:

$$p_r - \frac{E \cdot d}{R^2} \cdot w = EI \cdot \frac{d^4 w}{dx^4} \quad \text{or} \quad EI \cdot \frac{d^4 w}{dx^4} + \frac{E \cdot d}{R^2} \cdot w = p_r$$

In this way, the well-known equation of a beam of unit width resting on an elastic foundation with the subgrade modulus $k = (E \cdot d/R^2)$ can be directly recognised (see Section 8.3.3.1).

If the lower shell boundary is fixed, as previously stated, the required uniformly distributed radial forces and moments have to cancel not only the radial displacement, w_m, due to membrane action, but also the rotation of the generatrix at its base (see Figure 3.9). It is clear that the bending moment resulting at the lower boundary of the shell will be identical to the bending moment, M_e, at the fixed end of the corresponding beam on an elastic foundation (Timoshenko, 1956):

$$M_e = \frac{2}{c^2} \cdot EI \cdot \frac{p_r}{k}$$

where

$$c = \sqrt[4]{\frac{4 \cdot EI}{k \cdot 1}}$$

Physically, the quantity c, which is called the *characteristic length* (see Chapter 8, Section 8.3.3.1), relates to the distance $(c \cdot \pi/4)$ from the fixed end of the beam (equivalently from the end of the shell) over which the bending disturbance extends $(c = 0.76 \cdot \sqrt{R \cdot d})$, while the rest of the beam remains essentially free of bending stress due to the imposed constant displacement, w_m (see Figure 3.9).

It is interesting to realise that the above expression for M_e, after the necessary substitutions, yields the bending moment of a fictitious cantilever having a length, c, and bearing the load, p_r – that is,

$$M_e = p_r \cdot c^2/2 = 0.29 \cdot p_r \cdot R \cdot d \qquad \text{(see Figure 3.9)}$$

This suggests again a direct involvement with the elastic length of a beam resting on an elastic base, as examined in Section 8.3.3.1. The above bending moment obviously causes tension to the inner fibres of the cylinder (see Figure 3.9).

It is interesting to point out that the term p_r/k in the initial expression for M_e is equal to w_m and thus can also be written:

$$M_e = 0.29 \cdot E \cdot d^2 \cdot \frac{w_m}{R}$$

In the case in which the lower boundary of the shell is hinged, the maximum bending moment M_{max}, which applies tension to the external fibres, occurs, by analogy to the beam on an elastic foundation with one hinged end, at the characteristic distance ($c \cdot \pi/4$) from the hinged end (see Figure 3.9). This moment is given by (see Timoshenko, 1956):

$$M_{max} = \frac{0.64}{c^2} \cdot \frac{p_r}{k}$$

or

$$M_{max} = 0.092 \cdot E \cdot d^2 \cdot \frac{w_m}{R} = 0.092 \cdot p_r \cdot R \cdot d$$

3.3.2 Cylindrical tanks

Cylindrical tanks are usually under the action of a permanent internal hydrostatic pressure, p_r, perpendicular to their surface and that grows linearly with depth, H:

$$p_r = H \cdot \gamma$$

where γ is the specific weight of the stored liquid (see Figure 3.10).

Figure 3.10 Load-bearing behaviour of a cylindrical liquid tank

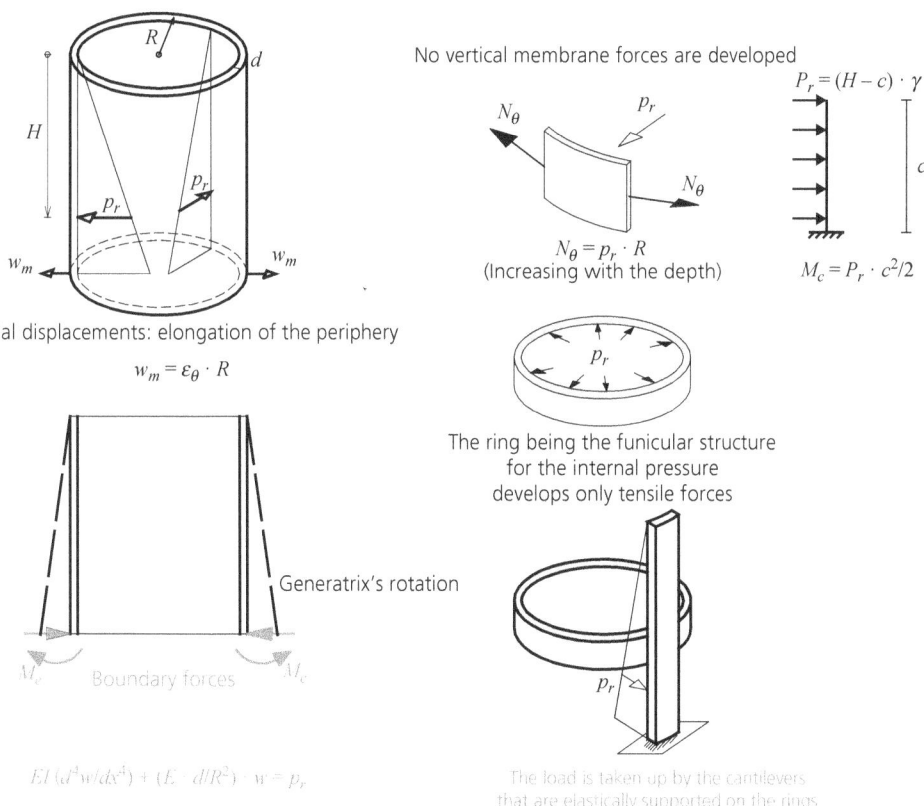

The bending moment, M_e, at the fixed boundary of the shell can be approximately assessed according to the previous section as the fixed-end moment of a fictitious cantilever having a length, c, and subjected to the uniform load $(H - c) \cdot \gamma$:

$$M_e = (H - c) \cdot \gamma \cdot c^2 / 2$$

The bending response in the shell becomes zero at a distance $(c \cdot \pi/4)$ from the bottom (see Figure 3.10).

In the case of a hinged base, the maximum bending moment occurs at a distance $(c \cdot \pi/2)$ with a value resulting from the equilibrium of the same fictitious 'short' beam as before but this time hinged at its lower end. This results in $maxM = (H \cdot \gamma) \cdot c^2/6$.

In both cases, the maximum tensile force, N_θ, occurs at a distance $x = c \cdot \pi/2$ from the bottom, thus resulting in $N_\theta = \gamma \cdot (H - x) \cdot R$ (Figure 3.10). ∎

Such a shell may also be loaded by the inward pressure of the possibly surrounding soil, while being empty inside (see Figure 3.11). This pressure, p_E, will correspond to the lateral earth pressure at rest, which also depends on possible acting live load, p, on the soil surface (see Chapter 8, Section 8.3.1). Thus,

$$p_E = \rho \cdot H \cdot 0.50 + p \cdot 0.50$$

where ρ is the specific weight of the soil and H is the depth, measured from the soil surface.

It is obvious that the lateral pressure, p_E, will produce a state of stress corresponding to the previously considered hydrostatic pressure and constant internal pressure acting in the opposite direction. It is clear that in the last expression, the value $0.50 \cdot \rho$ has been used instead of γ. ∎

However, the fact that the shell will be subjected to a ring compression raises the risk of buckling.

Figure 3.11 Cylindrical shell under lateral soil pressure

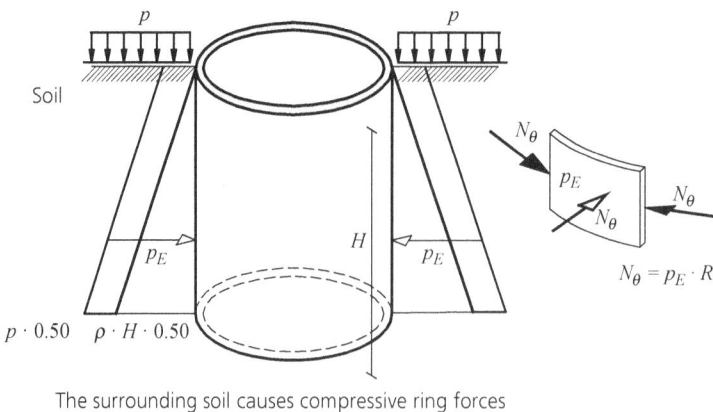

The surrounding soil causes compressive ring forces

The critical external pressure, p_D, of a closed cylindrical shell may be assessed from the relation (Pflüger, 1966):

$$p_D = 0.62 \cdot E \cdot \frac{R}{H} \cdot \left(\frac{d}{R}\right)^{2.50}$$

Any additional load acting along the longitudinal axis of the shell and producing a compressive stress, σ_x, must be taken separately into account.

The critical longitudinal compressive stress, σ_{xD}, may be assessed as (Pflüger, 1963):

$$\sigma_{xD} = \frac{0.48 \cdot E \cdot d/R}{\sqrt{1 + \dfrac{R}{100 \cdot d}}}$$

The coexistence of these two compressive states of stress can be regarded as secure if the following condition is satisfied (Flügge, 1960):

$$\frac{\sigma_x}{\sigma_{xD}} + \frac{p_E}{p_D} \leq 1 \qquad \blacksquare$$

The development of tensile ring stresses (N_θ/d) in a cylindrical tank may possibly be problematic for the design due to special requirements regarding leakage risk. Specifically, these stresses should clearly be lower than the concrete tensile strength, σ_{cz}. Thus, a peripheral prestressing is often applied, consisting either of circular tendons or high-strength wires tightened along the perimeter over the whole height of the shell (see Figure 3.12).

If Z is the prestressing force of a cable circling the shell, referring to the unit length of shell height and assuming its radius to be essentially equal to R, then the inward-acting deviation forces $u_p = Z/R$ cause the compressive ring force $N_{\theta P} = u_p \cdot R$ and, consequently, the developed compressive stress, σ_c, in the shell equals (see Figure 3.12):

$$\sigma_c = N_{\theta P}/d = Z/d$$

Figure 3.12 Peripheral prestressing of a cylindrical tank

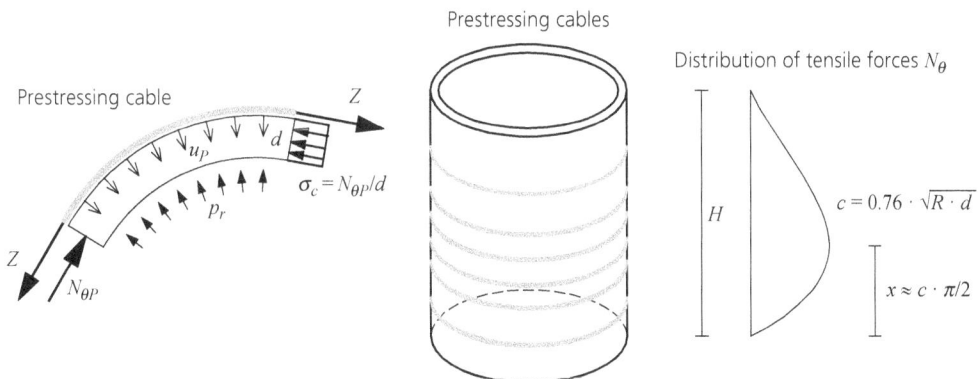

The prestressing force, which entirely takes over the internal pressure, p_r (i.e. $u_p = p_r$), equals $Z_p = p_r \cdot R$. Prestressing with this force will lead to approximately zero stresses in the shell wall. However, in order to take additional tensile stresses into consideration, due to the influence of concrete shrinkage as well as other secondary factors, a precompression, σ_c, of the order of 500–1000 kN/m² needs to be provided through additional prestressing. The required additional prestressing force, Z_d, for this purpose is, according to the above relation, equal to $Z_d = \sigma_c \cdot d$. Thus, the applied prestressing force, Z, in the circular cables – after losses – should be:

$$Z = Z_p + Z_d = p_r \cdot R + \sigma_c \cdot d$$

The circular cables should be placed in the external region of the wall thickness, which should in principle not be thinner than 25 cm. Moreover, the distribution of the prestressing force, Z, over the height should in principle follow the corresponding distribution of the tensile forces, N_θ (see Figure 3.12).

3.3.3 Silos

Silos also have a circular section and are used for storing granular material. Their height-to-diameter ratio is considerably greater than that for liquid tanks because in the latter, the internal pressure of the content increases linearly with depth, whereas in silos containing granular material with an angle of internal friction, φ, the internal pressure increases linearly only up to the depth (Ciesielski et al., 1970):

$$z = \frac{2 \cdot R}{4 \cdot \tan(0.6 \cdot \varphi)}$$

and then remains constant to the bottom (see Figure 3.13). The maximum internal pressure that thus can be developed is:

$$p_h = \gamma \cdot z = \frac{\gamma \cdot R}{2 \cdot \tan(0.6 \cdot \varphi)}$$

Figure 3.13 Distribution of internal pressure in a silo containing granular material

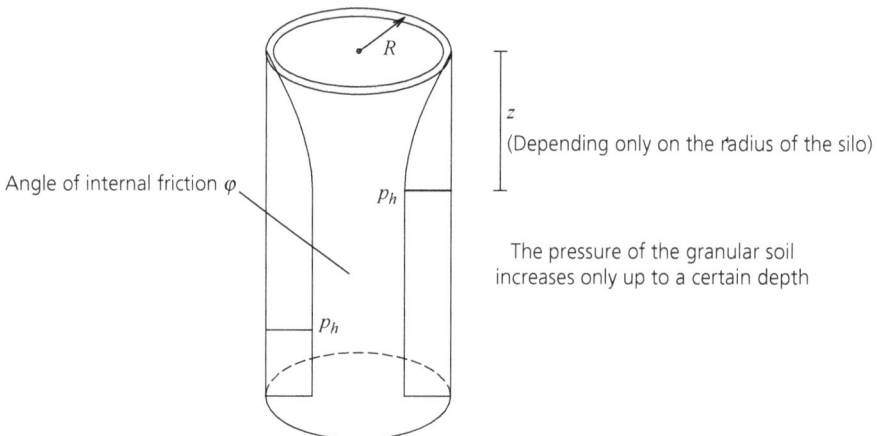

Angle of internal friction φ

p_h

z

(Depending only on the radius of the silo)

The pressure of the granular soil increases only up to a certain depth

p_h

R

3.3.4 Barrel shells

A long, thin cylindrical shell with a closed circular cross-section placed horizontally and supported only at its two extreme fronts in such a way that its profile remains undeformed (e.g. through an appropriate stiffener or a suitable circular wall) presents notable stiffness towards transverse loads. This stiffness makes this shell able to cover spans multiple times longer than its diameter by developing exclusively membrane forces (see Figure 3.14).

Assuming that the shell is under only the action of its self-weight, g, and considering the basic equation given in Section 3.2 that expresses how a load perpendicular to a shell surface can be taken over by its membrane forces, it can be deduced that, at the highest and lowest points of the cross-section, the developed ring forces, N_θ – tensile or compressive – are $N_\theta = g \cdot R$. These ring forces vanish at the ends of the horizontal diameter as the perpendicular component of g to the

Figure 3.14 Membrane load-bearing action of a closed beam-like cylindrical shell

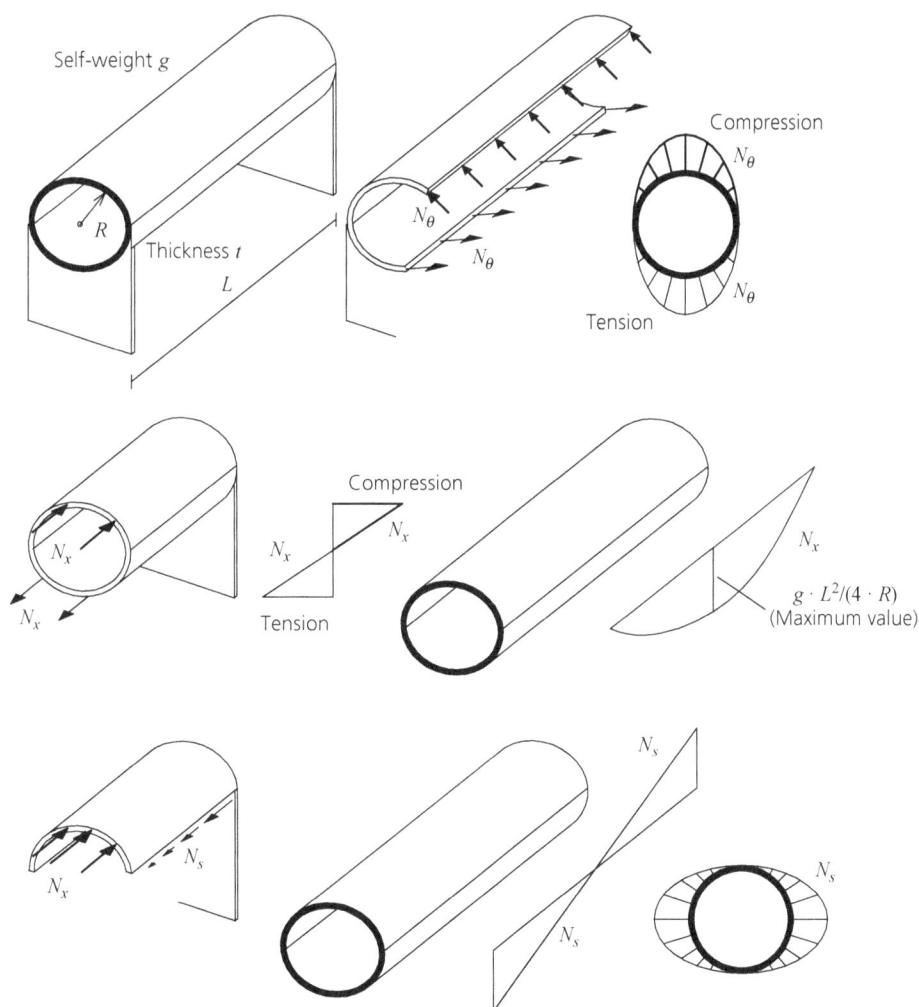

surface is zero there. Furthermore, it can be seen that the bending moment, M_0, over the span, L, which is equal to:

$$M_0 = (g \cdot 2R\pi) \cdot L^2/8$$

can be provided through longitudinal membrane forces, N_x, whose maximum value may result on the basis of an approximate value of moment of inertia, namely,

$$I = \pi \cdot d \cdot R^3$$

Thus

$$N_x = (M_0 \cdot R/I) \cdot d = g \cdot L^2/(4 \cdot R)$$

This maximum value appears as compression at the upper edge of the profile and as tension at the lower edge of the profile, while at the ends of the horizontal diameter N_x becomes null (see Figure 3.14).

It is now clear that longitudinal membrane shear forces are also developed along the generatrixes, arising from the longitudinal variation of N_x, which clearly vanish at the horizontal edges. Thus, by considering the horizontal equilibrium of, say, the cut of the upper half of the closed shell, membrane shear forces, N_s, must be developed in order to balance the compressive forces, N_x, acting over the cross-section at the middle (see Figure 3.14). These shear forces obtain their maximum value at the edges of the shell given that, acting along the periphery of the circular edges, they represent the only possibility of equilibrating the total vertical load on the shell. Following the shearing equilibrium requirements (*Cauchy's theorem*), the longitudinal shearing forces at the edges have the same value. This value is equal to

$$N_s = 2 \cdot g \cdot (L/2).$$

The above state of stress of the cylindrical shell is obviously a pure membrane state (see Figure 3.14). ∎

The form of a closed cylinder is certainly not suitable for covering an orthogonal area, unless it is cut by a horizontal plane and the resulting upper part is taken, again under the condition that its

Figure 3.15 Development of stiffness with the formation of a cylindrical cross-section

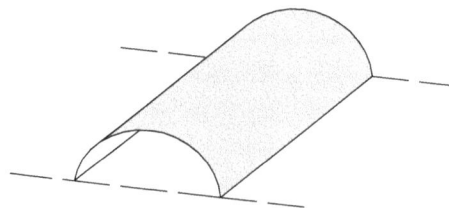

As shown also in folded slabs,
a sheet of paper acquires stiffness through its form

edges retain their cross-section – that is, are undeformable. A sheet of paper shaped appropriately illustrates this directly (see Figure 3.15).

In order now to examine whether it is possible for a pure membrane state of stress to be developed in such a structure under the action of gravity loads, those membrane forces that would be correspondingly developed in a closed cylinder should at first be applied along its free edges. If it is considered that the examined barrel constitutes a semicircle, given that the ring forces at this location are null, as previously stated, then the only forces that have to be offered externally are the longitudinal shearing forces, N_s, which may be offered by a horizontal beam (see Figure 3.16). However, such a beam acted on by the opposite longitudinal shearing forces, N_s, is tensioned and its elongation is not compatible with the absence of strain at the longitudinal fibres of the shell, given that $N_x = 0$, as previously explained. It can be seen that for the restoration of compatibility, additional longitudinal shearing forces should be

Figure 3.16 Consequences of membrane action in a barrel shell

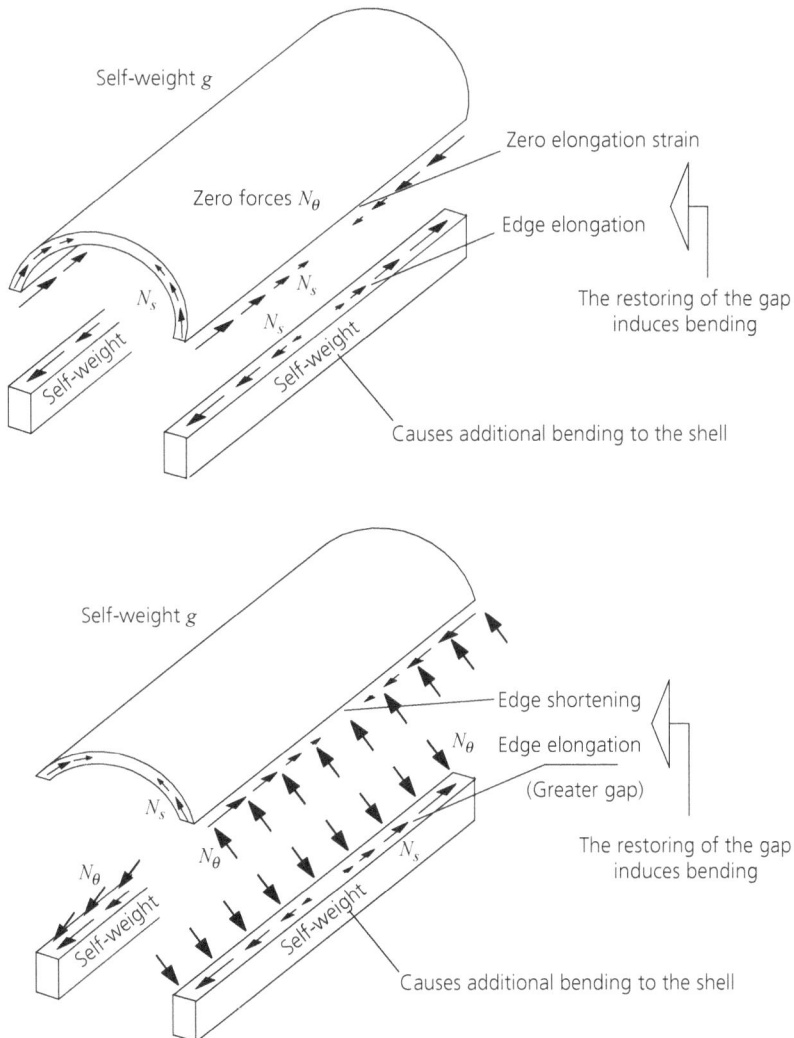

introduced at the free boundary of the shell – together, of course, with the corresponding opposite forces on the beam – which will obviously cause tension in the lower regions of the shell and that, together with the unavoidable action of self-weight of the beam along the free edge of the shell, will lead to a deviation from the pure membrane state – that is, they will cause (transverse) bending.

If the intersection of the full cylinder with the horizontal level is made at a higher level in order to maintain the membrane state, apart from the longitudinal shearing forces, N_s, of the free boundary, the ring forces, N_θ, should also be provided (see Figure 3.16). However, the gap in the longitudinal direction will now be greater due to the presence of the compressive force, N_x, and the subsequent shortening at the boundary of the shell, the beam having to provide the required inclined forces, N_θ, on the other side through its stiffness. Thus, with the additional exertion on the free edge of the shell of the self-weight of the beam, the deviation from the membrane condition will be even more intense.

Certainly, in order to take into account the introduced bending state of stress for the restoration of the gap, the shell and beam system may be designed by using appropriate computer software. However, it may be desirable to maintain, as far as possible, the membrane state of stress, and this can be done by prestressing the edge beam.

As shown in Figure 3.17, the compressive force of prestressing is able to eliminate the gap between the two longitudinal edges (between the shell and the beam), while, on the other hand, both the ring forces, N_θ, and the self-weight of the beam may be taken over by the cable deviation forces. ∎

Barrel shells can be divided into long and short ones, depending on the ratio of the length, L, to the width, b, of their ground plan.

Long barrel shells have a ratio L/b greater than 2 and may be designed at a preliminary stage as thin-walled beams, according to the classical theory of beams. On the basis of the maximum bending moment, M_0, the compressive force, N_x, occurring at the top of the middle section, can be determined from the relation $N_x = M_0 \cdot y_0 \cdot d / I$ (kN/m), whereas the tensile force, Z, which will be taken over by

Figure 3.17 Prestressing the edge beam of a barrel shell

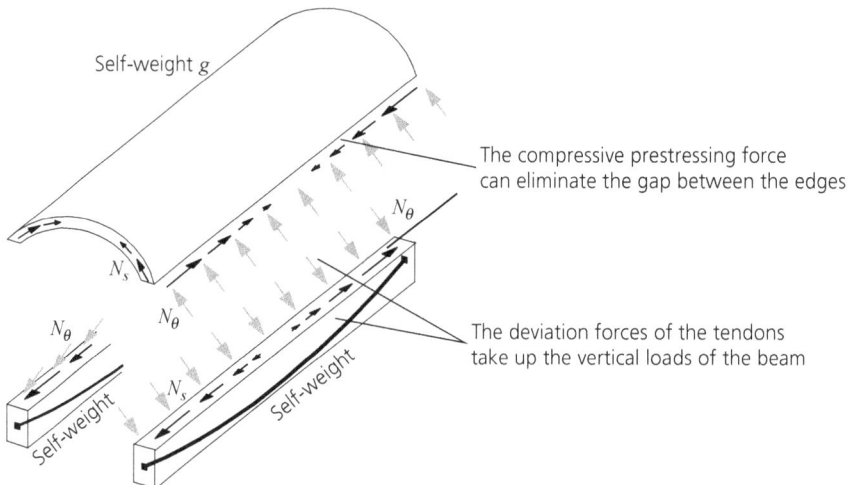

Self-weight g

The compressive prestressing force can eliminate the gap between the edges

N_θ

N_s

N_θ

N_θ

N_s

Self-weight

Self-weight

The deviation forces of the tendons take up the vertical loads of the beam

a corresponding reinforcement, may be considered equal to $Z = M_0/z$, where z is the lever arm of the internal longitudinal forces, estimated approximately as $z = 0.90 \cdot h$ (see Figure 3.18).

The shearing forces, N_s, which are applied to the end sections of the span and ensure equilibrium with the vertical loads, result from the vertical shear stresses (Stavridis and Georgiadis, 2025), Section 2.2.1) $f_s = (V \cdot S)/(I \cdot b)$ as $N_s = (f_s / \cos \alpha) \cdot d$. These forces show their maximum value at the centroidal axis of the section.

Of course, apart from these forces, ring forces, N_θ, are also developed, which have their maximum value at the top of the arch. This value, for preliminary design needs, may, according to the membrane state, be considered to be $N_\theta = g \cdot R$ (see Figure 3.18).

A possible additional support of the shell at an intermediate position implies a similar treatment to the continuous beam.

At this point, regarding the longitudinal bending response of the long barrel shell, an alternative approach can be considered, based on the results of Section 3.3.1. Specifically, by considering the vertical load on the barrel shell as playing, approximately, the role of constant internal pressure, but in the opposite sense to a closed cylindrical shell, direct use may be made of the relevant results.

Figure 3.18 Long barrel shells as beams with an additional transversal response

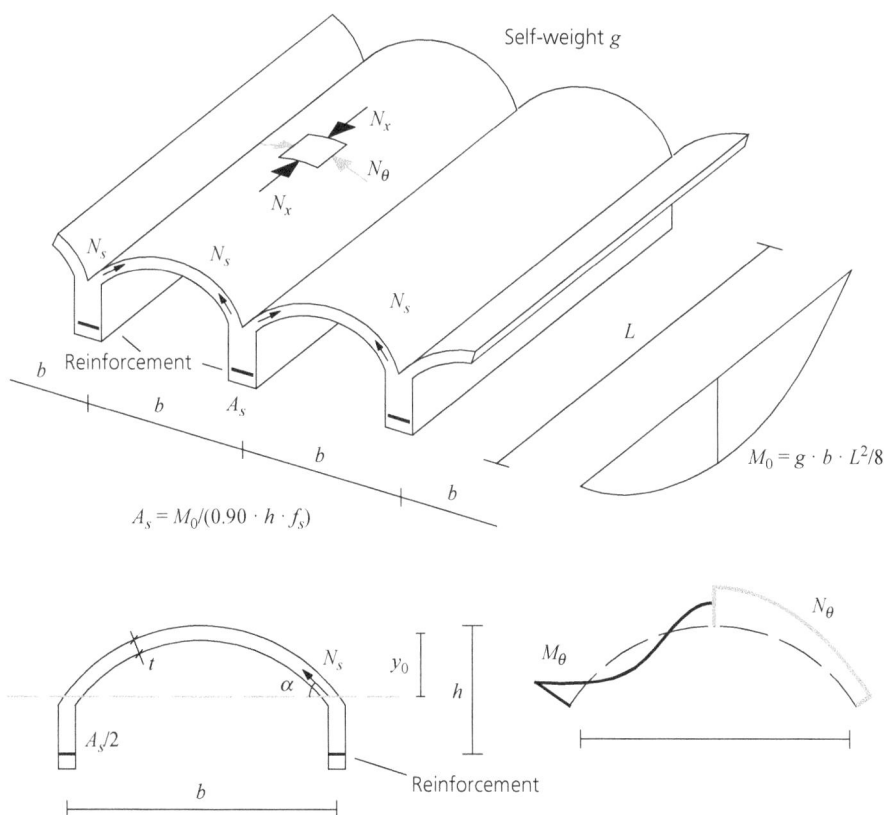

By considering the maximum value of N_θ equal to $g \cdot R$ and the corresponding compressive strain $\varepsilon_\theta = g \cdot R/E \cdot d$, the resulting inward displacement $w_m = \varepsilon_\theta \cdot R$ allows, by following Section 3.3.1, the assessment of the bending moments, M_e, at a clamped-end support, as well as of the maximum 'span' bending moment, M_{max}, in the case of a simply supported shell (both per unit length) according to the respective expressions:

$$M_e = 0.29 \cdot E \cdot d^2 \cdot \frac{w_m}{R} = 0.29 \cdot g \cdot R \cdot d \ \text{(tension outside)}$$

and

$$M_{max} = 0.092 \cdot E \cdot d^2 \cdot \frac{w_m}{R} = 0.092 \cdot p_r \cdot R \cdot d \ \text{(tension inside)}$$

It is clear that these values may be regarded only as approximate (see Section 3.3.1).

As explained above, the deviation from the membrane state of stress also means, for the shell, the development of transversal bending moments, M_θ, causing tension eventually at both the top and bottom fibres (see Figure 3.18). An approximate assessment of their values is given by the relation:

$$M_\theta = g \cdot (b/4)^2 / 8$$

However, these transversal moments may be very limited if care is taken to preserve the curved profile as undeformable through the insertion of a number of transverse ribs – that is, arches of significantly greater stiffness than that offered by the thickness of the cross-section of the shell only.

North light shells should also be included in the long cylindrical shell category, as shown in Figure 3.19, which, regarding lighting, show definite functional advantages compared with the full barrel shell when they are used for covering industrial areas.

Figure 3.19 Formation of north light shell beams for covering an orthogonal layout

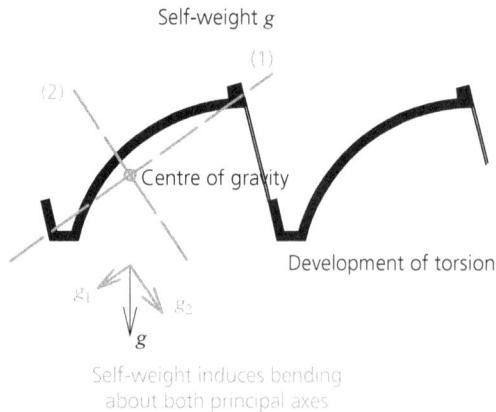

Self-weight g

(1)

(2)

Centre of gravity

Development of torsion

g_1

g_2

g

Self-weight induces bending
about both principal axes

These shells may be designed at a preliminary stage in the same way, namely as thin-walled beams according to theory, taking into account the fact that their principal axes are skew. These beams develop a certain torsional response, as will be explained in Chapter 5, but this is not of significance in their preliminary design. However, for the validity of the above approach, it is important to ensure the undeformability of the cross-section – for example, with transversal ribs. Moreover, the ring forces, N_θ, are necessary in order to maintain equilibrium and may be assessed as before.

In short barrel shells having a ratio L/b less than 2, the distribution of longitudinal membrane forces, N_x, cannot be based on the theory of bending. Of course, the free edges will develop tension, but tension may also occur in the top region. The ring forces, N_θ, can be considered, as in the case of long shells, and the transversal bending moments may also be assessed here according to the last expression. However, for the final design of these shells, appropriate computer software should be used. ∎

The risk of buckling should in no case be neglected, given that this is the factor that, in preliminary design, determines the thickness of the shell. Here, particular attention must be paid to both longitudinal and ring compressive stresses.

For the longitudinal (bending) compressive stresses, the following stress limit may be considered as critical (Seide, 1981):

$$\sigma_{xD} = \frac{0.58 \cdot E \cdot d/R}{\sqrt{1 + \dfrac{R}{100 \cdot d}}}$$

while for the transverse compression, the critical load, p_{cr}, can be considered, referring to that established for the closed cylindrical shell in Section 3.3.2, equal to:

$$p_{cr} = 0.6 \cdot E \cdot \left(\frac{d}{R}\right)^2$$

Both criteria should be examined. ∎

Cylindrical shells are particularly suitable for the aesthetically satisfactory covering of square as well as triangular ground plans by using a layout of intersected vaults.

Figure 3.20 shows the covering of a square ground plan with four intersecting barrel vaults, one based on each side. The arches created along the two diagonals receive the practically constant ring forces $N_\theta = g \cdot R$ from both their sides. The resulting V-shape cross-section of arches gives them a clearly greater bending stiffness than that which the shell thickness alone can afford.

The same logic is followed for covering a triangular ground plan, as depicted in Figure 3.21. The previous remarks are again valid.

The preliminary design of these shells can be made on the basis of the above-mentioned buckling criteria, as well as of the ring forces, N_θ, according to the previously exposed membrane state. However, it is worth repeating that, for ground plans of large dimensions, the buckling criterion prevails.

Figure 3.20 Intersecting barrel vaults for covering a square area (cross-vault)

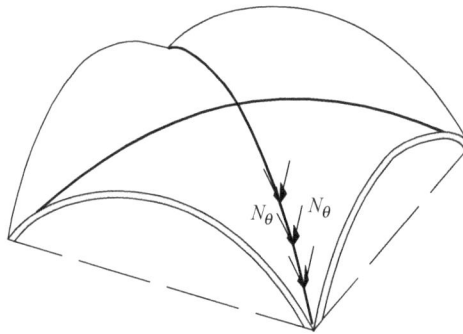

The arches take up the vertical components of N_θ

Figure 3.21 Intersecting barrel vaults for covering a triangular area

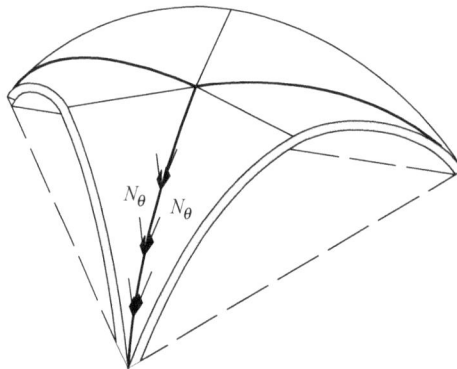

3.4. Dome shells
3.4.1 Non-shallow shells

With their middle surface being part of a spherical surface, dome shells are suitable for covering not only circular areas, but also various polygonal plans. The corresponding form is derived by cutting off the spherical surface with planes passing through every linear segment of the periphery of the base (see Figure 3.22). Of course, in the case of a circular base, the shell boundary will be a circle, but in all other cases it will consist of consecutive arch-like segments, representing the trace of the sphere on the previously mentioned planes. It is clear that these planes are not necessarily vertical, and a small outward slope leads to a more aesthetically appealing result.

Spherical shells are particularly suitable for carrying vertical distributed loads by membrane action, something that is always aimed for with their appropriate design. However, the degree to which this may be accomplished depends on the support conditions at the boundary of the shell, as will be examined later.

Figure 3.22 Design alternatives for covering areas with spherical shells

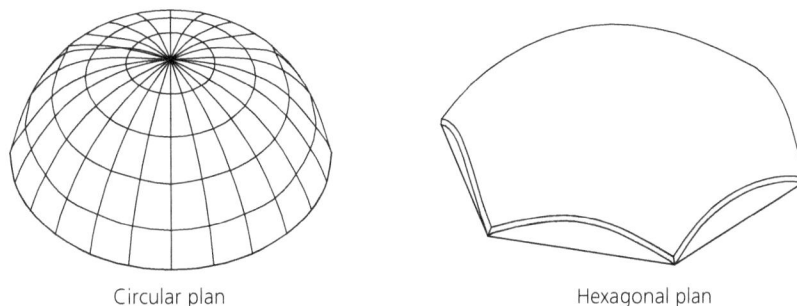

Circular plan   Hexagonal plan

Figure 3.23 Membrane forces in a spherical shell

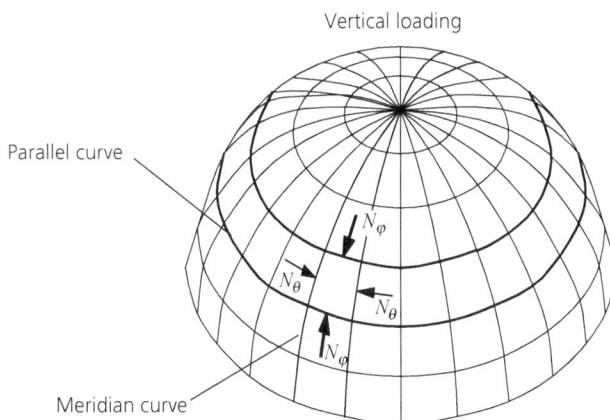

Vertical loading

Parallel curve

N_φ

N_θ

N_θ

N_φ

Meridian curve

The rotational symmetry of a shell suggests the consideration of the equilibrium of that surface element that is formed by two adjacent meridian and parallel curves (see Figure 3.23). The membrane forces, N_φ, and the membrane ring forces, N_θ, act along the parallel and meridian curves, respectively, and in a direction perpendicular to them. Moreover, if the vertical loading is also axisymmetric about the same axis as the shell, then it is easily concluded that the membrane shearing forces along the edges of the element, either meridian or parallel, vanish.

The membrane forces, N_φ, acting along a parallel curve can always be determined from the equilibrium condition in the vertical direction of the typical cut-off element (see Figure 3.24). If a represents the radius of the parallel circle, φ is the angle between a shell radius to the periphery and the vertical axis of symmetry, and V is equal to the total load applied to the upper shell portion being examined, then,

$$N_\varphi = V/(2\pi \cdot a \cdot \sin \varphi)$$

It is now assumed that the shell is subjected to a uniform load that is axisymmetric about the vertical axis. Application of the above relation for a certain central angle results in:

$$N_\varphi = p \cdot \pi \cdot a^2/(2\pi \cdot a \cdot \sin \varphi) = p \cdot R/2$$

However, this relation is not directly applicable for an element at the top of the shell under uniform loading, p – which may also represent the weight of the shell element – but instead the membrane equation of equilibrium in Section 3.2 may be applied. So, for the four-sided shell element with its centre located at the top of the shell, it holds that:

$$2 \cdot (N_\varphi/R) = p$$

Figure 3.24 Determination of the meridional membrane forces

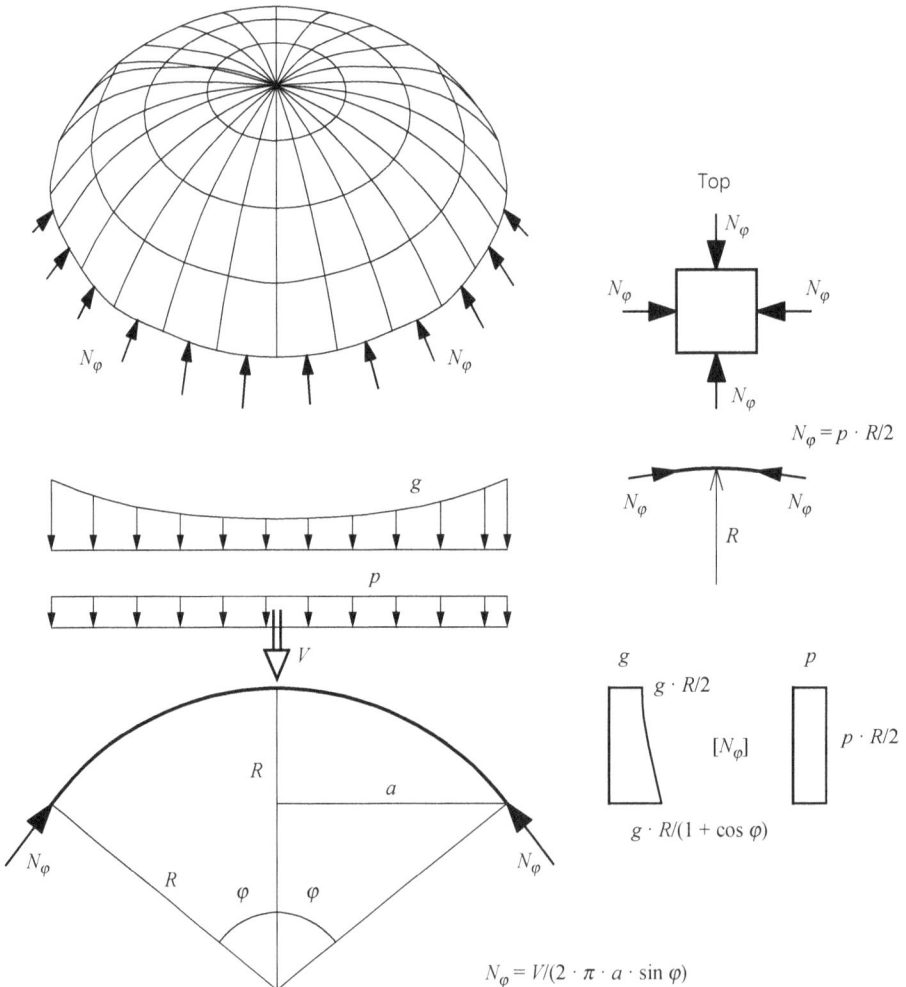

since in the absence of twist it is $S = 0$. The resulting value:

$$N_\varphi = p \cdot (R/2)$$

shows that this value of N_φ under a uniform constant load, p, is constant over the entire shell (see Figure 3.24).

Regarding now the self-weight, g, of the shell that is not uniformly distributed over the horizontal projection of the shell, its value increasing towards the edges, the meridional forces, N_φ, can be expressed, on the basis of the previous relation, as:

$$N_\varphi = R \cdot g \cdot \frac{1}{1+\cos\varphi}$$

this expression being obviously valid also for the top of the shell where $\varphi = 0$.

The ring force, N_θ, forming a right angle with N_φ is derived from the general membrane relation given in Section 3.2 by taking into account the corresponding principal curvatures at the point under consideration. It should be noted that, whereas the meridional curve of radius $R_\varphi = R$ refers to a principal curvature at that point, the corresponding parallel circle of radius, a, whose tangent at each point provides the line of action of the ring force, N_θ, does not represent the other principal curvature at the point under consideration (see Figure 3.24). This corresponds to a circle of radius $R_\theta = a/\sin\varphi = R$, which is a tangent to the parallel circle at that point, and N_θ belongs to both these planes as a common tangent to both circles (see Figure 3.23). Given that $S = 0$, the following relation can be written as:

$$\frac{N_\varphi}{R_\varphi} + \frac{N_\theta}{R_\theta} = p_r$$

where p_r is the component of the distributed load normal to the surface of the shell. The component p_r is equal to $(p \cdot \cos^2\varphi)$ or $(g \cdot \cos\varphi)$ for uniform load, p, and self-weight, g, respectively. So, the ring force, N_θ, may be expressed as:

$$N_\theta = R \cdot \left(p_r - \frac{N_\varphi}{R} \right)$$

The developed ring force, N_θ, due only to the self-weight, g, of the shell, is evaluated from the expression (Billington, 1965):

$$N_\theta = R \cdot g \cdot \left(\frac{1}{1+\cos\varphi} - \cos\varphi \right) \quad \text{(positive values refer to tension)}$$

In addition, for a uniformly distributed vertical load, p:

$$N_\theta = p \cdot R \cdot \cos(2 \cdot \varphi)/2$$

The development of ring forces, N_θ, can easily be understood by considering the equilibrium of a narrow ring cut off from the shell and subjected to the applied meridional forces, N_φ, along its upper and lower edges (see Figure 3.25).

Figure 3.25 Behaviour of the ring membrane actions

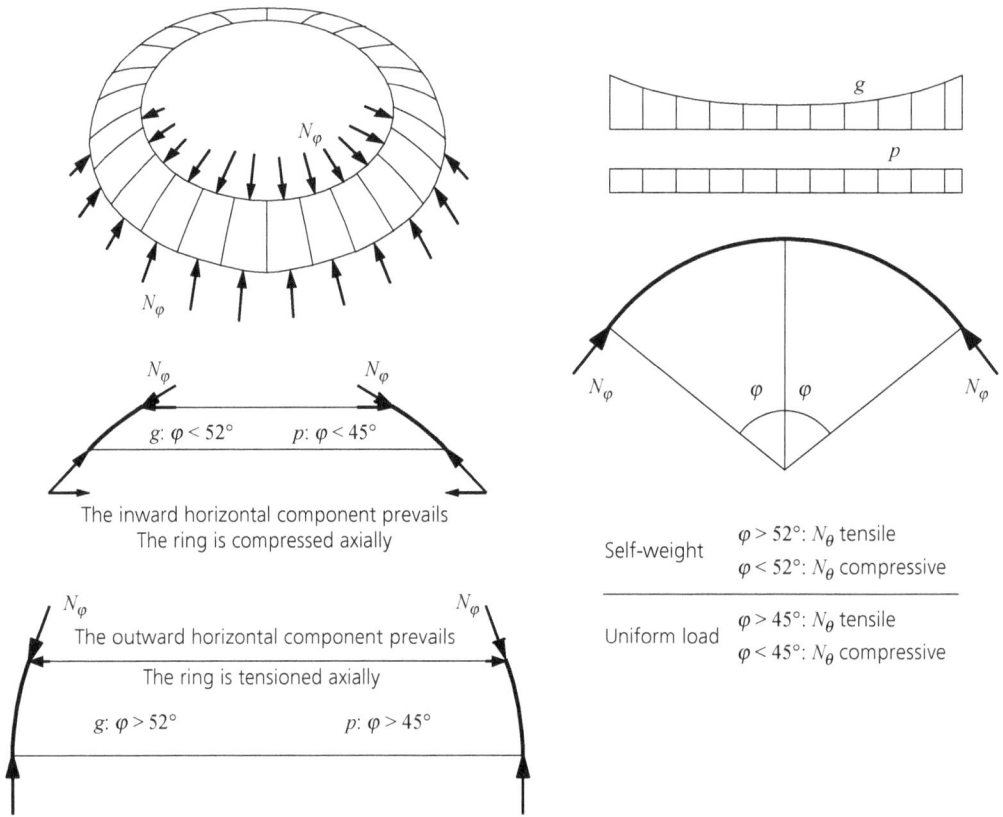

It is clear that as long as the ring belongs to the upper shell region, the inward horizontal component of N_φ at the lower edge prevails over the outward one of the upper edge, resulting in a compression of the ring (see Stavridis and Georgiadis, 2025, Section 2.2.8 and Figure 2.42). However, in the lower shell region, the outward component of N_φ at the upper edge dominates, therefore leading to tensile ring forces, N_θ. The exact value of the angle, φ, for which the transition from positive to negative values of N_θ takes place is derived from the above expressions for N_θ as $\varphi = 51.50°$ for the self-weight and $\varphi = 45°$ for the uniform load, p. This means that provided the opening angle, φ, is less than 45°, a spherical shell can carry any distributed vertical load with exclusively compressive forces. ∎

In the case in which the shell base is a horizontal circular plan, the meridional membrane forces, N_φ, at its boundary also have to act on the element that supports the shell, but in the opposite direction – that is, outwards (see Figure 3.26).

This element may normally be a circular beam (ring) that, under the action of the horizontal components of N_φ, will develop only tension without bending since it represents the funicular form for such a radial uniform loading (see Stavridis and Georgiadis, 2025, Section 2.2.8 and Figure 2.42). Consequently, its fibres tend to elongate and do not conform to the deformation of

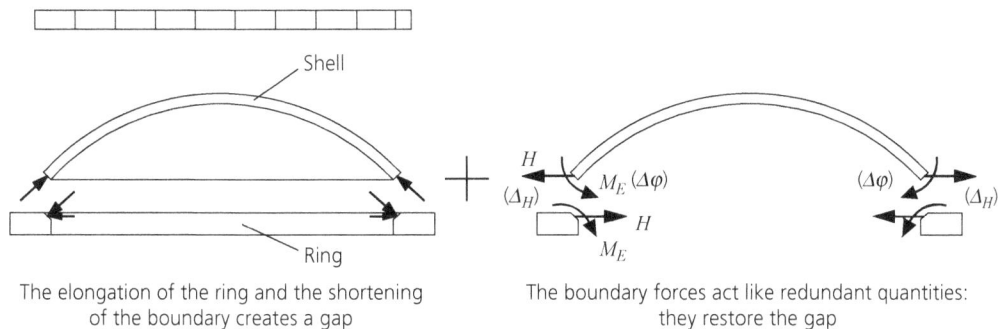

Figure 3.26 Interaction between the shell and the ring beam

The elongation of the ring and the shortening of the boundary creates a gap

The boundary forces act like redundant quantities: they restore the gap

the shell boundary as is implied by its ring forces N_θ. If the ring beam lies over the limit of the approximately 52° angle, the shell boundary will shorten ($N_\theta < 0$), whereas under this limit it will elongate ($N_\theta > 0$). Hence, in both these cases there will be a gap in the deformation of the two systems, although in the latter the gap will be less. Even in the case of a hemispherical shell, where the horizontal component of N_φ vanishes and the ring beam is not loaded within its plane, the existence of the tensile ring shell forces, N_θ, leads again to a gap between the ring beam and the shell.

Re-establishment of the compatibility of deformations will bring some bending along the shell boundary, which will be transmitted also to the ring beam but in the opposite sense. The ring will be subjected to an internal outward uniform pressure, H, as well as a uniformly distributed torsional load, M_E (see Figure 3.26). However, regarding the shell boundary, an analytical examination reveals that, due to the double curvature of the shell, this disturbance of the membrane state does not extend significantly into the body of the shell. Rather, it is restricted to its boundary region and thus does not have an essential influence on the preliminary design of the shell. Despite this, a detailed examination of the stress state and deformation of the boundary region is necessary in order to have a better understanding of its response.

3.4.1.1 Shell boundary

Generally, under the applied loads, the boundary of a shell exhibits a radial horizontal displacement, Δ_H, and a rotation of angle, $\Delta\varphi$, the influence of the latter in the overall response being practically negligible (see Figure 3.26). As it is clear that Δ_H is produced by the strain:

$$\varepsilon_\theta = N_\theta / E \cdot d$$

of the horizontal boundary ring having the radius $R \cdot \sin \varphi$, according to the relation:

$$\Delta_H = \varepsilon_\theta \cdot (R \cdot \sin \varphi)$$

and on the basis of the previous expression of N_θ for the permanent weight, g, it can be written as:

$$\Delta_H = \frac{R^2 \cdot g}{E \cdot d} \left(\frac{1}{1 + \cos \varphi} - \cos \varphi \right) \cdot \sin \varphi$$

It should be noted that Δ_H is inwards when $\varphi < 51.50°$ and outwards when $\varphi > 51.50°$.

On the other hand, for a constant vertical load, p, uniformly distributed all over the shell, according to the previous analysis it is:

$$\Delta_H = \varepsilon_\theta \cdot R \cdot \sin\varphi = \frac{N_\theta}{E \cdot d} \cdot R \cdot \sin\varphi = -\frac{R^2 \cdot p}{2 \cdot E \cdot d} \cdot \cos(2\varphi) \cdot \sin(\varphi)$$

In both the above expressions a positive result for Δ_H corresponds to an outward displacement and vice versa.

It should be noted also that a uniform horizontal force, H (kN/m), in the outward radial direction as well as a uniformly distributed bending moment, M_E, applying tension to the inner boundary fibres of the shell induce an outwards directed displacement, Δ_H, according to the following relations, respectively:

$$\Delta_H = \frac{2 \cdot R^2 \cdot \sin^2\varphi}{E \cdot d \cdot c} \cdot H$$

and

$$\Delta_H = \frac{2 \cdot R^2 \cdot \sin^2\varphi}{E \cdot d \cdot c^2} \cdot M_E$$

where $c = 0.76 \cdot \sqrt{R \cdot d}$, according to Section 3.3.1.

Here, it must be pointed out that the bending response of the shell edge is essentially identical to that of a cylindrical shell of the same thickness and radius, R, which circumscribes

Figure 3.27 Identical bending behaviour at the boundary of a spherical and a cylindrical shell

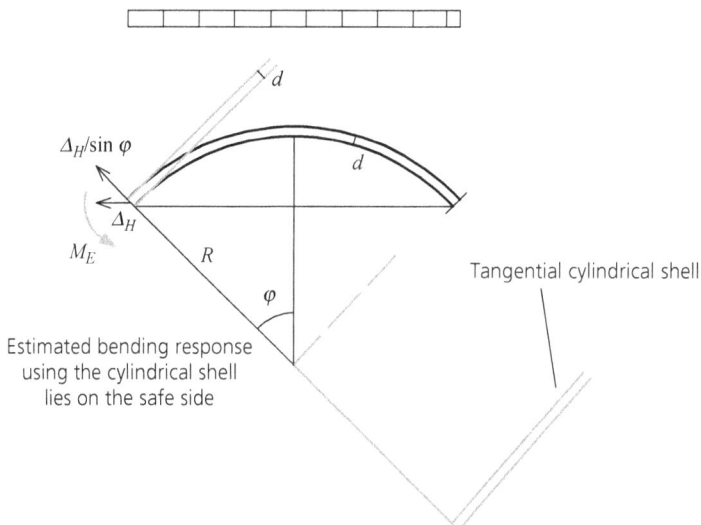

the spherical shell, being tangential to it at their common boundaries (see Figure 3.27). The involvement of the elastic length, c, in the above relations is thus evident according to Section 3.3.1 (Geckeler, 1926).

In this way, and for $\varphi > 20°$, the bending moment at the spherical shell boundary can be approximately determined according to the last equations of Section 3.3.1 through the respective cylinder displacement ($\Delta_H/\sin\varphi$) in its radial direction (see Figure 3.29):

$$M_E = 0.29 \cdot E \cdot d^2 \cdot \frac{\Delta_H}{R \cdot \sin\varphi}$$

while for a hinged boundary it is:

$$M_{max} = 0.092 \cdot E \cdot d^2 \cdot \frac{\Delta_H}{R \cdot \sin\varphi}$$

where Δ_H in the above equations takes the value that corresponds to the prevailing loading, which is either the self-weight or a uniform load.

Moreover, it should be emphasised that the above bending disturbance extends over a length not more than the corresponding one of the respective cylindrical shell – that is, practically equal to $\pi \cdot c/4$ according to Section 3.3.1. ∎

3.4.1.2   Ring beam

It is already known that a ring beam (with area, A_R, and second moment of area, I_R) subjected to a distributed radial force, H, develops only an axial force $N_R = H \cdot R$ and no bending at all, due to its funicular behaviour under that loading (see Stavridis and Georgiadis, 2025, Section 2.2.8 and Figure 3.28).

The strain in the ring beam is:

$$\varepsilon_R = N_R/(E \cdot A_R)$$

Figure 3.28 Behaviour of a ring beam under internal pressure

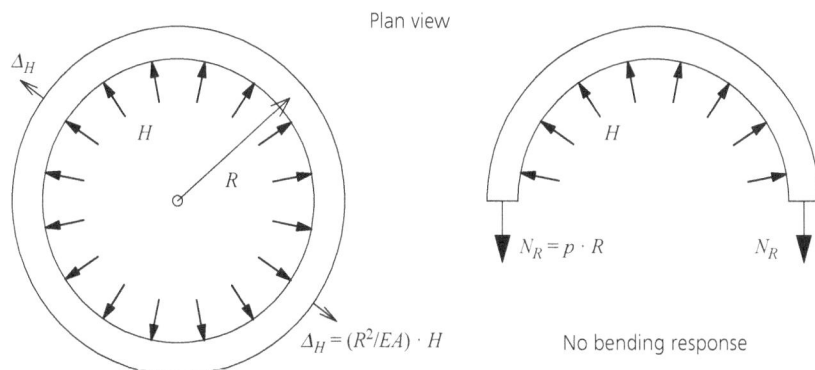

leading to an elongation of its circumference:

$$\varepsilon_R \cdot 2\pi R$$

and a radial displacement equal to:

$$\Delta_H = \varepsilon_R \cdot R = \frac{R^2}{E \cdot A_R} \cdot H$$

On the other hand, the application of a torsional load, M_E, uniformly distributed along the entire length of the periphery simply causes an equally directed rotation of its sections:

$$\Delta\varphi = \frac{R^2}{E \cdot I_R} \cdot M_E$$

and a constant bending moment:

$$M = M_E \cdot R$$

without creating any torsional moments, as shown in Figure 3.29. It should be noticed that under these conditions, the centroid of the beam section does not undergo any radial displacement, Δ_H. ∎

As has already been pointed out, the resulting bending disturbance of the pure membrane state in a spherical shell is, in any case, restricted to a relatively narrow region by its boundary, having a width of the order of magnitude of the characteristic length, c.

Prestressing of the ring beam can restore the incompatibility of the deformations. This additionally solves the problem of taking up the tensile stresses in the case of a concrete beam and also restricts the dimensions of its section (see Figure 3.30).

By placing a circular tendon prestressed with a force, P, in a concrete ring beam, the ring will be automatically subjected to an inward radial uniform pressure, P/R. If this pressure is greater than

Figure 3.29 Behaviour of a ring beam under a distributed torsional load

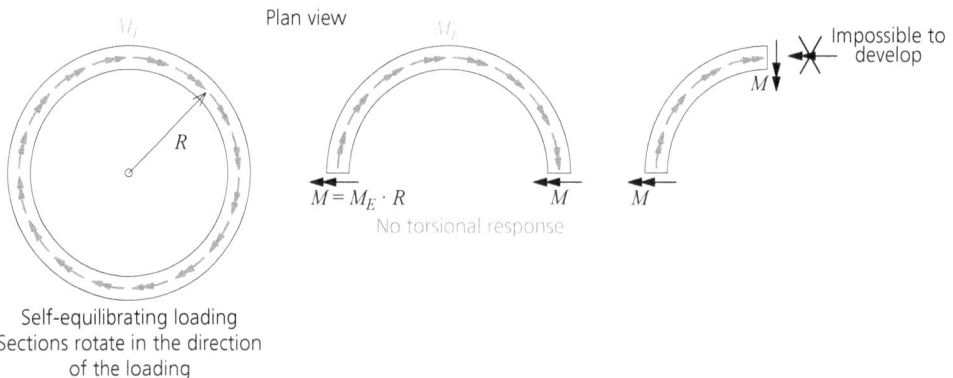

Plan view

Impossible to develop

M

$M = M_E \cdot R$ M M

No torsional response

R

Self-equilibrating loading
Sections rotate in the direction
of the loading

Figure 3.30 Prestressing of the ring beam

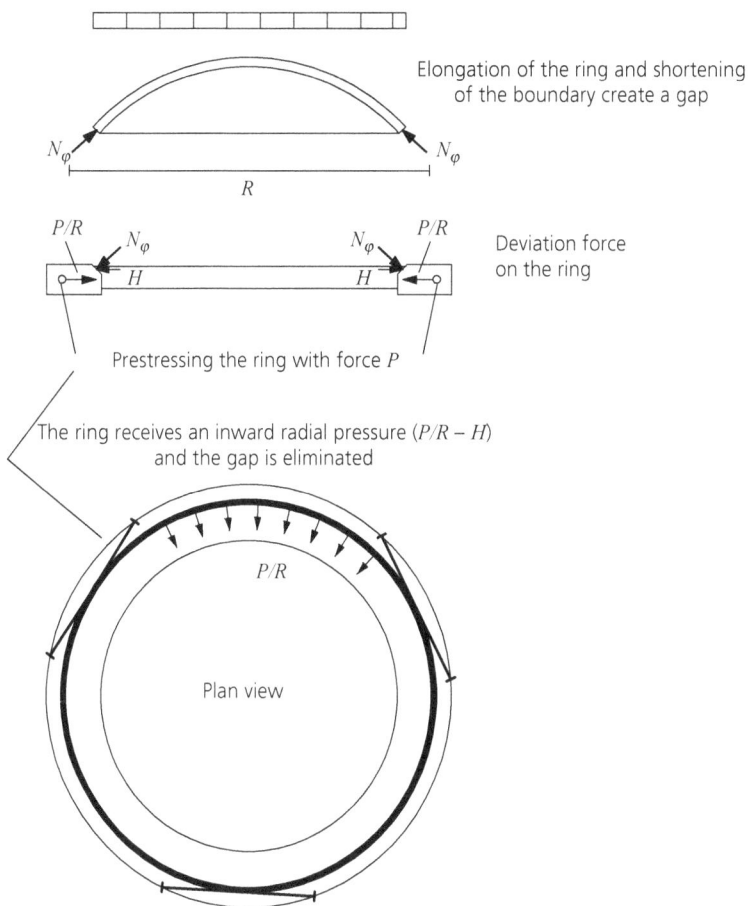

the already existing pressure, H, the ring will develop an axial compressive force $N_R = (P/R) \cdot R - H \cdot R = P - H \cdot R$ and, consequently, it will exhibit an inward radial displacement (see above):

$$\Delta_H = \frac{N_R}{E \cdot A_R} \cdot R = \frac{R^2}{E \cdot A_R} \cdot \left(\frac{P}{R} - H \right)$$

The prestressing force, P, can be selected in conjunction with the other structural parameters, so that, according to the previous analysis, the resulting gap between the ring beam and the shell boundary vanishes (see Figure 3.30).

3.4.2   Shallow shells

In order to confine the covered volume, as well as for functional or aesthetic reasons, a restriction of the dome height is often necessary, and thus the shell is characterised as shallow. This implies a decrease in curvatures and, consequently, an increase of the membrane compression forces. In such a shell, where the height, f, is less than about one-fifth of the covered span, a, the radius of curvature can be expressed as:

$$R = \frac{f^2 + (a/2)^2}{2f}$$

or approximately as:

$$R = \frac{(a/2)^2}{2.f}, \qquad \text{if } f < a/10 \qquad \qquad \text{(see Figure 3.31)}$$

Such a shell can cover an orthogonal, usually square, area, where the resulting boundary curves are sections of the spherical surface cut by vertical planes corresponding to the four sides of the shell base.

Due to the self-weight, g, the resulting meridional membrane forces, N_φ, along any 'parallel' are, according to the relations given in Section 3.4.1:

$$N_\varphi = \frac{\pi \cdot a^2 \cdot g}{2 \cdot \pi \cdot a \cdot \sin \varphi} = g \cdot \frac{R}{2}$$

So, it can be concluded that this force remains constant essentially over the entire shell surface. On the basis of this result and according to the analysis in Section 3.4.1, the ring force, N_θ, which is obviously compressive, is equal to the meridional one: $N_\theta = g \cdot (R/2)$.

In the case of an orthogonal base, the above result may also be confirmed by considering that the shallow shell consists of shallow arches that transfer their load in both directions through compressive forces. These compressive forces, N (per unit of width), are equal to:

$$N = \frac{g}{2} \cdot \frac{a^2}{8 \cdot f} = \frac{g \cdot R}{2}$$

according to the previous approximate expression for the radius of curvature.

Figure 3.31 Structural layout of a shallow shell

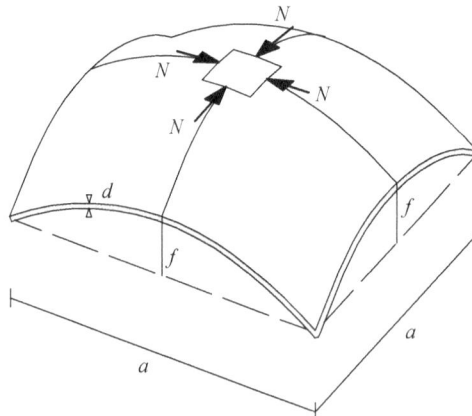

As may be seen from Figure 3.31, the shell boundaries can be formed by vertical arches. Although they cannot take up forces applied in a direction normal to the boundary and tangential to the shell, these arches can either offer to the shell membrane shear forces that should be applied along the shell boundary in order to maintain the equilibrium with the whole vertical load – as shown also in Figure 3.32 – or for the same purpose offer to the shell boundary vertical shear forces that will produce bending.

Thus, in such a shell, having essentially the form of a curved orthogonal slab, the bending response should be taken into consideration even during the preliminary design phase, especially when its boundary is restrained from rotation.

An estimation of the developed bending response, M_e, at the boundaries of a shallow shell with square base ($a \cdot a$) such as depicted in Figure 3.31 in the case in which these are considered as restrained from rotation, can be conservatively assessed by the expression $M_e = 25 \cdot p \cdot d^2$ (Stavridis and Armenakas, 1988) whereby, in contrast to slabs, the dependence of the bending moment on the shell thickness can be noticed. However, this bending response at the boundary causing tension on the upper fibres concerns only a perimetric band zone having a width of not more than $0.1 \cdot a$. The rest of the shell is subjected to a negligible bending of alternated sign, not exceeding one-fifth of the above value.

In the case in which a shallow shell is supported, as in Figure 3.31, not through tangential boundary shear but through vertical shear forces, a certain bending response at a distance $0.10 \cdot a$ from the supports is developed that can be conservatively estimated as $M = 15 \cdot p \cdot d^2$ (Stavridis and Armenakas, 1988), extending not more than $0.20 \cdot a$ from the boundary. ■

Figure 3.32 Load-bearing behaviour at the boundary of a shell with a polygonal base

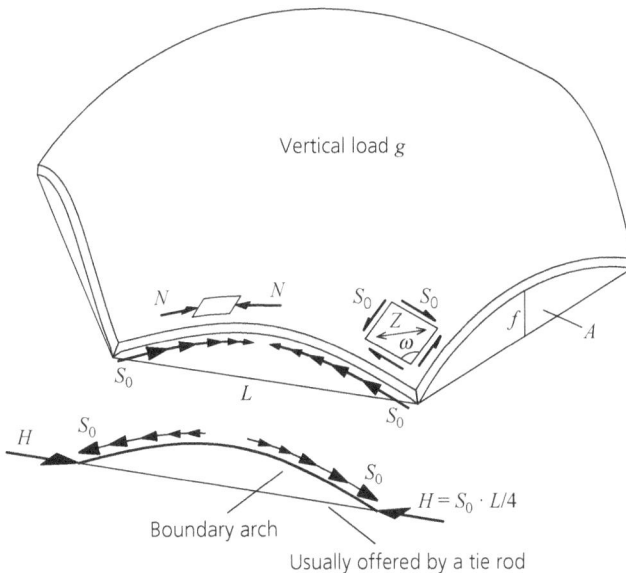

In a compressed dome, and particularly in a shallow dome, the risk of buckling deserves special attention. Of course, the existence of double curvature is a relieving factor that increases the critical load. This can be estimated using the relation (Schmidt, 1961):

$$p_D = 0.15 \cdot E \cdot (d/R)^2$$

whereby the application of a safety factor equal to 2 would be suggested.

From this relation, it is seen that an increase in the radius of curvature decreases the load-bearing capacity of the shell, whereas an increase in the shell thickness increases the critical load. One way to increase the critical load is to place transverse ribs along the mutually orthogonal directions, which increases the average shell thickness, as is often seen in older structures. ■

As previously mentioned, a spherical dome can cover a triangular, square or polygonal area. The sections of the shell surface on the corresponding planes passing through the sides of the polygonal base create arch-like openings with favourable lighting conditions (see Figure 3.32).

For the proper structural design of the shell boundary, it should be taken into account that forces cannot be applied in a direction that is normal to the boundary and tangential to the shell, while membrane shear forces should be applied along the boundary, taking up the whole vertical load on the shell, if equilibrium has to be maintained through only a membrane state of stress.

The fact that membrane forces, N_φ, transverse to the arch-like boundary cannot be developed means that, in order to achieve a membrane state of equilibrium, the vertical load in the boundary regions has be taken only by the membrane forces, N, which are parallel to the arch-like edges (see Figure 3.32). If R_b represents the radius of curvature of the corresponding arch, then according to the basic membrane relation given in Section 3.2

$$N = g \cdot R_b$$

Shear forces have to vanish along each axis of symmetry of the shell. So, in the arch-like boundary segment with span length, L, and height, f, they vary from zero at the crown to a maximum of S_0 at its ends. Clearly, the integral sum of their vertical components over all the boundary arches must equilibrate the total load $(g \cdot A)$ of the shell, where A denotes the area of the polygonal base. Assuming a parabolic distribution of the shear forces along each boundary arch, the maximum value can be conservatively obtained as:

$$S_0 = \frac{g \cdot A}{f \cdot k}$$

where k is the total number of boundary arches (see Figure 3.32) (Salvadori and Levy, 1967).

These shearing forces can be directly applied to the shell by an arch of suitable section, which may be adjusted to the shell boundary. Such an arch cannot transmit forces other than those lying in its own plane; therefore, it has to take on the shearing forces of the shell boundary along its axis and transfer them to the ends of the arch, which are also the support points of the shell. These forces are obviously directed towards the ends, and, consequently, produce an outward horizontal thrust, H. This thrust can be estimated on the basis of the adopted linear distribution of shearing forces as $H = S_0L/4$, and has to be borne either by a fixed support or through a tie connecting the ends of each boundary arch (see Figure 3.32).

At each shell corner, the developed membrane shear forces, S_0, produce a tensile force, Z, acting in a direction perpendicular to the angle and equal to:

$$Z = S_0/\tan(\omega/2)$$

This tensile force has to be taken up by an appropriate reinforcement.

3.5. Hyperbolic paraboloid ('hypar') shells
3.5.1 Overview

These shells belong to the translational shells category. Their midsurface is generated by a downwards curved parabola (generatrix), which 'slides' on another fixed parabola curved upwards (guide), keeping its plane always parallel to itself. The created surface has the form of a 'saddle' and, obviously, has a negative curvature, in contrast to the shells examined previously (see Figure 3.33).

Such shells, apart from their impressive aesthetics resulting from the alternating curvature, show a characteristic of particular constructional importance because of their negative Gaussian curvature – that is, they may also be generated using straight lines.

Thus, if a straight line, regarded as a generatrix, slides in space with both its edges, over another two fixed straight lines that belong to vertical parallel planes, keeping itself parallel to another fixed vertical plane on the one hand and with a constant rate of variation of its slope over the travelled distance by its sliding end on the other, then this line will produce the same surface as that previously described (see Figure 3.33). This will be explained in detail immediately below. What will, however, be pointed out first is that an understanding of this 'dual' nature of the hyperbolic paraboloid shell is of paramount importance for the comprehension of its load-bearing potential, as well as for the design of these shells in general, as they are able to cover orthogonal (and other) layouts with an unlimited variety of morphologies. In all cases, the possibility of creating the surface

Figure 3.33 Creation of hyperbolic paraboloid shells

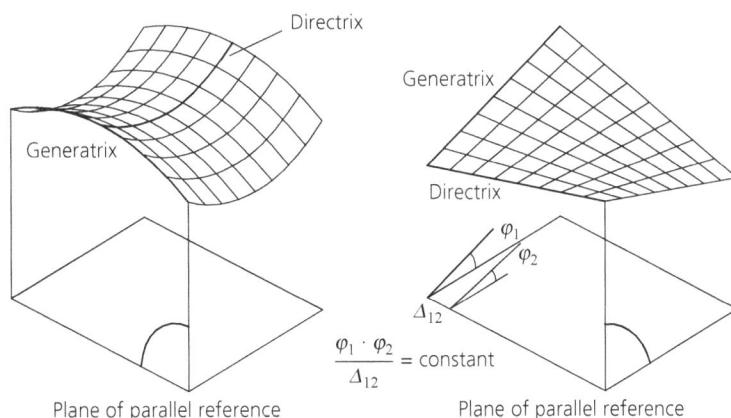

$$\frac{\varphi_1 \cdot \varphi_2}{\Delta_{12}} = \text{constant}$$

by using exclusively straight-line segments is a crucial constructional advantage that gives these shells a clear economic precedence.

As an understanding of the geometry of this shell is necessary for understanding its structural behaviour, it is appropriate first to examine it through the following analysis.

3.5.2   Perception of the geometry

The analytic description of the creation of the surface by the two parabolas is based on Figure 3.34.

In the considered orthogonal system $Oxyz$, the 'generatrix' parabola with a span and rise equal to $2a$ and f_1, respectively, has its plane parallel to the plane Oxz, while the 'guide' parabola curved in the opposite direction lies in the plane Oyz, having a span and rise (sag) equal to $2b$ and f_2, respectively. The coordinate z of the created surface is determined from the relation:

$$z = \frac{x^2}{2R_1} - \frac{y^2}{2R_2}$$

where R_1 and R_2 are the radii of curvature of the two parabolas, assumed to be constants, and equal to $R_1 = a^2/2 \cdot f_1$ and $R_2 = b^2/2 \cdot f_2$, respectively. It is assumed that both parabolas are 'shallow' in the sense described in Section 3.4.2.

It can be seen that the so-created surface covers a rectangular ground plan of dimensions $(2 \cdot a) \cdot (2 \cdot b)$. Every section of this surface with a vertical plane parallel to either x or y axis is a parabola, while a section with a horizontal plane produces a hyperbola. It is clear that within the considered orthogonal system Oxy, the twist of the surface everywhere, expressed as $t_{xy} = \dfrac{\partial^2 z}{\partial x \cdot \partial y}$, is equal to zero.

If a vertical plane is now considered whose trace on the horizontal plane Oxy forms an angle, ω, with the x axis such that $\tan^2 \omega = R_2/R_1$, then this plane intersects the shell along a straight line. In addition, if the symmetric of this plane with respect to the x axis is considered, its intersection with

Figure 3.34 Creation of a shell through a parabolic generatrix

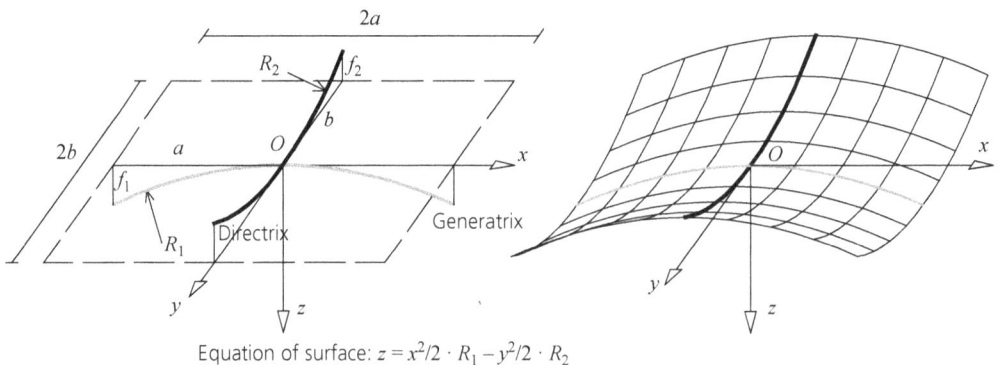

Equation of surface: $z = x^2/2 \cdot R_1 - y^2/2 \cdot R_2$

the shell is also a straight line. In other words, the x axis bisects the angle $2 \cdot \omega$ formed by the traces of the above two planes over the horizontal plane Oxy (see Figure 3.35).

If these two traces are considered as skew coordinate axes Ou and Ov, then it can be proved that the coordinates z of the surface may also be obtained from the expression:

$$z = k \cdot u \cdot v$$

where $k = 2/(R_1 + R_2)$ represents the existing twist of the surface within the skew axes of reference Ouv.

It is clear that the intersection of the surface with a vertical plane parallel to Ouz or parallel to Ovz is a straight line. The same shell can be assumed to be created if a straight line Oa, initially coinciding with axis Ou, is transposed in space while remaining parallel to the plane Ouz, with its edge always lying on the axis Ov. During this line movement, the ratio of $\Delta\varphi$, the change in the angle of its slope φ relative to the plane Ouv, to the corresponding distance travelled by its edge O on the axis Ov should be kept constant and equal to k (see Figure 3.35). It is clear that k represents the constant twist of the surface relative to the coordinate system Ouv considered:

$$k = d^2 z / \partial u \cdot \partial v \qquad \blacksquare$$

The transition from one geometry to the other may also take place.

Figure 3.35 Creation of a shell through a straight generatrix and the appearance of twist

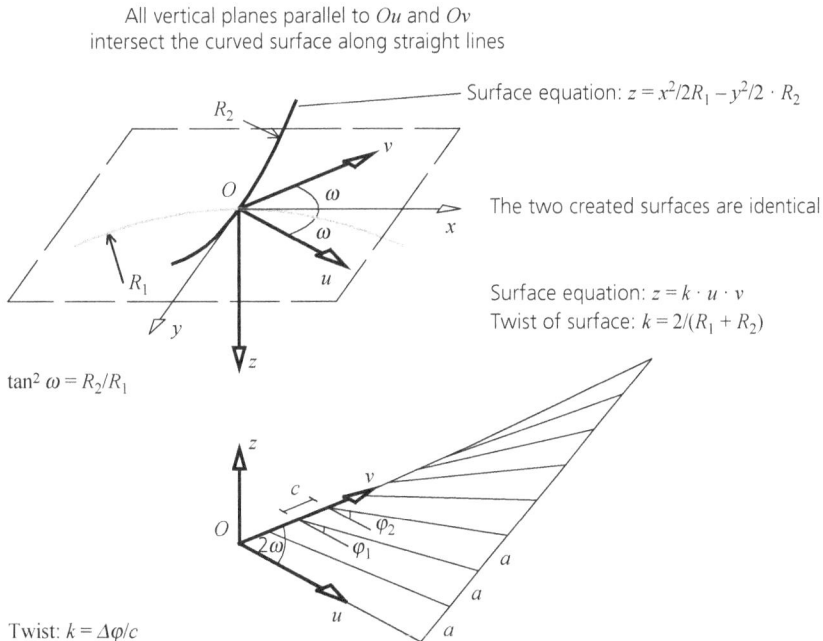

All vertical planes parallel to Ou and Ov
intersect the curved surface along straight lines

Surface equation: $z = x^2/2R_1 - y^2/2 \cdot R_2$

The two created surfaces are identical

Surface equation: $z = k \cdot u \cdot v$
Twist of surface: $k = 2/(R_1 + R_2)$

$\tan^2 \omega = R_2/R_1$

Twist: $k = \Delta\varphi/c$

Assume that the shell is intended to cover a parallelogram ground plan OACB with sides a, b enclosing an angle $2 \cdot \omega$, while the skew axes Ou and Ov coincide with the two sides of the parallelogram (see Figure 3.36).

Corresponding to the edge C with skew coordinates $u = a$, $v = b$, a point C' is defined with coordinate $z = f$ and connected through a line segment to the two other edges A and B. The hyperbolic paraboloid shell can be created if the segment OA moves parallel to the plane Ouz, inclined towards the lines OB and AC'. The coordinates z of the so-formed surface are $z = k \cdot u \cdot v$, where $k = f/a \cdot b$.

It is clear that any vertical plane parallel either to the axis Ou or Ov intersects this surface along a straight line, thus making it possible for the surface to cover any parallelogram ground plan.

This same surface may now be depicted through a system of mutually orthogonal parabolas with opposite curvatures in the following way (see Figure 3.37).

If the axis Ox is considered as the bisecting line of the angle $2 \cdot \omega$, and the axis Oy is drawn perpendicular to it, then each of the planes Oxz and Oyz intersects the surface along parabolas. The intersection with the plane Oxz shows an upward curvature with a radius:

$$R_1 = 2 \cdot \cos^2\omega / k$$

while the intersection with the plane Oyz shows a downward curvature with a radius:

$$R_2 = 2 \cdot \sin^2\omega / k$$

Thus, the coordinates z of this same surface referred to the orthogonal system Oxy can be expressed through the radii of curvature R_1 and R_2, as previously shown.

Figure 3.36 Creation of a hyperbolic paraboloid surface over a skew ground plan

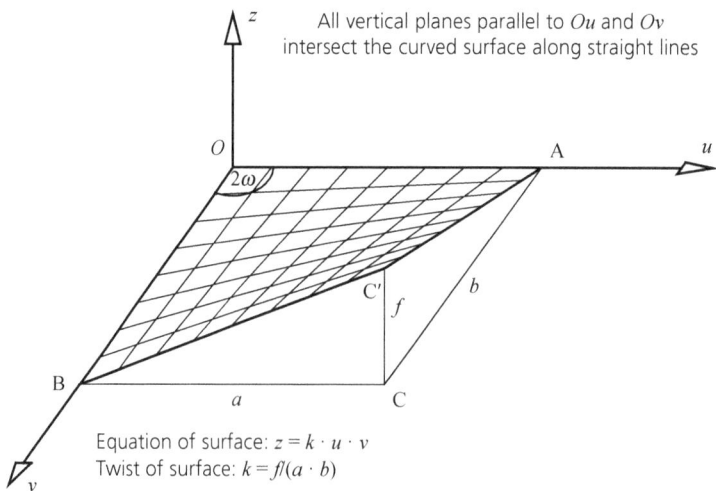

Figure 3.37 Restoration of parabolic generatrices in a shell over a rhomboid ground plan

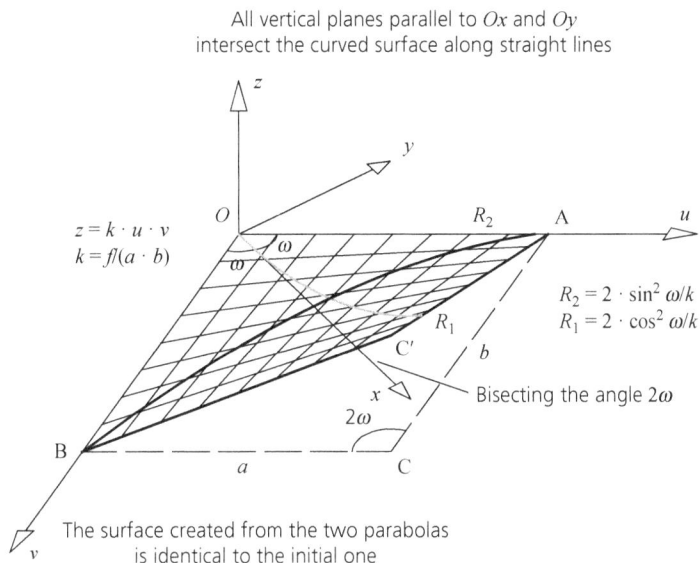

All vertical planes parallel to Ox and Oy
intersect the curved surface along straight lines

$z = k \cdot u \cdot v$
$k = f/(a \cdot b)$

$R_2 = 2 \cdot \sin^2 \omega/k$
$R_1 = 2 \cdot \cos^2 \omega/k$

Bisecting the angle 2ω

The surface created from the two parabolas
is identical to the initial one

Remark: In the case of rhomboid layout ($a = b$), the axis Ox coincides with the line OC,
but the two radii of curvature are not equal

If now the area to be covered is rectangular, meaning that the angle, ω, between the axes Ov and Ou is 90°, the resulting radii of curvature R_1 and R_2 referring to the orthogonal axes Ox and Oy produced as above are equal, as suggested from the last equations. Conversely, it is clear that a hyperbolic paraboloid shell with equal radii of curvature $R_1 = R_2 = R$, being referred to the orthogonal axes Oxy, can also be considered as created from the straight generatrices representing its intersections with the vertical and mutually perpendicular planes Ouz and Ovz that bisect the right angles of the axes Ox and Oy. The twist, k, of the surface corresponding to the equation $z = k \cdot u \cdot v$ is, according to the above, equal to $k = 1/R$.

In order to acquire a more direct understanding of the load-bearing behaviour, shells with equal radii of curvature in both the x and y directions will be further considered.

3.5.3 Consideration of equilibrium

In the orthogonal layout shown in Figure 3.38, $a^2/2 \cdot f_1 = b^2/2 \cdot f_2 = R$ (see Section 3.5.2). As can be seen from the figure, the shell extends between the four parabolas corresponding to the sides of the orthogonal base.

The twist, t_{xy}, of the surface with reference to the system Oxy is zero, as can be deduced from the expression $\partial^2 z/\partial x \cdot \partial y$, which implies that the tangential shear forces, S, in the directions x and y are also zero. The equilibrium of a curved quadrilateral element (see Figure 3.38) shows that the load, p, may be carried by the compressive forces, as well as by the tensile forces N_x and N_y. This is clear from the fact that the load, p, causes compression of downward-curved parabolas, making them act as *arches*, while simultaneously it applies tension to the upward-curved parabolas,

Figure 3.38 Load-bearing action of a hyperbolic paraboloid shell with parabolic edges

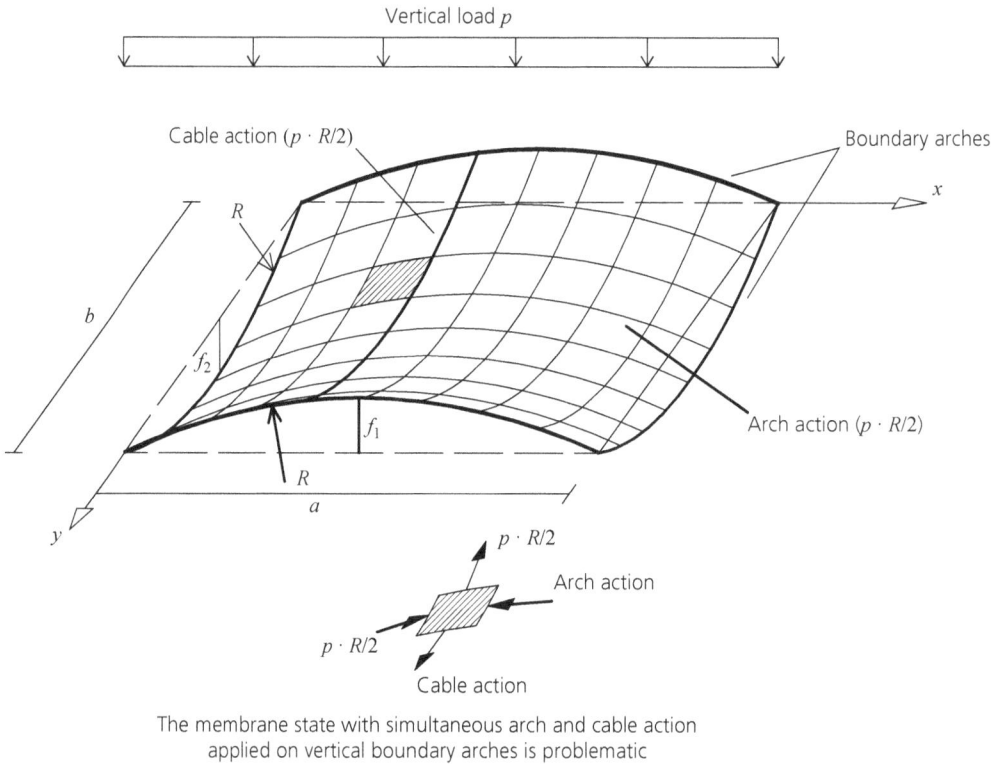

The membrane state with simultaneous arch and cable action
applied on vertical boundary arches is problematic

making them behave as *cables*. From the equilibrium equation in Section 3.2, it can be deduced that the membrane forces N_x and N_y have a common value equal to $p \cdot R/2$.

However, in order for this membrane state in the shell to exist, the above axial membrane forces should be appropriately introduced at the boundaries of the shell. If the shell is unyieldingly supported along its parabolic boundaries, then this is feasible. However, if the shell is supported on, for example, vertical parabolic arches, these are not able to carry forces perpendicular to their plane, and thus the purely membrane state of stress is impossible. The very fact, of course, that these arches are able to carry vertical loads means that vertical shear forces will necessarily be introduced in the shell, thus leading unavoidably to bending.

Another support possibility is the existence of unyielding supports along the ends of downward-curved (i.e. compressed) parabolic 'arches', while the edges of the upward-curved ones are free (see Figure 3.39). Then, the load, p, has to be supported entirely by the compressed parabolas, which will develop a compressive force, N_x, equal to $p \cdot R$, double the previous value (see Section 3.2).

It is now of interest to examine whether the above-discussed case of a shell bearing its load, p, through curved paths can be developed without transferring transversal loads to the edge-supporting arches, being previously described as problematic, while ensuring that a pure membrane state develops (see Franz and Schäfer, 1988). This is feasible on the basis of the fact that the cuts of the shell with vertical planes mutually perpendicular and with an angle of 45° to the axes x and y are straight lines, as mentioned previously (see Section 3.5.2).

Two cases are examined, one over a square plan and the other over an oblong plan with one side twice as long as the other.

In the first case depicted in Figure 3.40, the right-side boundary is assumed to represent an unyielding support bearing a compressive distributed load $(2 \cdot p \cdot R)$, while the boundary arches along the opposite boundaries are considered as offering tangential shear forces $(p \cdot R)$ all along their length. These shear forces are balanced (or offered) by the compressive and tensile forces equal

Figure 3.39 Feasible membrane equilibrium using arch action and free opposite boundaries

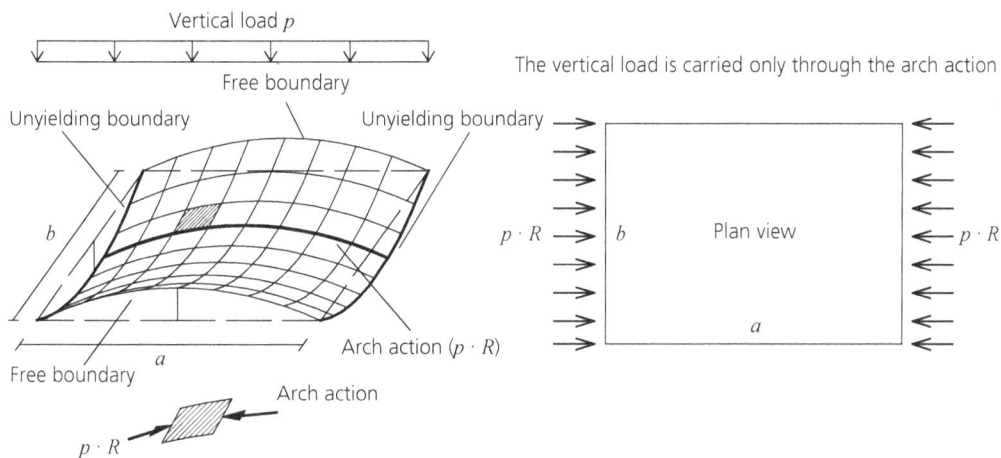

Figure 3.40 Feasible membrane equilibrium in a square layout with a free edge

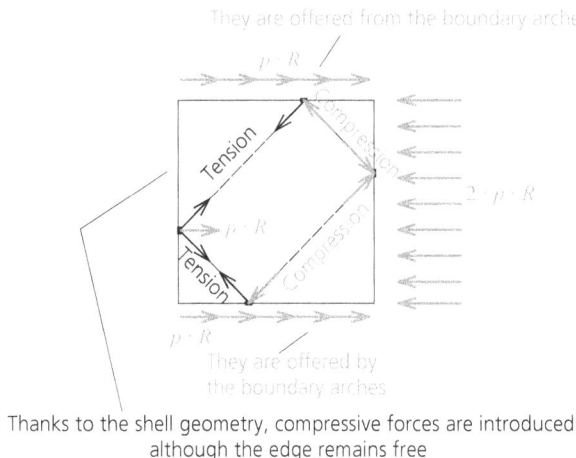

Thanks to the shell geometry, compressive forces are introduced although the edge remains free

Figure 3.41 Feasible membrane equilibrium in an oblong layout with two free edges

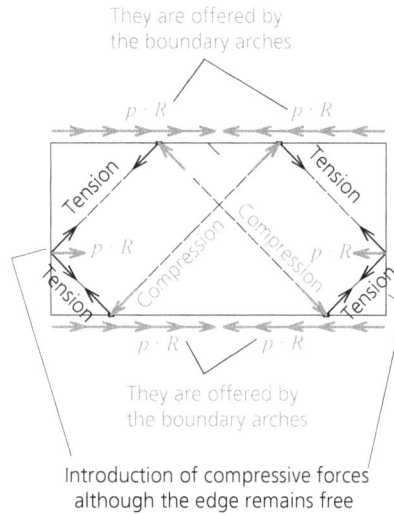

to $(0.707 \cdot p \cdot R)$ acting at $45°$ on both the opposite boundaries, while the tensile forces acting on the left boundary offer the required compressive force $N_x = p \cdot R$ to be developed there. The whole acting forces layout satisfies the global equilibrium of the shell, while, thanks to the existing recti-linear tracks in the shell, it offers the required internal forces for both the curved arch- or cable-like paths so that the vertical load, p, can be undertaken exclusively by membrane forces.

In the second case depicted in Figure 3.41, the short sides are considered free, whereas the long sides are supported by arch-like boundaries offering to the shell a self-equilibrating system of shear forces as shown, which is equilibrated by a system of tensile and compressive forces acting on the boundary at $45°$. This, in turn, offers through the concurring tensile forces acting on the short sides the compressive force required to ensure the equilibrium along the arch-like trajectories of the shell. In this way, the shell can take up the gravity load, p, by developing only membrane forces.

3.5.4   Hyperbolic paraboloid shells with straight edges

To cover orthogonal areas, *hypar* shells are usually formed with straight boundaries. The shell is necessarily referred to a system of axes Oxy comprising two adjacent sides of the orthogonal base $OABC$ (see Figure 3.42). With the origin O kept fixed, the three other edges can be moved verti-cally at any distance, as shown in Figure 3.42. The shell is created by sliding of the segment OA' along the guiding sides OB' and $A'C$, staying always parallel to the plane Oxz, until it reaches the position $B'C'$. The twist, k, of the created surface expresses the total change of the slope of the generatrix OA' (equal to $(f_1/a) + (f_2 - f_3)/a$), which takes place over the moving distance b – that is,

$$k = \frac{f_3 - f_1 - f_2}{a \cdot b}$$

The equation of the surface is expressed as

$$z = k \cdot x \cdot y + (f_1/a) \cdot x + (f_2/a) \cdot y$$

According to Section 3.5.2 and given that $2 \cdot \omega = 90°$, both surface curvatures (i.e. '$1/R$') are equal to k. These curvatures appear understandably along the skew directions at an angle $(2 \cdot \omega)/2 = 45°$ relative to the axes Ox and Oy, respectively.

Regarding now the equilibrium of an orthogonal shell element in the system Oxy under the vertical load, p, it can be seen that the only membrane mechanism through which this load may be carried is the system of the tangential shearing forces, S, acting along the sides of the element (see Figure 3.42). This is so because the shell element presents null curvatures about the two orthogonal axes x and y, and, consequently, any internal forces N_x or N_y that are developed, being co-linear for any corresponding pair of opposite sides, cannot offer a component that resists the load. Thus, it follows that the above forces cannot be developed, so that the basic equation for membrane equilibrium in Section 3.2 yields the following constant value for S:

$$S = \frac{p}{2 \cdot k} = \frac{p \cdot R}{2}$$

Indeed, the difference in tilt at opposite sides of the element allows the forces, S, to offer an upward component, which equilibrates the vertical load, p (see Figure 3.42).

These forces, S, must necessarily be offered at the straight edges of the shell too, which is possible only if edge beams are arranged along each boundary.

It should be noted here that the previous conclusion about zero membrane forces N_x and N_y is strictly valid only if the load, p, acts perpendicularly to the shell surface. Since the acting load is vertical, it is clear that for small slopes of the surface the above conclusion is essentially valid. 'Small slopes' are identical to the existence of 'small curvatures', which, in turn, means that the shell should be 'shallow', and this is the case if the edge beams have a slope not exceeding 18°. For steeper slopes, the vertical load, p, has a non-negligible component in the direction of the straight generatrices, leading to the development of axial forces in their direction and, consequently, the membrane shearing forces are no longer constant. However, for most cases of roofing, the hypar shells used can be considered 'shallow'.

Figure 3.42 Load-bearing shear mechanism of a hypar shell with straight edges

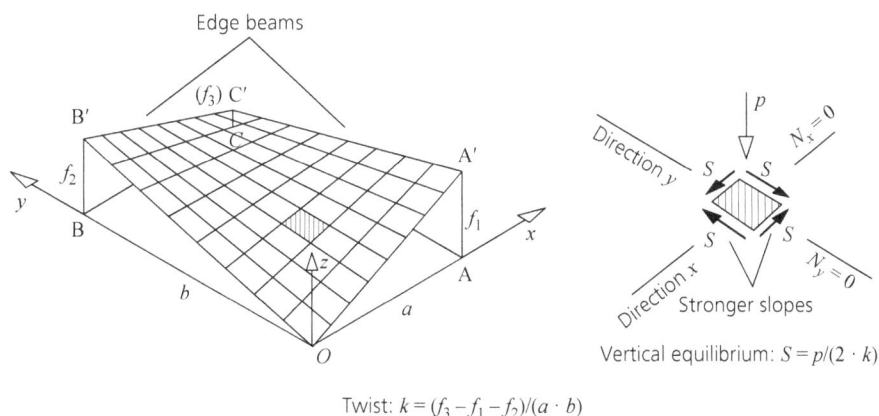

Twist: $k = (f_3 - f_1 - f_2)/(a \cdot b)$

Of course, irrespective of the above impact of the equilibrium consideration for the x and y axes, it should be noted that axial compressive and tensile forces, N, are developed due to the existing curvatures along both oblique directions of $45°$.

So, from the equilibrium of a cut-out part, as shown in Figure 3.43, for the compressive and tensile membrane forces N_{arch} and N_{cable}, respectively it can be written as:

$$S \cdot 1 \cdot \frac{\sqrt{2}}{2} \cdot 2 = N_{arch} \cdot \sqrt{2} \quad \text{or} \quad S \cdot 1 \cdot \frac{\sqrt{2}}{2} \cdot 2 = N_{cable} \cdot \sqrt{2}$$

and thus, according to the previous equations:

$$N_{arch} = N_{cable} = N = S = p/(2 \cdot k) = p \cdot R/2$$

that is, the normal forces N_{arch} and N_{cable} are equal to the membrane shear forces, S (see Figure 3.43).

Thus, the shell always works as a group of compressed 'arches' and suspended 'cables' both developing the same axial force. These forces reaching the boundary may be considered as being applied to the vertical sides of unit length of the orthogonal triangular element, as shown in Figure 3.43, the hypotenuse of which is $\sqrt{2}$. The equilibrium of this element, given that the component of forces, N, perpendicular to the boundary cancel each other, confirms the necessity for the membrane shearing force to act along the boundary and to be equal to N. This is exactly the shearing force found above, running through the whole shell, which should therefore be offered to the shell at the boundary by the edge beams being themselves subjected to the opposite forces.

Figure 3.43 Load-bearing mechanism through the arch and suspension action of a shell

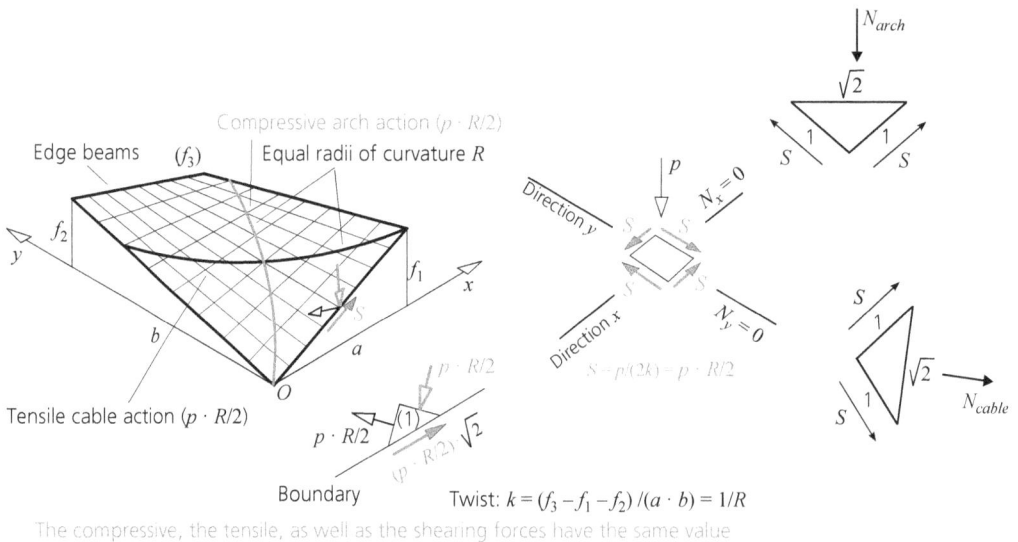

Twist: $k = (f_3 - f_1 - f_2)/(a \cdot b) = 1/R$

The compressive, the tensile, as well as the shearing forces have the same value

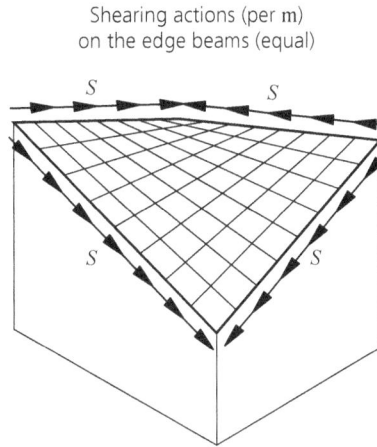

Figure 3.44 Forces acting on the edge beams of straight boundaries

Shearing actions (per m)
on the edge beams (equal)

The shearing actions are always summed up on the lowest corners

As shown in Figure 3.44, the forces on the edge beams accumulate to their lowest points, where they are taken up by appropriately designed columns. Thus, the edge beams develop a progressively increasing compression towards their lower supports, and in all cases care must be taken to carry their self-weight also, perhaps by providing a continuous support. ∎

A new question arises, however, of whether the edge beams that will carry the above shearing forces can be designed free of intermediate support and, moreover, if it is possible to design a layout of boundary beams supported (fixed) only at their corresponding lower ends and, remaining free all along their length, working as large cantilevers.

The fact is that, although according to the membrane state these cantilevers do not carry any part of the shell vertical loads but receive only longitudinal ones, they have nevertheless a significant self-weight, which they are not able to carry as long they are acting as cantilevers. The question, therefore, which essentially arises is whether the shell can participate in bearing the self-weight of the edge beams and, if so, with what consequences for its state of stress.

The answer to this question is in principle positive, and is based on the fact that there are straight generatrices between the edge beams that can act as inserted cables. These 'cables' can act against the tendency of the cantilevers to deform by applying transversally acting tensile forces along the boundary beam, which will relieve the cantilever from the response due to its self-weight. These forces are carried essentially unchanged from a given boundary edge to the opposite one (Schlaich, 1970).

As shown in Figure 3.45, the system of the two adjacent 'cantilevers' (OA) and (OB), the sides (AC) and (BC) considered unmovably fixed, is subjected to the vertical loads of their self-weight, as well as to the forces of the tensioned cables at every discrete level. The tendency of the cantilevers to tilt about the axis AB by an acting overturning moment, M_0, is counteracted by the cable force, Z, which acts above the base line AB.

Figure 3.45 Self-weight of the edge beams is taken up by 'cable action'

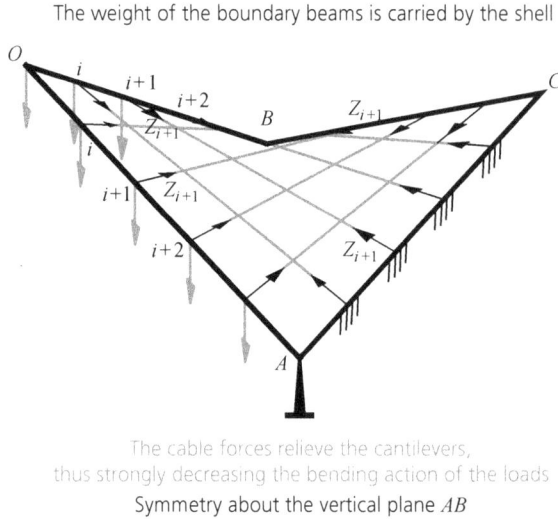

The weight of the boundary beams is carried by the shell

The cable forces relieve the cantilevers,
thus strongly decreasing the bending action of the loads
Symmetry about the vertical plane AB

These cable forces, Z, counteract the tendency of the 'cantilevers' to tilt about the axis AB as they cause an opposite bending moment to M_0 due to their self-weight, thus relieving them considerably. For simplicity, the structure is assumed to be symmetrical about the vertical plane passing through the axis AB.

An approximate estimation of the tensile cable forces can be made by assuming that the cantilevers are elastically undeformable and behave like rigid bodies rotating freely about the hinged axis of their support AB (see Figure 3.46). It can be obviously considered that the system of cables actually may consist of prestressed tendons anchored at the opposite edge beams.

Thus, for each level i having a distance l_i from the corresponding hinged end of the respective boundary, the forces Z_i are determined on the basis of the difference Δz_i, of their ends their projection length, Δx, on the vertical plane OC; the corresponding cable length, L_i; the distance, e_i, of the corresponding resultant R_i of forces Z_i at each node from the support axis AB; the inclination γ of the boundary beam and, finally, the cross-sectional area, F_i, of the 'cables', which may correspond to the tributary shell strips between them.

Thus, according to Schlaich (1970):

$$Z_i = M_0 \frac{K_i}{2 \cdot \cos(\omega/2)} \cdot \sum_i \frac{1}{K_i \cdot e_i} \quad (\omega = 90)$$

where

$$K_i = \frac{F_i \cdot l_i}{L_i^3}(\Delta x \cdot \sin\gamma + \Delta z_i \cdot \cos\gamma)$$

Figure 3.46 Assessment of the required tensile (prestressing) cable forces

Cable forces acting on the part AOB

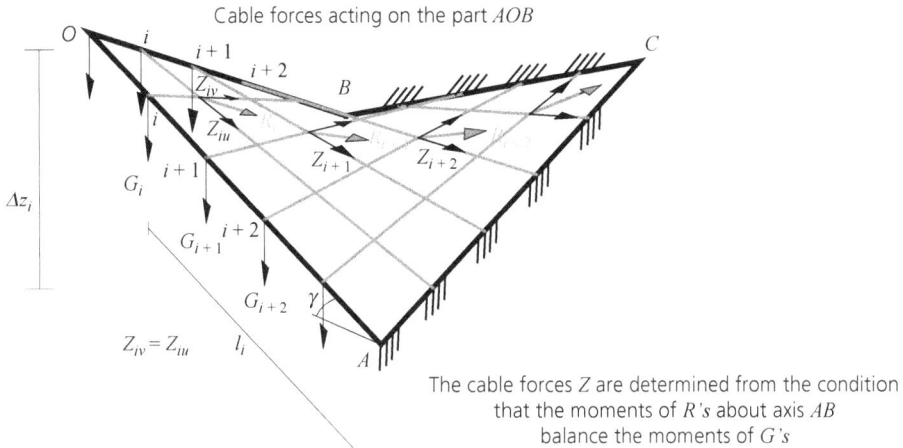

The cable forces Z are determined from the condition
that the moments of R's about axis AB
balance the moments of G's

The cable forces have to be equal in each direction
due to symmetry about the vertical plane AB

$R_i = 2 \cdot Z_i \cdot \cos(\omega/2)$

$\omega = 90°$

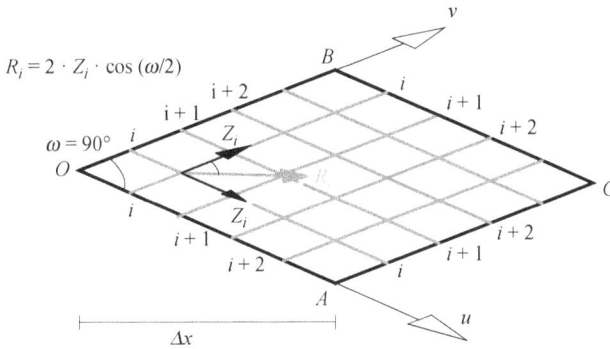

and

$$\Delta x = \frac{(AB)}{2 \cdot \tan(\omega/2)} \quad (\omega = 90)$$

As mentioned previously, these tensile forces, Z_i, should be taken up by prestressing (see Figure 3.47). By introducing the appropriate prestressing forces along the straight generatrices, a purely membrane state in the shell may be feasible in practice, given that each compressive force applied at the shell boundary is not spread out but remains in the 'strip' of the corresponding generatrix. In addition, the absence of curvature in the tendons essentially eliminates the friction losses. ∎

The synthesis of individual shells within a square plan in wider forms through appropriate juxtaposition of shell parts makes possible the covering of larger areas with a small number of supports.

Figure 3.47 Realisation of cable action through prestressing of a shell

The required tensile forces are provided through prestressing

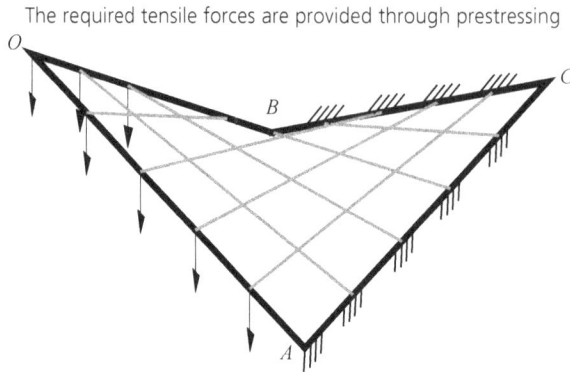

Figure 3.48 Layout of four hypar shells to cover a square ground plan

The horizontal acmes (beams) carry the self-equilibrating actions from the adjacent shells

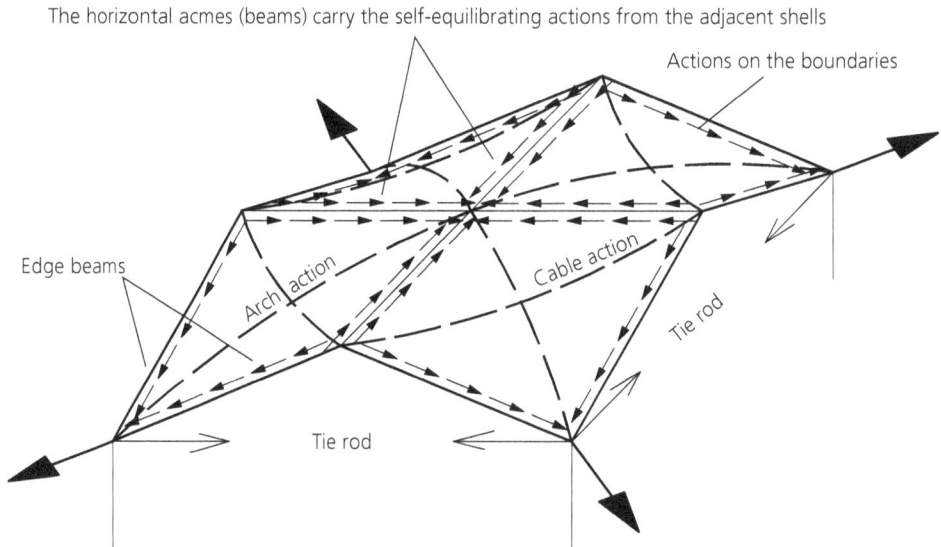

The covering of a square ground plan using four individual shallow hypar shells is shown in Figure 3.48. Each shell develops a shearing state of stress in the direction of its generatrices, which, as explained above, is caused by the simultaneous presence of the equal tensile and compressive forces of the 'arch' and 'suspension' action, respectively.

It can be seen that the two imaginary cut strips along the top horizontal edges are themselves in equilibrium under the opposite acting shearing forces, as shown in Figure 3.48. Thus, the required shearing forces along the interior boundary of each shell are provided by the adjacent shell. However, the required shearing forces at the 'external' boundaries have to be provided by edge beams that, being acted on by the opposite forces, have to give up their resultants to the corner columns.

These resultants have a horizontal component acting along the diagonals of the ground plan, according to the arch action indicated in Figure 3.48 (see Stavridis and Georgiadis, 2025, Section 8.1). The equilibrium of the column top necessitates the provision of a horizontal diagonal action in order to avoid an overstressing of both the columns and the edge beams and this is accomplished by providing peripheral ties that are prestressed accordingly. ∎

Using the same juxtaposition of four hypar shells as above and applying the previously described load-bearing possibilities, the covering of a similar orthogonal ground plan may be achieved, but this time with cantilevered perimetral edges, as shown in Figure 3.49. The prestressing cables may be inserted along the respective straight generatrices of each hypar shell and anchored all along both the perimetral and internal beam shell boundaries. With the horizontal components of the acting anchor forces on both sides of the internal beams being self-balanced, the vertical forces are taken up by the two transverse frames with inclined columns that, by having their geometry closer to the corresponding funicular parabolic form for these loads, exhibit a relatively moderate bending response.

Of course, the strong horizontal thrusts developed at the base of the inclined columns may necessitate the implementation of appropriate prestressed ties placed underneath the ground level (see Stavridis and Georgiadis, 2025, Section 6.2.1 and Section 1.4). ∎

Despite the fact that the preliminary design of hypar shells may be based on the assumption that the loads are carried by membrane action, it should be noted that various incompatibilities arising from this consideration may also affect the bending action of the shell. Thus, it is noted, among other things that, for example, non-uniform loadings may cause bending in the shell, or even that the mechanism through which the self-weight of the straight edge beams is carried through membrane action may never be complete.

Figure 3.49 Layout of hypar shells to cover a square ground plan with unsupported edges

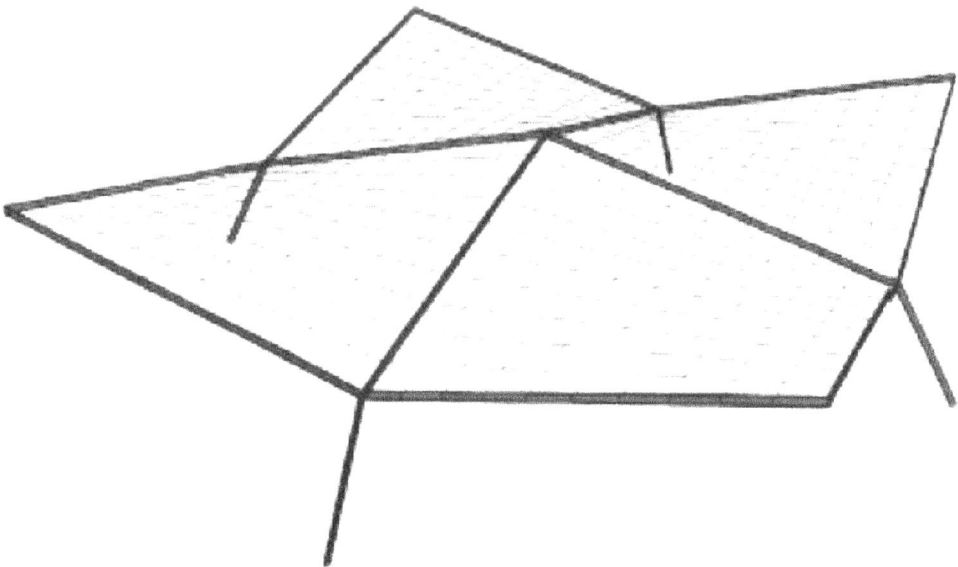

As has been emphasised elsewhere, in order to determine the final state of stress of the shell, the use of appropriate software is necessary so that all details of its constructional formation can be taken into account.

3.5.5   Elastic stability
The presence of compressive forces in hypar shells constitutes a danger of buckling that should be evaluated sufficiently during their preliminary design.

In the case of shells with parabolic boundaries, it is clear that the presence of tensile forces along the upward-curved parabolas clearly contributes to the stability of the embedded 'arches'. Thus, the use, in this case, of the expression for the critical load of a cylindrical shell having the same dimensions in plan as well as the same curvature will definitely lead to safe results (see Section 3.3.2).

In the case of hypar shells with straight boundaries that are rigidly supported, having ground plan dimensions a and b ($a \cdot b$), a height difference, f, and a thickness, d, the critical load, p_K, may be assessed from the relation (Beles and Soare, 1972):

$$p_K = 0.40 \cdot E \cdot \frac{f^2 \cdot d^2}{a^2 \cdot b^2}$$

If there are also edge beams having a moment of inertia equal to I, the critical load is greater. This can be assessed from the relation:

$$p_K = 2 \cdot E \cdot \frac{f^2 \cdot d^2}{a^2 \cdot b^2} \sqrt{\frac{2}{3} \cdot \left(1 + \frac{24 \cdot I}{d^3 \cdot a}\right)}$$

3.6.   Conoidal shells
Along with the translational shells are also classified those shells that, in order to cover a rectangular ground plan between two parabolic fronts over two opposite sides of the rectangle, are formed through the segment of the other side as a strait generatrix by using the two parabolic edges as 'guides'. This generatrix moves parallel to itself, sliding between the two parabolic guides and thus offering obvious advantages regarding the construction of the shell formwork (see Figure 3.50). One of the two parabolic fronts is usually made higher than the other, which could also simply be a straight one.

The preliminary design of such a shell can be based on considering it as an arch having a height, f_m, equal to the average height of the two front parabolas. If L is the span of the parabolic arch, the compressive force per unit metre of the shell due to the vertical load, g, is:

$$N_d = g \cdot L^2 / (8 \cdot f_m)$$

In addition, the critical buckling load can be assessed on the basis of the average radius of curvature $R = L^2 / 8 \cdot f_m$ from the corresponding relation for cylindrical shells (see Section 3.3.2).

In the case in which the shell is not directly supported along its straight edges of length, B, it can be assumed that the above compressive 'arch' forces of the shell are carried by the deep beams,

Figure 3.50 Load-bearing action of conoidal shells

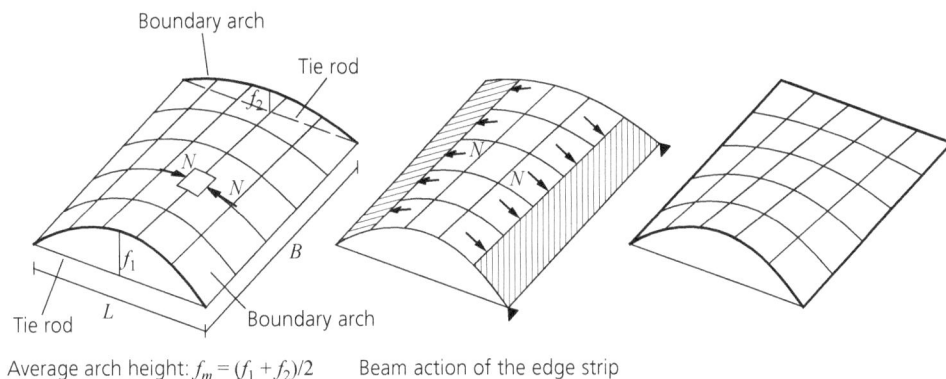

Boundary arch
Tie rod
f_2
N
N
f_1
B
Tie rod
L
Boundary arch
N
N

Average arch height: $f_m = (f_1 + f_2)/2$ Beam action of the edge strip

which are formed along the edge region with an assumed structural height of $B/4$ acting as simply supported beams (see Figure 3.50).

A not unusual use of conoidal shells is that of a shelter in the form of a cantilever, fixed at its parabolic edge and having all its three other straight edges free (see Figure 3.50).

Its structural function may be conceived as that of a cantilever of length, B, having a parabolic cross-section, which presents an advantageous increase in its moment of inertia towards the fixed end, where the maximum tensile and compressive stresses are developed.

According to the above approach, the maximum tensile force, N_z, which results under a vertical load, g, is (Kollar, 1984):

$$N_z = \frac{B^2 \cdot g/2}{0.265 \cdot f} = 1.90 \cdot \frac{B^2 \cdot g}{f}$$

and is developed at the top of the supported parabolic edge.

Beyond this 'membrane' load-bearing mechanism with regard to the cantilever moment $M_e = g \cdot B^2/2$, a bending moment, M_b, of much less intensity also has to be considered, being assessed as $M_b = (d/f)^2 \cdot M_e$.

Furthermore, a bending moment, M_{tr} in the transverse direction has to be taken into account, which causes tension at the top fibres since it increases the transversal curvature of the shell. Its value at the middle of span, L, may be assessed as:

$$M_{tr} = 0.13 \cdot g \cdot L^2$$

In the case in which such shells are arranged monolithically connected in order to shelter larger spaces (see Figure 3.51), the transversal moments are limited to about 25% of the above value, with alternating signs across each field, causing a tension at the bottom fibres over the top of the connected 'arches'.

Figure 3.51 Structural action of a conoidal shell as a cantilever

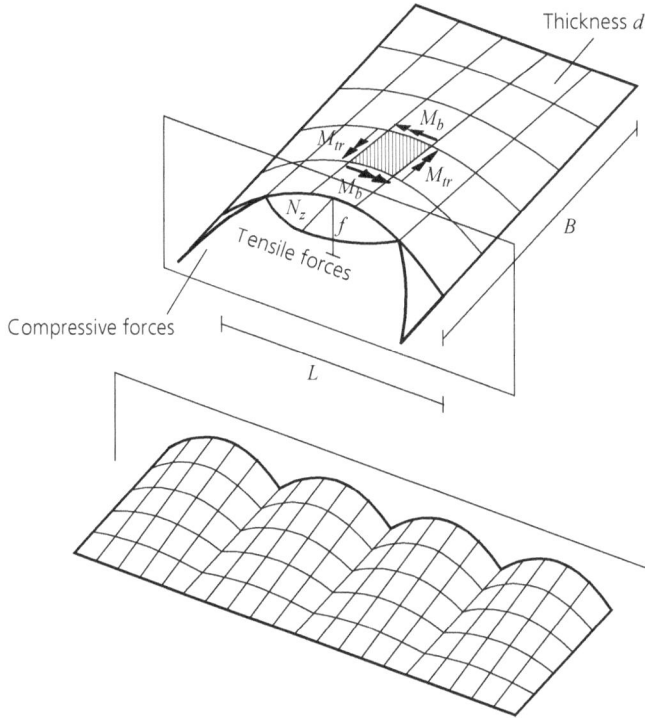

3.7. Worked examples

Example 1

A continuous concrete barrel shell with two spans $L = 16.0$ m, $R = 4.0$ m and thickness $d = 13$ cm carries, besides its own self-weight ($\gamma = 25$ kN/m³), a snow load equal to $q = 1.4$ kN/m². The barrel shell is continuous over three stiffeners (see Figure 3.52).

The total uniform vertical load acting on the shell is:

$$w = \pi \cdot R \cdot (d \cdot \gamma) + q \cdot 2 \cdot R = 52.0 \text{ kN/m}$$

and produces a maximum hogging bending moment over the middle support equal to:

$$M = w \cdot L^2/8 = 52.0 \cdot 16.0^2/8 = 1664 \text{ kNm}$$

The maximum compressive and tensile unit forces due to M are:

$$N_{x,max} = (M/(0.83 \cdot d \cdot R^2)) \cdot d = (1664.0/(0.83 \cdot 0.13 \cdot 4.0^2)) \cdot 0.13 = 35 \text{ kN/m (compression)}$$

and

$$N_{x,max} = (M/(0.47 \cdot d \cdot R^2)) \cdot d = (1664.0/(0.47 \cdot 0.13 \cdot 4.0^2)) \cdot 0.13 = 221 \text{ kN/m (tension)}$$

Figure 3.52 Concrete barell shell

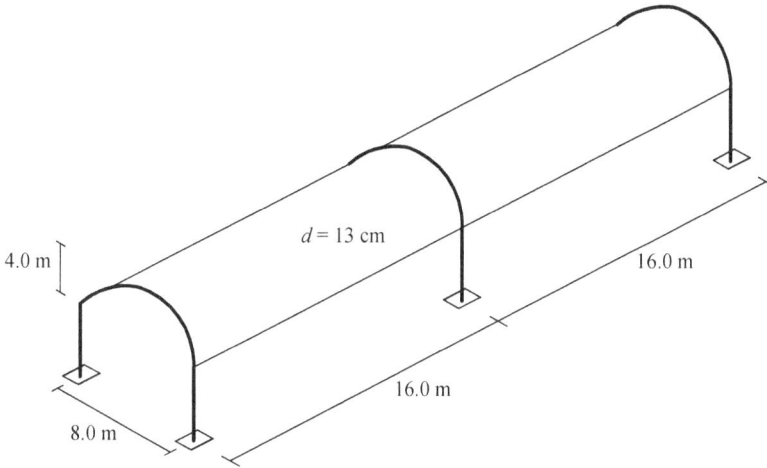

Over the middle support where the barrel shell is fixed, the bending moment that is developed is (see Section 3.3):

$$M_e = 0.29 \cdot g \cdot R \cdot d = 0.29 \cdot (0.13 \cdot 25 + 1.4) \cdot 4.0 \cdot 0.13 = 0.70 \text{ kNm/m},$$

extending over a distance of:

$$c = 0.76 \cdot \sqrt{R \cdot d} = 0.76 \cdot (4.0 \cdot 0.13)^{0.5} = 0.20 \text{ m}$$

As this distance is very small, the shell takes up the vertical load mainly through membrane forces.

Example 2

A spherical concrete dome has a radius of $R = 15.0$ m and, besides its own self-weight ($\gamma = 25$ kN/m³), is loaded by a snow load of $q = 1.4$ kN/m². The shell has a thickness $d = 0.15$ m and is considered fixed at the boundary (see Figure 3.53).

Figure 3.53 Spherical concrete dome

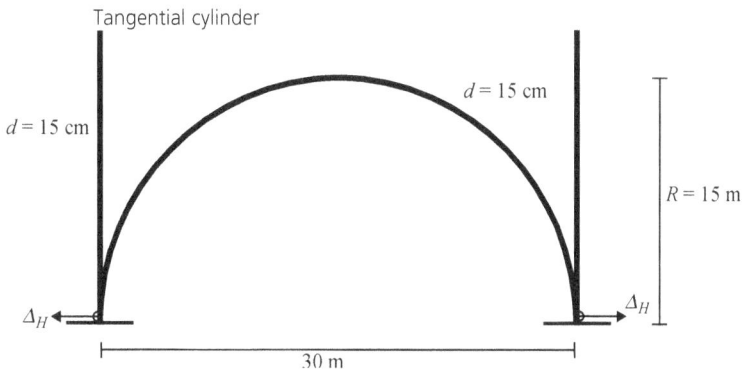

Self-weight: $g = d \cdot \gamma = 3.75$ kN/m^2

Tensile ring force near the base: $N_\theta = R \cdot g \cdot (1/(1 + \cos 90) - \cos 90) - p \cdot R \cdot \cos (2 \cdot 90)/2 = R \cdot g - p \cdot R \cdot \cos (180)/2 = 15.0 \cdot 3.75 + 1.4 \cdot 15.0/2 = 66.75$ kN/m

Required reinforcement: $A_s = 6675/2000 = 3.5$ cm^2/m

Elongation strain: $\varepsilon_\theta = 66.75/(0.15 \cdot 3 \cdot 10^7)$

Radial outward displacement for the equivalent tangential cylinder according to Geckeler (1926): $\Delta_H = \varepsilon_\theta \cdot R = 0.00022$ m

Bending moment at fixed base $M_e = (2/c^2) \cdot EI \cdot p_r/k = 0.29 \cdot \Delta_H \cdot k \cdot R \cdot h = 2.87$ kNm/m

Example 3
A concrete spherical cap with a radius of $R = 15.0$ m and an opening angle of $30°$ carries, besides by its own self-weight, a snow load of 1.4 kN/m^2. The shell is considered fixed at its circular boundary (see Figure 3.54).

Self-weight: $g = d \cdot \gamma = 3.75$ kN/m^2

Compressive ring force near the base: $N_\theta = R \cdot g \cdot (1/(1 + \cos 30) - \cos 30) - p \cdot R \cdot \cos (2 \cdot 30)/2 = -23.82$ kN/m

Figure 3.54 Concrete spherical cap

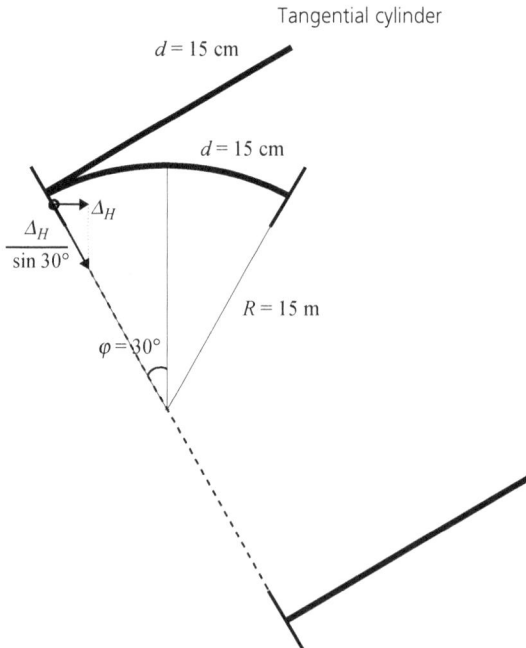

Shortening strain: $\varepsilon_\theta = 23.82/(0.15 \cdot 3 \cdot 10^7) = -0.000079$

Radial inward displacement: $\Delta_H = \varepsilon_\theta \cdot R = 0.00119$ m

Equivalent displacement of the tangential cylinder according to Geckeler (1926): $\Delta_H/\sin 30 = 0.001374$ m

Bending moment at fixed base: $M_e = (2/c^2) \cdot EI \cdot p_r/k = 0.29 \cdot \Delta_H \cdot k \cdot R \cdot h = 2.87$ kNm/m

Example 4

A shallow translational concrete elliptic paraboloid shell with a total rise of 4.0 m and a thickness of $d = 15$ cm covers a square area 20 m by 20 m resting on four vertical arches with a rise $f = 2.0$ m (see Figure 3.55).

Due to the low value of the ratio $2.0/20.0 = 0.1$ the shell is characterised as shallow. The radius of curvature is:

$$R = 20.0^2/(8 \cdot (4.0 \cdot 2.0)) = 25.0 \text{ m}$$

and the self-weight can be considered uniform over the whole area and equal to:

$$w = d \cdot \gamma = 0.15 \cdot 25.0 = 5.25 \text{ kN/m}^2.$$

The developed compressive forces are estimated as:

$$N_x = N_y = 5.25 \cdot 20.0^2/(8 \cdot 2.0) = 131.25 \text{ kN/m}$$

Figure 3.55 Shallow translational concrete elliptic paraboloid shell

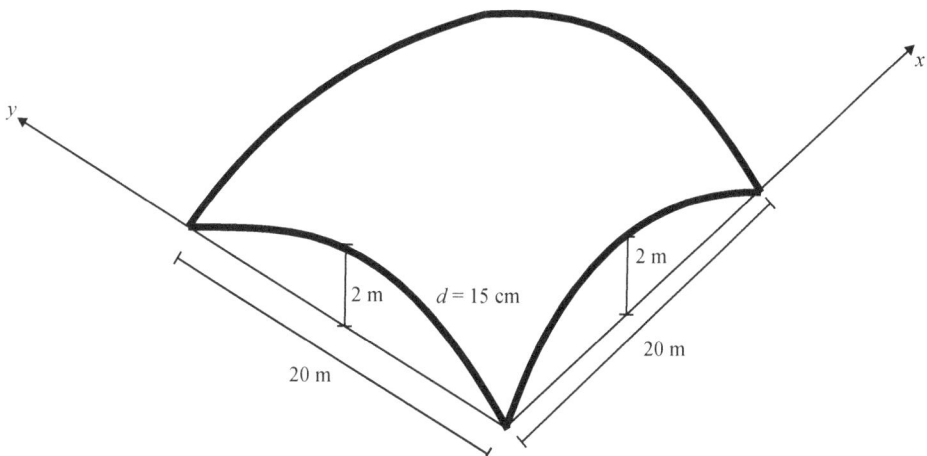

The total load is taken up by the vertical components of the boundary shear forces acting along the parabolic boundary arches of the shell. The maximum value of shear force at the shell corner is:

$$S = w \cdot L^2 / f = 5.25 \cdot 20.0^2 / 2.0 = 1050 \text{ kN/m}$$

Thus, a reinforcement, A_s, is required to cover the induced equal tensile force $Z = 1050$ kN/m at the corner. It is:

$$A_s = 1050/180000 = 58 \text{ cm}^2/\text{m}$$

Example 5

A concrete roof is composed of four identical hypar shells with a thickness $d = 3$ cm and adjointed as shown in Figure 3.55, covering a square area of 24 m by 24 m and bearing, besides its self-weight, a live load equal to 10.0 kN/m². The edge beams are provided with tie rods in order to take up the resulting thrust. The hypar shell surface exhibits a twist, k, equal to $k = (4.0/3.0)/3.0 = 0.028/\text{m}$ (see Figure 3.56).

The total load is $q = 0.3 \cdot 25.0 + 10.0 = 13.75$ kN/m² and is considered uniformly distributed. It induces equal membrane shear and normal forces $S = N = 13.75/(2 \cdot 0.028) = 245$ kN/m giving a compressive and tensile stress of $\sigma = 245.0/0.3 = 2042$ kN/m² along the diagonal arch and cable action, respectively. The edge beams undergo a maximum compressive force, N, equal to $N = 245.0 \cdot (3.0/\cos 18.4°) = 3099$ kN leading to a horizontal thrust $H = 3099.0 \cdot \cos 18.4 = 2940$ kN. Thus, a tie cross-sectional area $A_s = 2940.0/150\,000$ – that is, 196 cm² is required.

Figure 3.56 Concrete roof made of four hypar shells

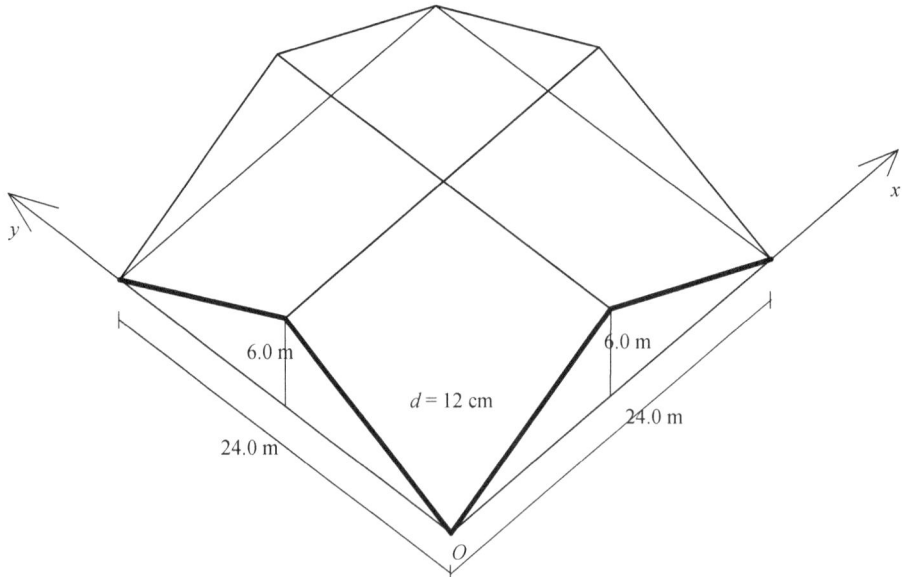

REFERENCES

Beles AA and Soare MV (1972) *Berechnung von Schalentragwerken*. Bauverlag, Wiesbaden, Germany. (In German.)

Billington D (1965) *Thin Shell Concrete Structures*. McGraw-Hill, New York, NY, USA.

Ciesielski R, Mitzel A, Stachurrski W, Suwalski J and Zmudzinski Z (1970) *Behälter, Bunker, Silos, Schornsteine, Fernsehtürme und Freileitungsmaste*. Wilhelm Ernst, Berlin, Germany. (In German.)

Flügge W (1960) *Stresses in Shells*. Springer, Berlin, Germany.

Franz G and Schäfer K (1988) *Konstruktionslehre des Stahlbetons, Band II, A*. Springer, Berlin, Germany. (In German.)

Geckeler J (1926) Über die Festigkeit achsensymmetrischer Schalen. *Forschungsarbeiten des Vereines deutscher Ingenieure 276*, Berlin, Germany. (In German.)

Kollar L (1984) *Schalenkonstruktionen – Beton – Kalender 1984 – Teil II*. Wilhelm Ernst, Berlin, Germany. (In German.)

Pflüger A (1963) Zur praktischen Berechnung der axial gedrückten Kreiszylinderschale. *Stahlbau* **32**: 161–164. (In German.)

Pflüger A (1966) Zur praktischen Berechnung der axial gedrückten Kreiszylinderschale unter Mantel-druck. *Stahlbau* **35**: 249–252. (In German.)

Salvadori M and Levy M (1967) *Structural Design in Architecture*. Prentice-Hall, Englewood Cliffs, NJ, USA.

Schlaich J (1970) Zum Tragverhalten von Hyparschalen mit nicht unterstützten Randtragern. *Beton u. Stahlbetonbau* **3**: 54–63. (In German.)

Schmidt H (1961) Ergebnisse von Beulversuchen mit doppeltgekmmmten Schalenmodellen aus Aluminium. *IASS Symposium Delft*, pp. 159–181.

Seide P (1981) Stability of cylindrical reinforced concrete shells. *American Concrete Institute ACI – SP* **67**: 43–62.

Stavridis LT and Armenakas AE (1988) Analysis of shallow shells with rectangular projection: analysis. *Journal of Engineering Mechanics ASCE* **114(6)**: 943–952.

Stavridis L and Georgiadis K (2025) *Understanding and Designing Structures without a Computer: Plane structural systems*. Emerald Publishing, Leeds, UK.

Timoshenko S (1956) *Strength of Materials, Part II*. Van Nostrand, Princeton, NJ, USA.

emerald
PUBLISHING

ice

Leonidas Stavridis and Konstantinos Georgiadis
ISBN 978-1-83662-945-0
https://doi.org/10.1108/978-1-83662-942-920251004
Emerald Publishing Limited: All rights reserved

Chapter 4
Thin-walled beams

4.1. Introduction

The concept of thin-walled beams is applied to a wide variety of structural forms, mainly regarding bridges, having, however, a field of application in buildings too. The cross-section of a thin-walled beam is formed by using thin surface elements – mainly plane elements – that are monolithically connected to each other at common edges, thus shaping oblong members of beam type (see Figure 4.1). This configuration results in a drastic material saving compared with the corresponding solid cross-section with the same height and width. It is for this reason that bridge superstructures are generally formed from thin-walled beams.

As in beams with a solid cross-section, in thin-walled beams the dimension of length, L, is clearly greater than the maximum transverse dimension, B, with a length-to-width ratio (L/B) higher than 5, while the thickness, t, of the plane elements is thin enough so that it is $t/B \leq 0.1$ (see Figure 4.1). Thus, the beam is formulated as a folded structure according to Chapter 2, Section 2.6 and can basically be analysed by using suitable computer software. However, the main goal of the present chapter is the examination of this structural type as a linear beam element so that its load-bearing action may be appraised and understood accordingly.

The main difference between a thin-walled and a solid beam, which causes the greatest difficulties in analysing a thin-walled beam as a beam element, lies in the way the torsion is taken up. In thin-walled beams under torsion, not only shear but also longitudinal normal stresses are developed, a fact that is not foreseen in the so-called *classical technical theory* of solid beams. Torsional stresses develop in thin-walled beams in rectilinear bridges due to the eccentricity of traffic loads with respect to the cross-sectional axis. They also develop in curved bridges where the self-weight alone leads to a torsional response, as well as in rectilinear beams with a cross-section without a vertical axis of symmetry.

Thin-walled cross-sections are distinguished as *open* or *closed* sections, each category having its own load-carrying characteristics. Closed cross-sections are also called *box sections*. ∎

The comparison between a closed and an open double T-section, as used in bridge design with respect to the evaluation of their structural behaviour, shows clear advantages of the former, without forgetting at all the effectiveness of the latter, in a number of cases.

Thus, in a beam with a double T-section, the ability of developing compressive forces in the bottom region over the supports, either in a continuous system or in a cantilever (see Figure 4.2), is certainly limited in comparison with a box section with the same depth (see Figure 4.1), where the bottom slab of the box can fully participate in taking up the compressive force.

This limitation of the double T-section is translated as a restriction of the tensile force the top slab can carry; thus, the respective bending resistance of the cross-section is also limited. ∎

Figure 4.1 Formation of thin-walled beams

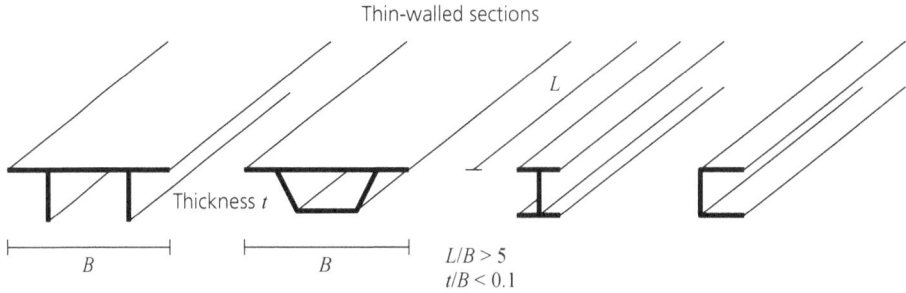

Thin-walled sections

Thickness t

B

B

L

$L/B > 5$
$t/B < 0.1$

Figure 4.2 Need for closed cross-section formation

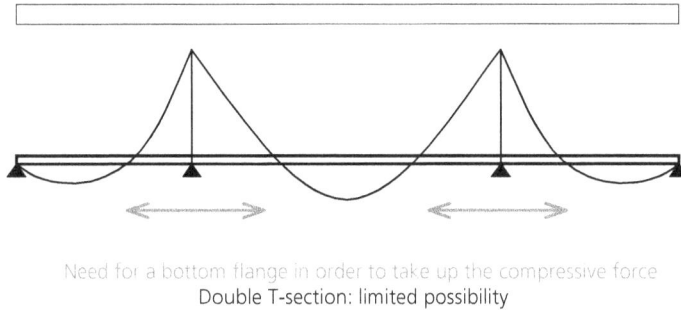

Need for a bottom flange in order to take up the compressive force
Double T-section: limited possibility

Apart from the fact that the bending resistance of closed sections is clearly greater compared with open sections, another unfavourable characteristic of open sections is related to their use in service conditions due to the combination of a smaller core depth and a higher positioned centroidal axis, as explained below.

A simply supported prestressed beam is considered and the midspan section is examined. The prestressing compressive force, P, that is initially applied to the cable trace is finally shifted upwards to a distance $a = M_g/P$ due to the self-weight bending moment, M_g, according to Stavridis and Georgiadis (2025, Section 4.3.1) (see Figure 4.3). However, the position of the applied force, P, in the open section is closer to the bottom line of the central kern than that corresponding to the closed section. Hence, the diagram of compressive stress, σ, shows a greater difference, $\Delta\sigma$, between the extreme fibres in the open cross-section than that corresponding to the closed cross-section. Now, as mentioned in Stavridis and Georgiadis (2025, Section 4.1.1), the curvature $(1/r)$ of the beam is expressed as $|\varepsilon_o - \varepsilon_u|/h$ – that is, $|\Delta\sigma|/(E \cdot h)$. Thus, the open section presents greater deformations due to prestressing than the closed section, and it can be clearly seen that this difference becomes two to three times greater under the influence of creep (see Stavridis and Georgiadis, 2025, Section 1.2.2.1) (see Figure 4.3). ∎

The last and perhaps most important difference between the two types of cross-section (open and closed) is related to torsion.

Figure 4.3 Influence of type of section on the stress variation along the depth due to prestressing

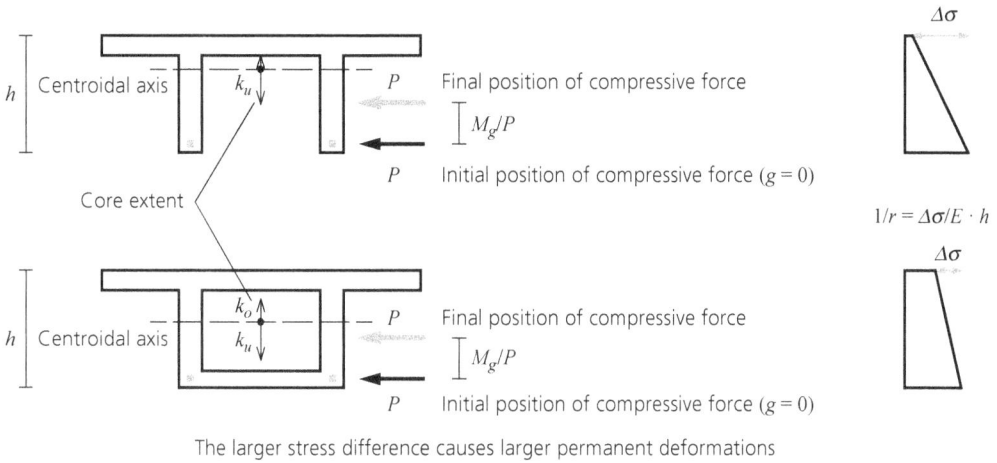

P	Final position of compressive force
P	Initial position of compressive force ($g = 0$)

$$1/r = \Delta\sigma/E \cdot h$$

P	Final position of compressive force
P	Initial position of compressive force ($g = 0$)

The larger stress difference causes larger permanent deformations

A torsional moment, M_T, applied to the open section is taken up according to the classical techni-cal theory by the three oblong orthogonal parts through the development of shear stresses with maximum values, τ_{max}, occurring at the external longitudinal sides (see Figure 4.4), as explained in Chapter 1, Section 1.1.4. Thus, it is $\tau_{max} = (3 \cdot M_{T,i})/(B_i \cdot t_i^2)$ where B_i and t_i are the length and the thickness, respectively, of each oblong side and $M_{T,i}$ is the portion of M_T that is carried by each part and offered naturally by the couple of the contrary directed stresses, τ. It is pointed out that the distribution of M_T to the parts is proportional to their torsional inertia $I_T = B_i \cdot t_i^3/3$ (see Chapter 1, Section 1.1.4). However, as a matter of fact, due to the small lever arm (roughly $2 \cdot t/3$) of the above couple, the ability to take up a torsional moment through this mechanism is always limited (see Chapter 1, Section 1.1.4).

In a closed section, on the other hand, shear stresses of the same magnitude with a constant flow all over the profile have the ability to offer a much greater torsional moment due to the relatively large lever arm between the shear forces. So, in each wall of the closed section, uniformly distributed

Figure 4.4 Basic torsional behaviour of open and closed cross-sections

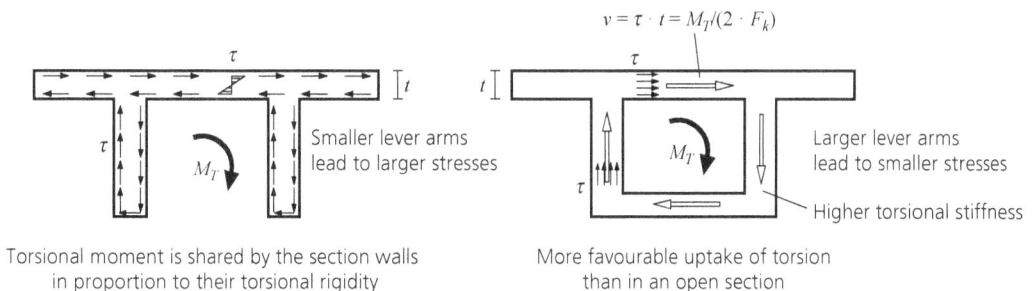

$$v = \tau \cdot t = M_T/(2 \cdot F_k)$$

Smaller lever arms lead to larger stresses

Larger lever arms lead to smaller stresses

Higher torsional stiffness

Torsional moment is shared by the section walls in proportion to their torsional rigidity

More favourable uptake of torsion than in an open section

shear stresses τ_1, τ_2, τ_3 and τ_4, are developed, presenting a shear flow $v = (\tau \cdot t)$ that according to Bredt's formula has the constant value:

$$v = \frac{M_T}{2 \cdot F_k}$$

This means that $v = \tau_1 \cdot t_1 = \tau_2 \cdot t_2 = \tau_3 \cdot t_3 = \tau_4 \cdot t_4$, thus representing for each wall a constant shear force per unit length (see Figure 4.4). In the above relation, which is generally applicable to any closed section, F_k is the area encompassed by the mid-line of the cross-section (see Chapter 1, Section 1.1.4).

Certainly, the torsional rigidity of the closed section is much greater than that of the open section with the same cross-sectional area, since for constant thickness, t, the torsional inertia, I_T, of a closed section is equal to $4 \cdot t \cdot F_k^2 / L_\Omega$ where L_Ω is the perimeter of the mid-line of the cross-section (see Chapter 1, Section 1.1.4). This generally means smaller torsional deformations for the closed section. ∎

However, the ability of open sections to undertake torsional moments is not exhausted in the classical shear stress mechanism mentioned above, also known as the *St Venant mechanism*. In addition to this, there also exists another mechanism that is not predicted by the classical technical theory and that helps the St Venant mechanism to take up the torsional moment. Despite this, the ability of open sections to carry torsional moments remains limited compared with closed ones.

Lastly, the participation of the two webs in receiving an eccentrically applied load in open and closed cross-sections will be examined. Both types of cross-sections have a width and a height equal to b and h, respectively (see Figure 4.5).

A concentrated load, Q, is considered, applied at the mid-section of a simply supported beam, placed precisely on the left web of each section. The load is analysed as a symmetric and an antisymmetric part, as shown in Figure 4.5. The symmetric loading causes bending and shearing to the beam, whereas the antisymmetric loading causes torsion.

Examining, first, the beam with open section at a position on the left of its midpoint, it may be concluded that for the symmetric loading, the shear force $Q/2$ acts through the downward forces $Q/4$ at each web, while the slab shear flows have the directions shown in Figure 4.5, vanishing at the midpoint of the section. In the antisymmetric loading, the torsional moment $(Q \cdot b/4)$ is counterbalanced by a couple of opposite shear forces in the webs being equal to $(Q \cdot b/4)/b = Q/4$, while the shear flow in the section is as shown in Figure 4.5. At this point, it should be noted that the directions of the shearing stresses generally correspond to those followed by an analogous hydraulic flow, regardless of whether they emanate from shear or torsion. Of course, the presence of shear forces from torsion implies bending of the left and the right web, downwards and upwards, respectively; consequently, this opposite bending of the two webs will cause such rotations at their ends where Bernoulli's law states that plane sections will remain plane is automatically ruled out.

This is a rough description of the bending mechanism of an open cross-section mobilised in addition to the classical St Venant shear stress mechanism in order to take up a certain torsional moment. However, this leads to the conclusion that the left web receives all the shear force $Q/2$ while the right web remains unstressed (see Figure 4.5), which strictly speaking is not quite true. This will be explained in the following sections of the present chapter through a more detailed examination of the additionally developed bending mechanism in an open cross-section when subjected to a torsional moment.

Figure 4.5 Uptake of eccentric load by a beam having an open and a closed cross-section

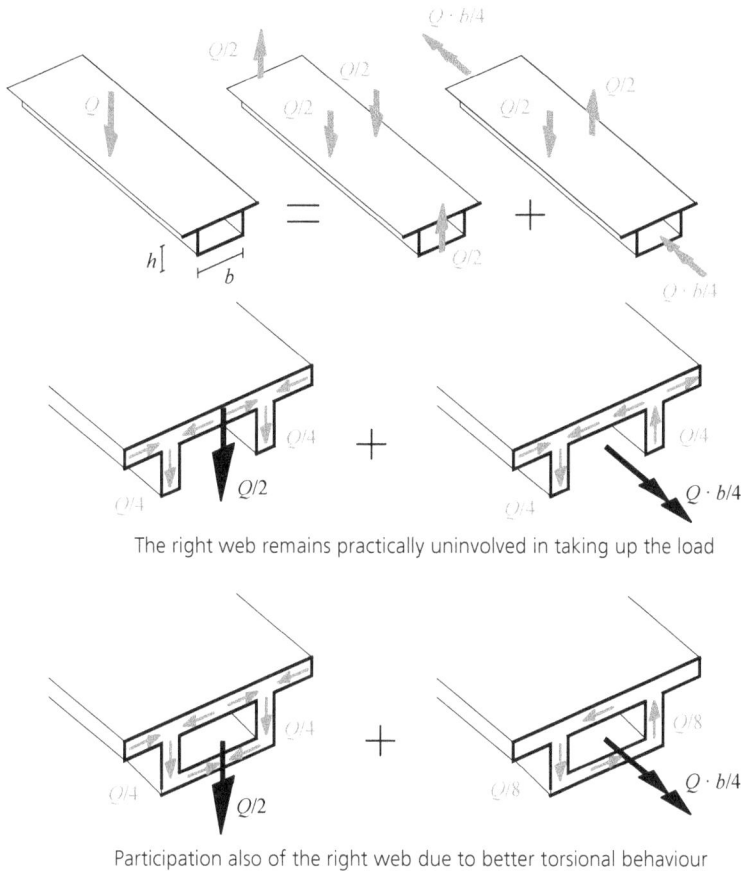

The right web remains practically uninvolved in taking up the load

Participation also of the right web due to better torsional behaviour

On the other hand, a beam with closed cross-section shows better behaviour. Again, a cross-section on the left of the beam midpoint is examined. In the symmetric loading, the shear force $Q/2$ causes again the same shear forces $Q/4$ in the webs, and the shear stresses at the top and bottom slab are according to the above rule of *hydraulic flow*, as shown in Figure 4.5. In the antisymmetric loading, however, the torsional moment $(Q \cdot b/4)$ causes a peripheral shear flow according to Bredt's formula (see Figure 4.5) that develops in each web a shear force equal to:

$$\frac{Q \cdot b}{4(2bh)} h = \frac{Q}{8}$$

instead of $(Q/4)$, as in the open cross-section. Thus, the left web receives a shear force $3 \cdot Q/8$ (i.e. 75% of the total shear force $Q/2$) while the remaining 25% representing a shear force $Q/8$ goes to the right web. It is clear that the better behaviour of the closed cross-section in the distribution of an eccentrically applied load is owing to a more favourable uptake (i.e. with smaller shear forces) of the torsional moment $Q \cdot b/4$ (see Figure 4.5).

After the above introductory presentation of thin-walled cross-sections, the basic principles governing their behaviour will be examined in a more systematic way.

4.2. The basic assumption of non-deformable cross-section

A thin-walled beam with open cross-section that is subjected to a torsional moment presents a perceptible distortion of each cross-sectional plane (see Figure 4.6). More specifically, the longitudinal fibres of the beam are deformed along their length so that, generally, a cross-section does not remain plane and thus is subjected to *warping*. The reason for this will be explained later. What, however, becomes acceptable in this respect is that the new outline of the cross-section, mapped over its initially single plane, gives a form that is rotated with respect to its initial position, not presenting any deformation relative to its initial form (see Figure 4.6). This is the so-called assumption of *non-deformable cross-section*. Despite the fact that this assumption can be implicitly satisfied, the placement of transverse diaphragms at certain distances over the length of the beam may be additionally required, which, given the negligible rigidity transversely to their plane, do not affect the deformability of the longitudinal fibres of the beam – i.e., its warping.

The distortion of an open cross-section due to a torsional moment that is developed in a supported beam implies, apart from shear flow (as mentioned in Chapter 1, Section 1.1.4 and Section 4.1), the development of a certain bending of its constituent thin-walled elements, as shown for example in Figure 4.6. This involves, of course, bending rotation of the elements' sections within their plane; therefore, the *synthesis* of these bending rotations of elements leads to the warping of the cross-section. The presence of warping, in any case, should not be confused with the requirement of non-deformability of the cross-section.

Regarding beams with closed cross-section, it can be proved that the magnitude of their warping is very limited. It should be noted, however, that the non-deformability assumption for closed cross-sections which will be adopted throughout this chapter may potentially be abandoned for constructability reasons, with corresponding consequences in their structural behaviour, as will be examined in Chapter 5.

Figure 4.6 Warping and undeformability of cross-section

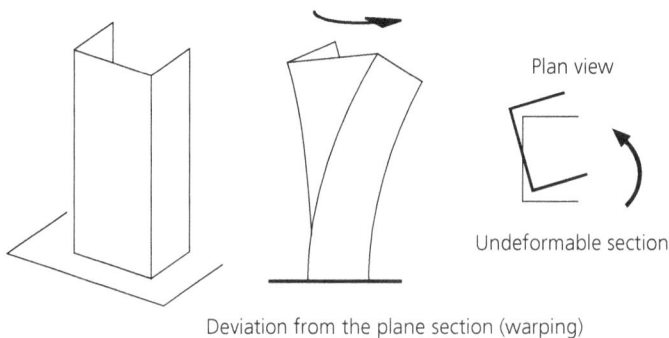

Plan view

Undeformable section

Deviation from the plane section (warping)

4.3.   Shear centre

The concept of shear centre has fundamental importance for the behaviour of thin-walled cross-sections in torsion.

In the particular example shown in Figure 4.7, a thin-walled beam with a channel section is fixed at its left end and is subjected to a vertical distributed load positioned over the web. This load causes at the fixed end, besides the bending moment, a shear flow over the cross-section of the beam.

Apart from the hydraulic flow perception mentioned previously, the way to determine the directions of shear forces in the flanges is now explained. A cut-out part of the top and bottom flange

Figure 4.7 The shear centre in a thin-walled beam

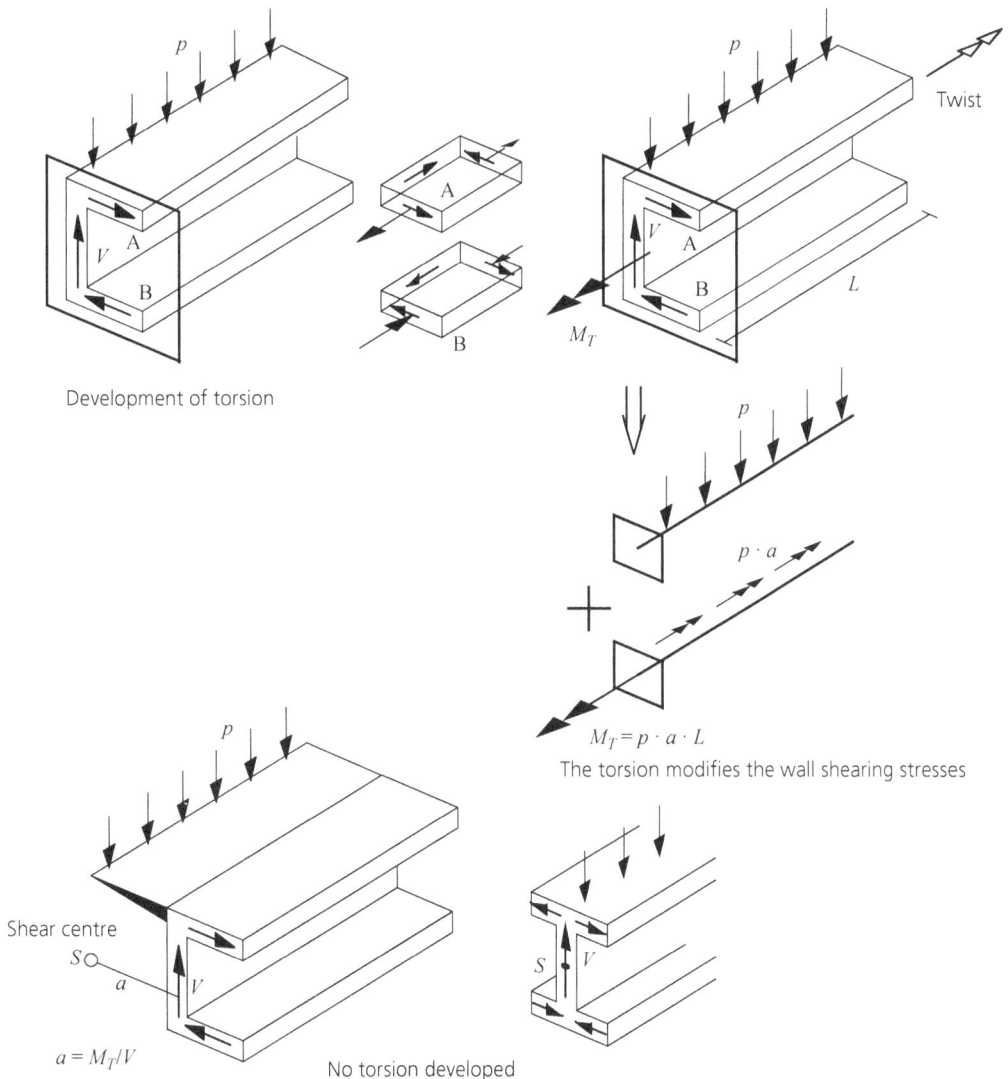

Development of torsion

Twist

M_T

$p \cdot a$

$M_T = p \cdot a \cdot L$

The torsion modifies the wall shearing stresses

Shear centre
S
a
V

$a = M_T / V$

No torsion developed

that is under different longitudinal normal forces requires a shear force on its internal face for the establishment of the longitudinal equilibrium (see Figure 4.7). It is clear that the direction of the shear stresses on the plane of the cross-section is derived according to the shear equilibrium condition (Cauchy's theorem) (see Stavridis and Georgiadis, 2025, Section 4.1.1).

If, now, the beam is considered as a free body under the distributed uniform load, the shear flow forces and the normal stresses on the fixed end section, then the equilibrium is not satisfied since the transverse shear forces of the two flanges produce an unbalanced moment. It is clear that for the establishment of the equilibrium a torsional moment (M_T) must be applied from the fixed end on the beam, with a vector showing towards the fixed end (see Figure 4.7). The twisting angle of each cross-section is shown by a vector with the opposite sign.

In order to eliminate the above torsion, the plane of the distributed load must be shifted to a suitable distance, a, on the outside, thus passing from the point, S, as shown in Figure 4.7. With this loading arrangement, the three applied shear forces on the walls at the fixed cross-section will be in equilibrium with the loads in the vertical direction and the moments that are developed with respect to any longitudinal axis (as the one passing from point, S), so that no torsional moment is required by the fixed end to maintain equilibrium.

Thus, it is obvious that the beam does not twist. The point, S, is called the *shear centre* of the cross-section and it is always the point to which torsion is referred (see Figure 4.7). Any load that passes through the shear centre does not cause torsion but only bending deformation, whereas the action of each load, acting at a distance, e, from it, is equivalent to that of the same load shifted to the shear centre thus producing only bending plus a torsional moment. This moment, in the case in which the load, p, is a distributed one, constitutes a distributed torsional load $m_D = p \cdot e$ along the beam axis, whereas in the case of a concentrated force, P, it constitutes a concentrated torsional load ($P \cdot e$). It is obvious that the development of a torsional reaction on the beam will cause an additional shear flow in the walls of the cross-section.

It is clear that if a thin-walled section has one axis of symmetry, then its shear centre lies obligatorily on this axis, whereas if the section is double symmetric, the shear centre coincides with the centroid of the section.

It should also be pointed out that in any case the shear centre represents the point around which any cross-section rotates, with this point remaining obviously immovable.

The above qualitative search of the shear centre (i.e. through the development of shear flows) can in principle be applied to any cross-section (Zbirohowski-Koscia, 1967). Of course, its general analytical determination is cumbersome and should be performed only by suitable computer software. In Figure 4.11, analytical results for the location of the shear centre of some typical cross-sections are given, while in Figure 4.12 is shown the position of shear centre in some special cases.

4.4.   Warping of thin-walled beams and the stress state due to its prevention

The application of a torsional moment to a thin-walled beam causes warping of its cross-section, as mentioned in Section 4.2. However, the mechanism in which open and closed sections undertake torsion is different, as shown below.

Figure 4.8 Compulsory warping of a free beam under torsion

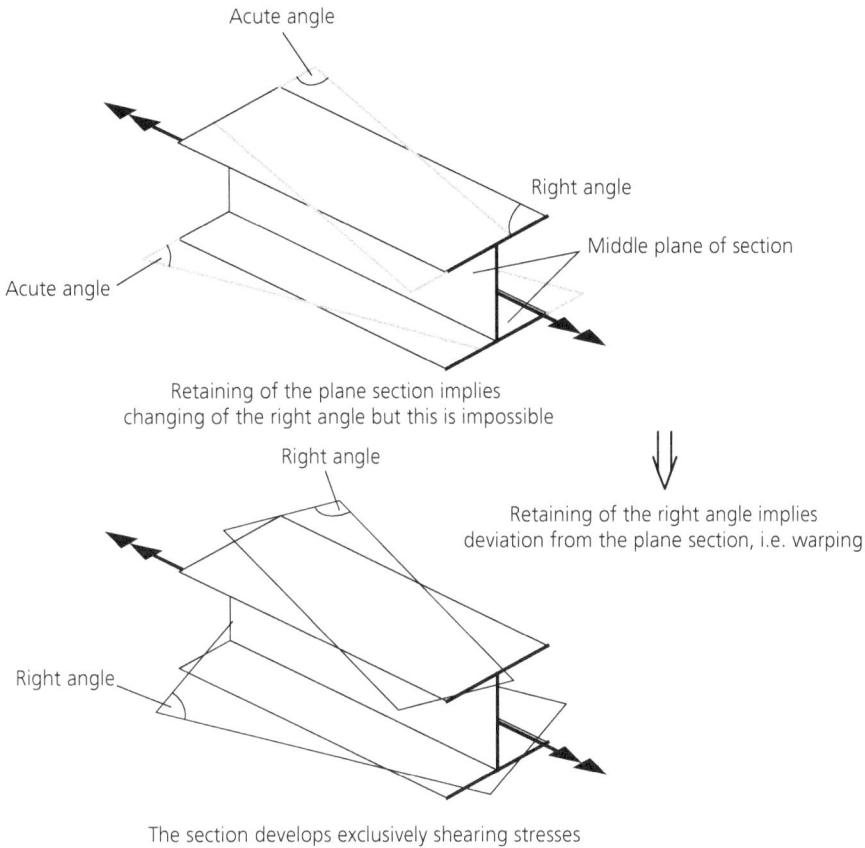

Acute angle

Right angle

Middle plane of section

Acute angle

Retaining of the plane section implies
changing of the right angle but this is impossible

Right angle

Retaining of the right angle implies
deviation from the plane section, i.e. warping

Right angle

The section develops exclusively shearing stresses

4.4.1 Open sections

The I-section beam shown in Figure 4.8 is examined. This beam is considered to be in free body equilibrium under two equal and opposite torsional moments, as shown in Figure 4.8.

It is clear that the constant torsional moment at each cross-section of the beam is taken up by the shear stresses developed in each wall of the cross-section. These shear stresses vanish, of course, at the middle of each wall thickness, so that the mid-plane of each wall is free from shear stresses (see Chapter 1, Section 1.1.4).

Under the influence of the torsional moments, every cross-section of the beam, and hence the end ones as well, will rotate about their centre of symmetry (shear centre), which means that the section flanges will move in opposite directions. As shown in Figure 4.8, the assumption that the extreme cross-sections will maintain their initial plane leads to a certain shear deformation of the beam flanges, which naturally is not possible, given the stressless state of the midplane of the section. Hence, it is concluded that the development of the above shear stresses in the section walls to undertake torsion involves the warping of the cross-section. It is clear that no other stress needs to be developed in the examined element. ∎

The above stress state changes drastically if one end of the beam is fixed – say the left one – while at the free end the previous torsional loading is maintained. It is clear that at the fixed end the cross-section cannot warp and it obviously remains plane (see Figure 4.9).

For the stress state prevailing at the free end cross-section, what was mentioned previously is still valid. The applied torsional moment, M_T, causes a turning around shear flow in each wall. The totally developed torsional contribution of the walls according to the St Venant theorem is equal to $M_{T,S}$. Obviously, at the free end cross-section it is $M_T = M_{T,S}$; however, this will be different for any other position in the beam.

Because of the developed rotation of the cross-sections, θ, which naturally varies along the length of the beam, the two section flanges are bent within their plane in opposite directions – the top flange is bent to the left and the bottom flange to the right. This bending of flanges and the developed curvature is naturally accompanied by longitudinal stresses, σ_ω, that vary along the length of the beam, since the curvature also varies with maximum value at the fixed end and zero value at the free end, where obviously normal stresses cannot be developed. Moreover, it is clear that this variation of longitudinal normal stresses, σ_ω, leads, from the equilibrium, to the development of shear stresses, τ_ω, as shown in Figure 4.9, both for the top and the bottom flange.

Thus, at each position of the top or bottom flange, a bending moment is developed corresponding to a linearly varying diagram of longitudinal normal stresses, as well as a transverse shear force in each flange for each position. It is clear from Figure 4.9 that, while the longitudinal normal stresses, σ_ω, in the two flanges are self-equilibrating – that is, they do not create an axial force in the cross-section – the shear forces in the two flanges offer a torsional moment, $M_{T,\omega}$. This moment along with moment, $M_{T,S}$, constitute the imposed torsional moment, M_T. It is:

$$M_T = M_{T,S} + M_{T,\omega}$$

It is obvious that the contribution of $M_{T,S}$ to the total moment, M_T, in the cross-section decreases as one moves away from the free end, where, of course, it constitutes 100%.

It becomes clear that both σ_ω and τ_ω along the length of the beam owe their existence to the prevention of cross-sectional warping at the fixed end. Generally, any warping prevention of the cross-section in places where torsion develops causes the above stresses. One might think that at first the axial stresses, σ_ω, must be applied in order to 'remove' or prevent the warping. As these axial stresses necessarily vary along the length of the beam, they give rise to the development of shear stresses. It is clear that the higher this distortion, the greater the stresses, and vice versa. A detailed examination of warping is therefore absolutely necessary, since the consequences of its prevention constitute the only reason for which the technical theory of torsion (St Venant) needs to be revised. As a matter of fact, the latter predicts the sectional distortion, as is clear from the above explanation, but it does not predict the stresses σ_ω and τ_ω that develop due to its prevention.

4.4.2   Closed sections

Before examining warping in detail, it is important to point out the fact that the above contribution of shear stresses, τ_ω, to the total torsional moment, M_T, concerns only open sections. The reason is that in closed sections the developing τ_ω constitutes a self-equilibrating system of stresses (in the same manner as stresses, σ_ω) and, consequently, they do not contribute to the uptake of the torsional moment of the section. Thus, in closed sections the developed shear flow, according to

Figure 4.9 Development of longitudinal stress state due to restrained warping

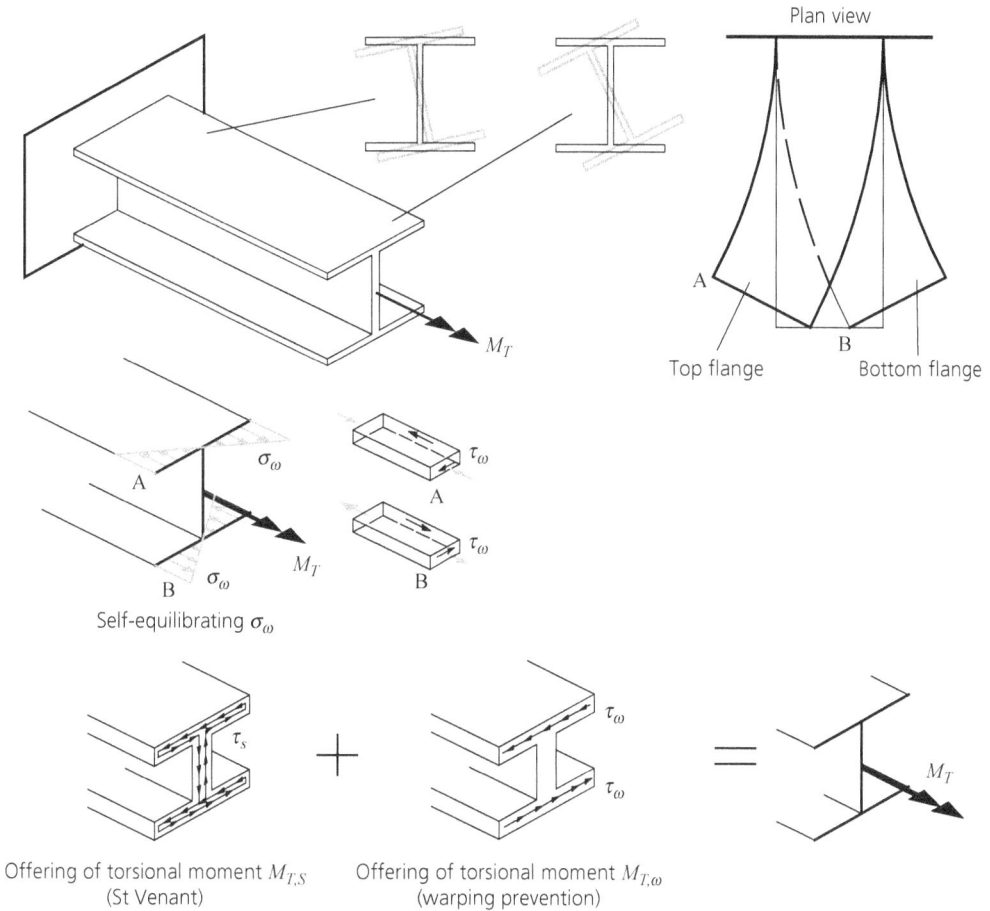

Plan view

Self-equilibrating σ_ω

Offering of torsional moment $M_{T,S}$ (St Venant)

Offering of torsional moment $M_{T,\omega}$ (warping prevention)

Bredt's formula, constitutes the only mechanism of taking up the applied torsional moment, as will be explained below.

As in the case of open sections, the torsion mechanism of closed sections is shown in a cantilever beam (see Figure 4.10). The applied torsional moment is considered to be introduced through a transverse diaphragm that is supposed to be very stiff in its own plane. However, it does not offer any resistance to imposed displacements that are perpendicular to its own plane, as those caused by warping of the cross-section. This diaphragm transmits (according to Bredt's formula) a peripheral shear flow to the walls of the closed section (see Section 4.1). It is obvious that the cross-section rotates about the shear centre due to the applied torsional moment by an angle that reduces to zero at the fixed support.

In order to understand the deformed picture of the beam, the top slab is first considered cut out from the webs, having the midpoint, M, of its left side simply pinned against horizontal displacement (see Figure 4.10).

Figure 4.10 Restrained warping and longitudinal stress state in a closed cross-section

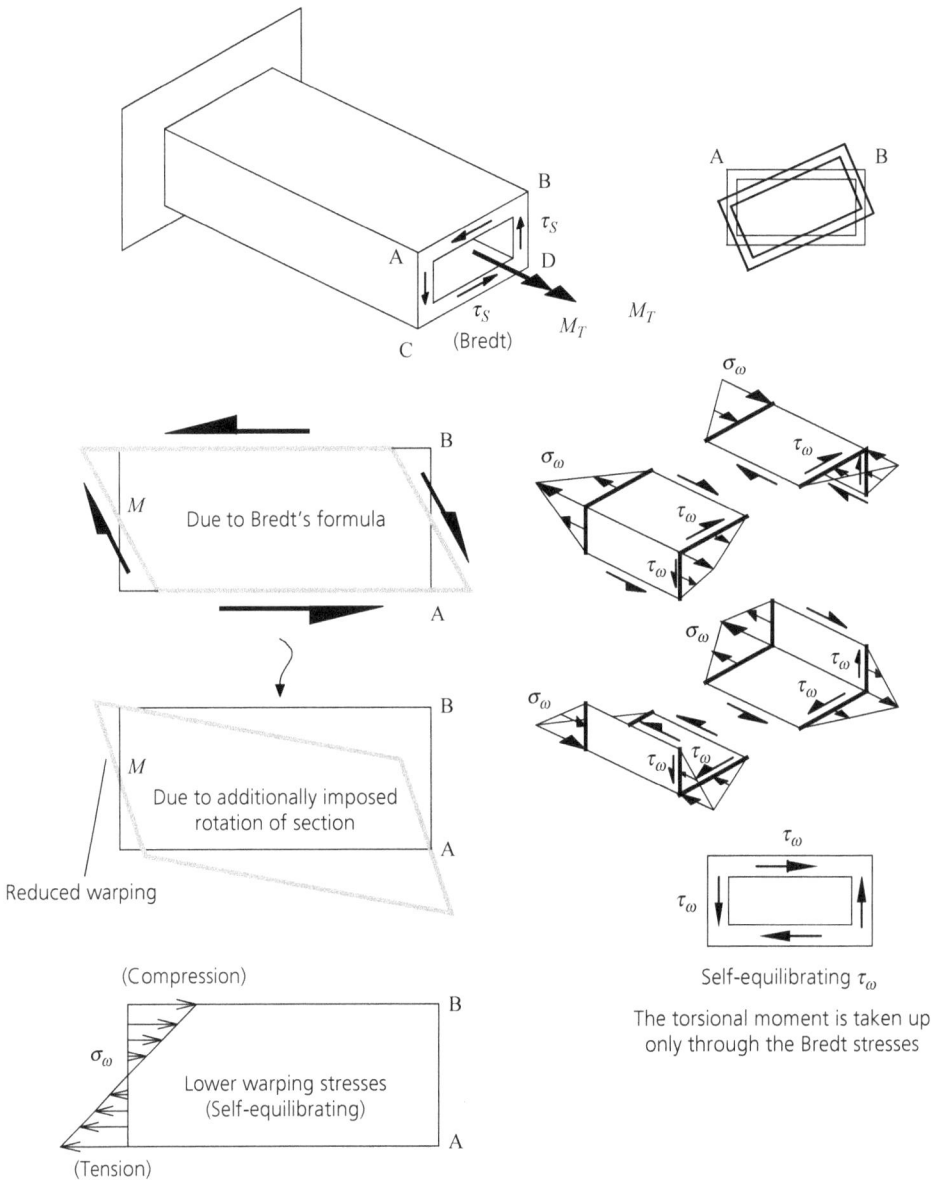

Due to Bredt's formula

M

Due to additionally imposed rotation of section

M

Reduced warping

(Compression)

Lower warping stresses (Self-equilibrating)

σ_ω

(Tension)

Self-equilibrating τ_ω

The torsional moment is taken up only through the Bredt stresses

The slab is in equilibrium under the shear actions (according to Bredt's formula) at both ends as well as the shear forces (according to Cauchy's theorem) on its longitudinal sides. Under the above forces, the slab is deformed in shear and the two end sides acquire a slope with respect to their initial (undeformed) positions, thus causing a deviation from the cross-sectional plane that leads them to warping. Certainly, this deviation (warping) is reduced due to the imposed movement of the right side of the slab 'downwards', with the point M being fixed, due to the anticlockwise rotation of the cross-section of the loaded end (see Figure 4.10). Thus, it is clear that the developed warping in a closed cross-section is definitely smaller than in an open one.

The presence of fixity does not allow the left end side to develop this slope – that is, to warp – therefore, in order to maintain the plane section, longitudinal stresses, σ_ω, must be developed, as shown in Figure 4.10, all over the perimeter of the section, since the same logic applies to each individual wall of the cross-section. As these imposed longitudinal stresses induce bending, it becomes clear that each wall is subjected to bending. The developing longitudinal stresses are smaller than the ones of an analogous open section because the exhibited warping of the closed section is reduced. The stresses, σ_ω, constitute here also a self-equilibrating system and the fact that they obviously vanish at the free end means that they keep reducing all along the length of the beam, thus giving rise to the development of shear stresses, τ_ω.

Indeed, by cutting the top and bottom slabs, as well as the two webs at their mid-width along some length of the beam, the need for development of longitudinal shear stresses is apparent from the longitudinal equilibrium of the free cut-out part, which in turn (according to Cauchy's theorem) produce corresponding shear stresses, τ_ω, on the cross-section itself. However, as can be easily understood from Figure 4.10, the total τ_ω on the cross-section is self-equilibrating and thus does not contribute to taking up the torsional moment of the section, although it modifies, even to a small degree, the existing shear stresses by Bredt's formula. Hence, in closed sections it may be written as:

$$M_T = M_{T,S}$$

The length for which stresses σ_ω and τ_ω are extended is in the order of the cross-sectional width, given the self-equilibrating character of σ_ω and the non-deformability of the cross-section, according to a corresponding conclusion of the elasticity theory, known as the St Venant theorem.

The above additional stresses, σ_ω and τ_ω, since they do not particularly modify the stress state obtained by the technical theory, may generally be omitted in the preliminary design stage, provided that the non-deformability of the cross-section is ascertained.

4.4.3   Analysis of warping

If each point of the cross-section outline is characterised by its measured distance, s, on it from a certain constant point, then the longitudinal displacement of a point (s) of the section that is found on the abscissa, x, of the beam may be represented by the function $u(x,s)$, while $\theta(x)$ is the twisting angle at the specific position (x) of the beam (see Figure 4.11).

As is clear from the above, the warping of a cross-section at position x results from the new positions $u(x,s)$ of its points (x,s) after the longitudinal deformation of the beam fibres. It is found that (Zbirohowski-Koscia, 1967):

$$u(x,s) = \frac{d\theta}{dx} \cdot \omega(s)$$

where the *warping function* $\omega(s)$ is referred at each point of the cross-section outline and is determined on the basis of its shear centre position. Thus, each thin-walled cross-section has its own characteristic diagram of the warping function $\omega(s)$. For each point of the cross-section, this diagram shows how much this point will be shifted along the beam axis due to a unit change $(d\theta/dx)$ of the twisting angle of the beam. The diagram $\omega(s)$ (m^2) always satisfies the relation:

$$\int \omega(s) \cdot t \cdot ds = 0$$

where t is the thickness of the section wall.

Figure 4.11 Warping behaviour of cross-sections

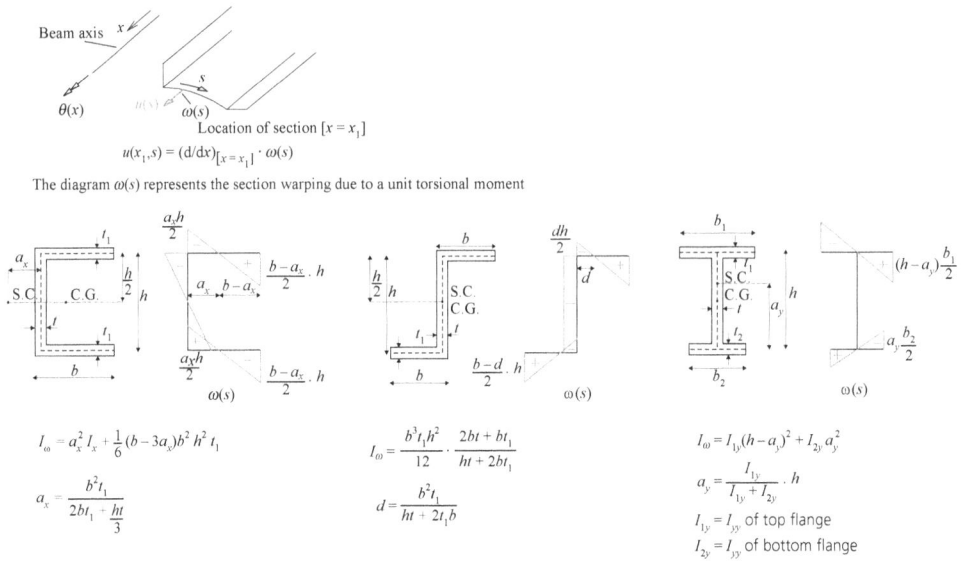

Beam axis

$\theta(x)$ $u(s)$ $\omega(s)$

Location of section $[x = x_1]$

$u(x_1,s) = (d/dx)_{[x=x_1]} \cdot \omega(s)$

The diagram $\omega(s)$ represents the section warping due to a unit torsional moment

$$I_\omega = a_x^2 I_x + \frac{1}{6}(b - 3a_x)b^2 h^2 t_1$$

$$a_x = \frac{b^2 t_1}{2bt_1 \cdot \frac{ht}{3}}$$

$$I_\omega = \frac{b^3 t_1 h^2}{12} \cdot \frac{2bt + bt_1}{ht + 2bt_1}$$

$$d = \frac{b^2 t_1}{ht + 2t_1 b}$$

$$I_\omega = I_{1y}(h - a_y)^2 + I_{2y} a_y^2$$

$$a_y = \frac{I_{1y}}{I_{1y} + I_{2y}} \cdot h$$

$I_{1y} = I_{yy}$ of top flange

$I_{2y} = I_{yy}$ of bottom flange

The determination of the warping diagram $\omega(s)$ for a cross-section is a cumbersome process that can generally be carried out using special software. In Figure 4.11, the warping diagrams of some characteristic cross-sections are shown. ∎

It should be pointed out here that there are some thin-walled cross-sections that do not develop any warping under the torsional response and, consequently, are never subjected to the stresses σ_ω and τ_ω due to warping restraint. The following cross-sections belong to this category (see Figure 4.12):

- Cross-sections having a centre of symmetry. The shear centre coincides with this point.
- Open cross-sections having walls passing through a common point. The shear centre always coincides with this point.
- All the closed cross-sections that can be circumscribed to a circle, having constant wall thickness (triangular etc.).
- More generally, if in a closed cross-section the wall thicknesses are considered as vectors and the resultants of these vectors at each corner pass through a common point, then the cross-section does not suffer warping (see Figure 4.12).

4.4.4 Longitudinal stresses due to prevention of warping

It is again pointed out that the longitudinal normal stresses, σ_ω, due to the restraint of warping are not predicted by the classical technical theory of beams since the only normal stresses appearing there refer either to bending moments or to axial forces. As the stresses, σ_ω, are self-equilibrating, they do not contribute to the global equilibrium of the structure. However, it has been found that they may be expressed in the same way as the normal stresses in technical theory through a sectional magnitude, M_ω, called *bimoment* (see Figure 4.13).

Figure 4.12 Cross-sections that do not develop warping

Sections that do not warp under a torsional moment

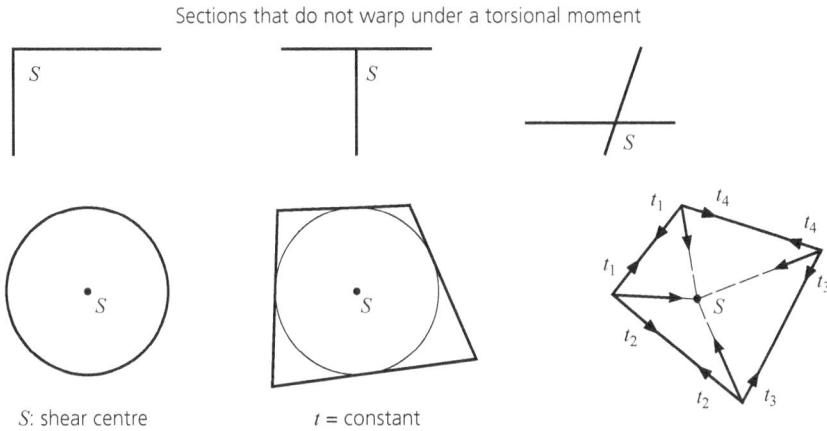

S: shear centre t = constant

Figure 4.13 Size and significance of the bimoment

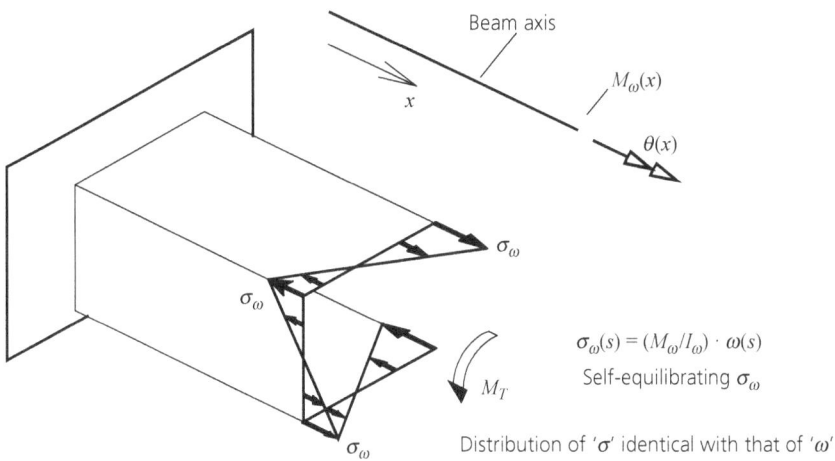

Beam axis

$M_\omega(x)$

$\theta(x)$

σ_ω

σ_ω

M_T

$\sigma_\omega(s) = (M_\omega/I_\omega) \cdot \omega(s)$

Self-equilibrating σ_ω

σ_ω

Distribution of 'σ' identical with that of 'ω'

Bimoment

$M_\omega = \int \sigma_\omega(s) \cdot \omega(s) \cdot dA = -(EI_\omega) \cdot (d^2\theta/dx^2)$
$I_\omega = \int \omega \cdot dA$ Much less in closed than in open sections
The bimoment represents quantitatively the self-equilibrating longitudinal normal stresses

In absolute analogy with the technical theory of bending, it is found that for each point of the cross-section, the longitudinal warping stress can be expressed as (Vlasov, 1961):

$$\sigma_\omega = \frac{M_\omega}{I_\omega} \omega(s)$$

where,

$$I_\omega = \int_A \omega^2 \cdot \mathrm{d}A$$

The magnitude I_ω (m^6) is called the *warping constant*, whereby integration is performed on each elementary part of surface $\mathrm{d}A$ of the cross-section. The similarity of the above equation with the equation of pure bending is obvious. It is observed that for a given value of bimoment, the distribution of σ_ω is identical to the distribution of $\omega(s)$. Certainly, the analytical determination of I_ω is very cumbersome and can generally be performed through special computer software. The values of I_ω of some characteristic cross-sections are shown in Figure 4.11.

The bimoment itself (kNm^2), the physical meaning of which will be discussed in the next two sections, is defined through the following expression, in an analogous way to the bending moment of technical theory:

$$M_\omega = \int_A \sigma_\omega(s) \cdot \omega(s) \cdot \mathrm{d}A$$

If $\theta(x)$ is the function of twisting angle along the length of the beam, then it is:

$$M_\omega = -EI_\omega \cdot \frac{\mathrm{d}^2\theta}{\mathrm{d}x^2}$$

whereby the similarity with the relation between the bending moment, M, and the beam deflection, w, is obvious (see Stavridis and Georgiadis, 2025, Section 2.3.6).

As the closed cross-sections exhibit a clearly smaller warping compared with open ones, they also have a much more 'moderate' warping function, $\omega(s)$; consequently, their warping constant, I_ω, is comparatively much smaller. This fact certainly leads to limited normal stresses, σ_ω, compared with open cross-sections.

4.5.   The concept of bimoment

The concept of bimoment in a thin-walled beam is directly related to the self-equilibrating system of normal stresses aroused by the prevention of cross-section warping under a torsional moment. It should be clear that a constructional warping prevention of even one single section (e.g. a beam's end) is enough to exert a restraining effect in the warping of all other sections of a beam. It should also be understood that the aforementioned prevention activates the bending rigidity of the section walls and automatically causes longitudinal normal stresses that are not negligible, being in any case obliged in an inherent self-equilibrium along the longitudinal direction, given that no external axial force is present. It may be considered that these normal stresses represent the bimoment itself, with the latter expressing actually the induced bending in each section wall.

The fact that a torsional moment generally causes bending, thus giving rise to the concept of bimoment, besides what was examined in Section 4.4.1, is also explained in the system in Figure 4.14.

The vertical beam with an open profile is fixed at its bottom end, while the top end is rigidly connected to a horizontal diaphragm, non-deformable within its own plane and without any transverse bending rigidity. A torsional moment, M_T, is applied to the diaphragm.

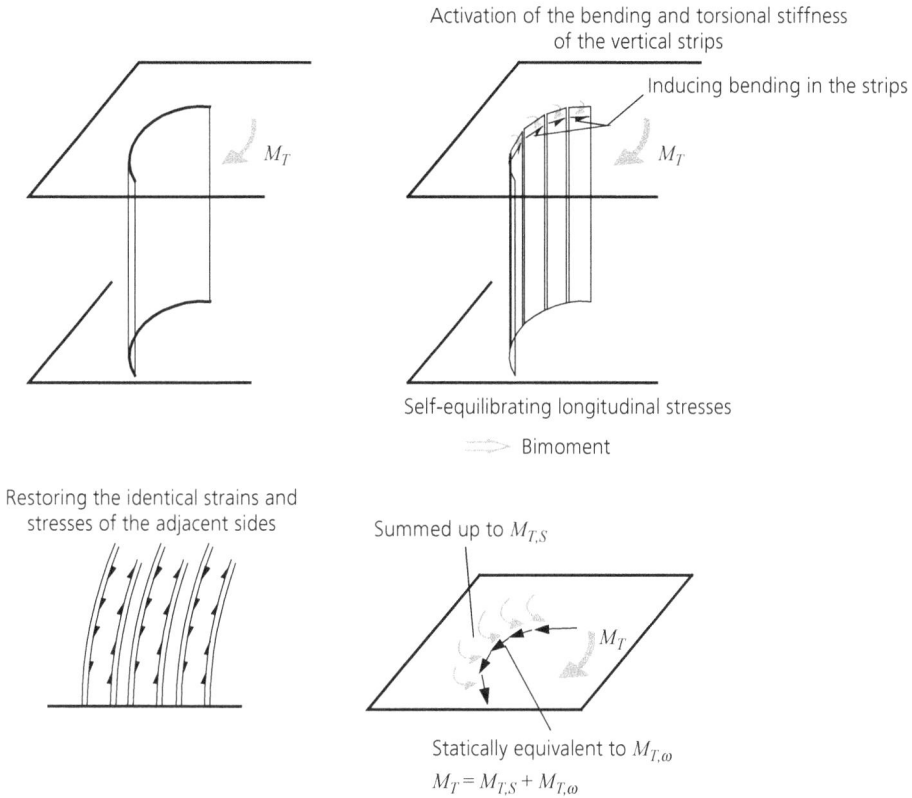

Figure 4.14 Example of physical perception of bimoment

The beam may at first be considered as constituted from mutually independent vertical plane strips, exhibiting a bending rigidity only within their plane, while possessing a torsional stiffness according to the St Venant shear stress mechanism.

The forced movement of the horizontal diaphragm under the torsional moment, consisting of a displacement and a rotation around the shear centre of the beam's section, imposes displacements and rotations to each wall segment that mobilise their bending and shear response on one hand and their torsional response on the other, according to its own bending and torsional stiffness, respectively. The shear and torsional shear stresses are constant along the height of the segment. However, the development of a shear force at the top of each wall segment causes bending with axial stresses that keep increasing towards its fixed end, constituting always a self-equilibrated system in the vertical direction. These stresses are obviously self-equilibrating – that is, they have no resultant along the longitudinal direction since no axial force is acting on the segment, thus being the outcome of a bimoment that is developed along the whole beam.

This situation will not change even with the self-equilibrating system of longitudinal shear forces (see Figure 4.14) imposed along the free sides of the wall segments in order to restore the incompatible deformations occurring there. It is clear that the finally resulting set of self-equilibrating axial stresses ascertains the existence of a corresponding bimoment, M_ω, of the whole section.

The horizontal diaphragm is in equilibrium under the externally applied torque, as well as the reacting shear forces and torsional moments of the walls, acting in the opposite sense. It is clear that these shear forces will amount only to a total moment, $M_{T,\omega}$, which, together with the summation of St Venant moments, $M_{T,S}$, will balance the moment, M_T. It should be clear that the absence of any applied horizontal force on the diaphragm ensures that the above shear forces do not have any resultant force acting on the same plane of the diaphragm. Hence, it is:

$$M_T = M_{T,\omega} + M_{T,S}$$

proving that a torsional moment, M_T, is constituted of a St Venant moment, $M_{T,S}$, and a warping moment, $M_{T,\omega}$. The relation between these two contributions depends not only on the cross-section itself. but also on the length of the beam. As the length increases, for example, the ratio of bending stiffness $[3 \cdot EI/(\text{length})^3]$ to torsional stiffness $[GI_T/(\text{length})]$ with respect to horizontal displacement and rotation angle at the top, respectively, for each wall segment decreases and, consequently, the contribution percentage of $M_{T,\omega}$ with respect to $M_{T,S}$ in taking up the torsional moment, M_T, decreases too.

4.6.   Two theorems regarding the bimoment
4.6.1   Theorem 1
A force, P, parallel to the beam axis being applied to a point of the cross-section having a warping value, ω, causes a bimoment equal to $M_\omega = P \cdot \omega$ (see Figure 4.15). The compressive loads are considered negative.

In the beam with an 'I' cross-section, where three self-equilibrating forces are applied ($N = 0$), a bimoment equal to $M_\omega = 2 \cdot (P/2) \cdot \omega_1$ is developed, obviously causing axial stresses according to the basic relation mentioned in Section 4.4.4, contrary to what would be predicted by the classical technical theory.

The equal and opposite compressive forces, P, on the beam that can be considered as prestressing forces cause a positive bimoment $M_\omega = P \cdot \omega_1$, ($P < 0$, $\omega_1 < 0$) and, consequently, tensile normal stresses at the ends of the cross-section ($\omega > 0$), contrary to what would be concluded according to the classical technical theory. However, a uniform compressive stress, p_0, applied to the entire cross-section does not cause a bimoment since:

$$M_\omega = \int p_0 \cdot dA \cdot \omega = p_0 \cdot \int \omega \cdot dA = 0 \qquad \text{(see Section 4.4.3)}$$

and, therefore, the whole section is under the above compressive stress.

4.6.2   Theorem 2
A moment, M, applied on a plane parallel to the beam axis at a distance, e, from the shear centre causes a bimoment $M_\omega = M \cdot e$. The bimoment is considered positive if the vector of the applied moment, M, is directed to the shear centre.

According to the example in Figure 4.15, even though the two externally applied moments, M, on the beam are in equilibrium, they cause a bimoment $M_\omega = -M \cdot (h/2) - M \cdot (h/2) = -M \cdot h$ and, accordingly, a longitudinal stress response.

Figure 4.15 Development of a bimoment without the presence of a torsional load

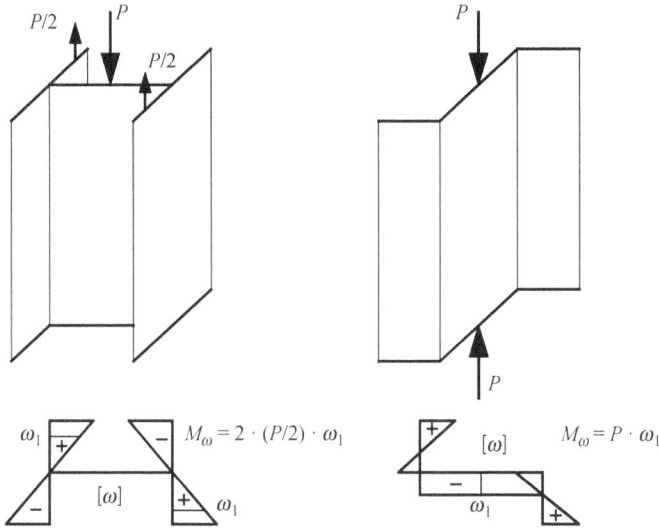

$M_\omega = 2 \cdot (P/2) \cdot \omega_1$

$M_\omega = P \cdot \omega_1$

The presence of bimoment means development of self-equilibrating stresses

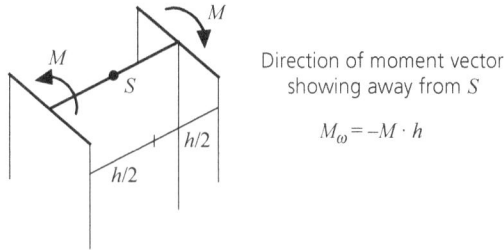

Direction of moment vector showing away from S

$M_\omega = -M \cdot h$

4.7. Warping shear stresses

As explained previously, warping shear stresses, τ_ω, are required in order to ensure equilibrium, given the variation of the normal stresses, σ_ω, along the length of the beam. As has been asserted, their distribution over the cross-section is not self-equilibrating, as happens with σ_ω, but leads to the warping torsional moment, $M_{T,\omega}$. It should be noted, however, that this particular effect appears only in open sections. According to Section 4.4.2, the stresses, τ_ω, in closed sections constitute a self-equilibrating system, which certainly affects the distribution of shear stresses according to Bredt's formula. However, it does not contribute to the uptake of the torsional moment, M_T. This is taken up exclusively by the mechanism of Bredt's peripheral shear flow.

It should be mentioned here that in closed cross-sections a warping torsional moment, $M_{T,\omega}$, is also present, leading, however, to self-equilibrating shear stresses, τ_ω, as previously explained.

The analytical determination of the distribution of stresses, τ_ω, in the cross-section (open or closed) based on the moment, $M_{T,\omega}$, may be considered according to the formula:

$$\tau_\omega = \frac{M_{T,\omega}(x) \cdot S_\omega(s)}{I_\omega \cdot t}$$

Figure 4.16 Shear stresses due to distortion prevention

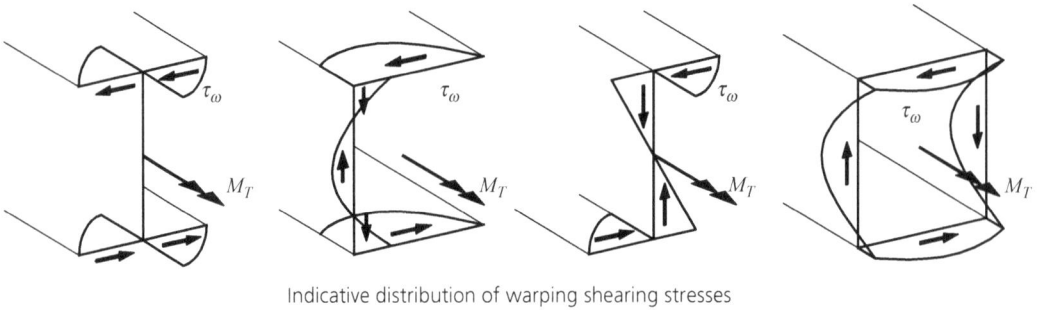

Figure 4.16 Shear stresses due to distortion prevention

Indicative distribution of warping shearing stresses

where,

$$S_\omega(s) = \int_0^S \omega(s) \cdot t \cdot \mathrm{d}s$$

with s measured along the section profile from a certain point on it (see fig 4.11) but it is generally cumbersome. For this reason, the use of special computer software is required. The distribution of warping shear stresses, τ_ω, for some typical cross-sections is shown in Figure 4.16.

4.8. The governing equation of torsion and its practical treatment

A suitably supported rectilinear thin-walled beam is considered, which is subjected to a distributed torsional load, m_D, and, possibly, to a concentrated torque, T, as well. The beam is referred to a respective axis with abscissa, x, as it has been shown in Figure 4.13.

The characteristic deformation magnitude on the basis of which all the torsional magnitudes of the beam are expressed is the rotation $\theta(x)$. Thus, it is found that the warping torsional moment, $M_{T,\omega}$, is expressed as:

$$M_{T,\omega} = -EI_\omega \cdot \frac{\mathrm{d}^3\theta}{\mathrm{d}x^3}$$

while the torsional moment according to St Venant, $M_{T,S}$, according to Section 1.1.6, is:

$$M_{T,S} = GI_D \cdot \frac{\mathrm{d}\theta}{\mathrm{d}x}$$

Moreover, since the bimoment, M_ω, in Section 4.4 is expressed as $M_\omega = -EI_\omega \cdot (\mathrm{d}^2\theta/\mathrm{d}x^2)$, the following relation between bimoment and warping torsional moment is directly obtained from the above:

$$M_{T,\omega} = \frac{\mathrm{d}M_\omega}{\mathrm{d}x}$$

Now, according to the above expressions, the relation established in Section 4.4.1:

$$M_T = M_{T,\omega} + M_{T,S}$$

may be written as:

$$M_T = GI_D \cdot \frac{d\theta}{dx} - EI_\omega \cdot \frac{d^3\theta}{dx^3}$$

and, given the equilibrium relation (Section 1.1.6):

$$\frac{dM_T}{dx} = -m_D$$

the last equation takes the form (Vlasov, 1961):

$$EI_\omega \cdot \frac{d^4\theta}{dx^4} - GI_D \cdot \frac{d^2\theta}{dx^2} = m_D$$

It should be pointed out that this equation is valid both for open and closed cross-sections, despite the fact that, as previously mentioned, in the closed cross-sections the magnitude $M_{T,\omega}$ leads to self-equilibrating shear stresses, τ_ω. It is, however, possible that the last result may also be derived from the relation:

$$\tau = \tau_{T,\omega} + \tau_{T,S}$$

that is valid both for open and closed cross-sections.

The above equation is the fundamental relation that governs the torsional behaviour of thin-walled cross-sections. In this equation, the pertinent boundary conditions in terms of the function $\theta(x)$ are required to be satisfied. Thus, the following boundary conditions are distinguished.

- In the case of a fully fixed cross-section at position x, it should be:

$$\theta = 0 \text{ and } (d\theta / dx) = 0$$

- In the case in which twisting of the cross-section is prevented but not its warping, which means that in the considered cross-section longitudinal warping stresses, σ_ω, cannot be developed, it should be $\theta = 0$ and $M_\omega = 0$ – that is, on the basis of the previous equations:

$$\theta = 0 \text{ and } (d^2\theta / dx^2) = 0$$

- In the case of a completely free cross-section, it will be $M_\omega = 0$, $M_T = 0$; therefore, on the basis of previous equations it is:

$$\left(d^2\theta / dx^2\right) = 0 \text{ and } GI_D \cdot \frac{d\theta}{dx} - EI_\omega \cdot \frac{d^3\theta}{dx^3} = 0$$

Based on the above, the boundary conditions for other cases can also be worked out. Thus, for example, when, in the place of internal support of a continuous beam where two beams are linked, the arranged bearings – with or without a diaphragm placed above the support – prohibit the end rotation, it will be $\theta_{(1)} = \theta_{(2)} = 0$, while due to the continuity of normal stresses, σ_ω, the equality of bimoments $M_{\omega(1)} = M_{\omega(2)}$ must also be required at the same position. ∎

Although it is in principle feasible to find the solution to the above basic equation that governs the torsional stress state of a beam through the classical analytical treatment of differential equations, it is very unpractical, especially in more complex cases.

As a matter of fact, the form of the equation in question is precisely the same as that of a tension beam under vertical loading according to the second-order theory (Roik, 1983), as examined in Stavridis and Georgiadis (2025, Section 7.4). This equation is:

$$EI \frac{dw^4}{dx^4} - H \frac{dw^2}{dx^2} = p(x)$$

This observation is of crucial importance as it establishes an absolute analogy between the two problems. This fact allows the treatment of the torsional problem using the conceptually much more amenable behaviour of beam bending. This analogy, with due allowance to the corresponding parameters, is depicted in Table 4.1, as well as in Figure 4.17.

According to the above analogy and following the result from Stavridis and Georgiadis (2025, Section 7.4), where:

$$w = w_1 \cdot \frac{1}{1 + \xi}, \quad (\xi = P / P_{cr}) \tag{a}$$

Table 4.1 Analogy between magnitudes in bending and torsional response

Beam in bending	←	→	Beam in torsion	
Displacement	w	←→	Rotation angle	θ
Tensile force	H	←→	Torsional rigidity	$G \cdot I_T$
Moment of inertia	I	←→	Warping constant	I_ω
Uniform load	p	←→	Uniform torsional moment	m_D
Concentrated transverse load	P	←→	Concentrated torque	T
Vertical component of the tensile force	H_V	←→	Torsional moment according to St Venant	M_{TS}
Bending moment	M	←→	Bimoment	M_ω
Shear force (vertical)	Q	←→	Torsional moment	M_T

$\overrightarrow{M_{T,\omega}} = \overrightarrow{V} - \overrightarrow{H_V}$ (see Figure 4.17)

Figure 4.17 Analogy between torsional and bending stress states of a tension beam

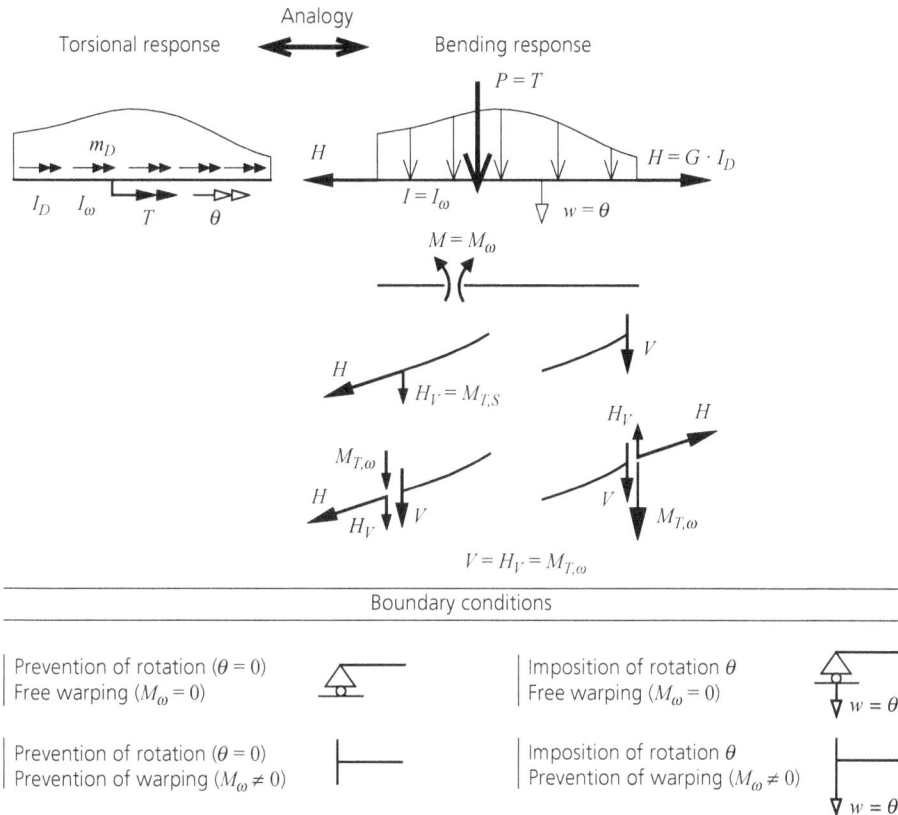

In the case of a simply supported beam subjected to either a uniformly distributed load or a concentrated load at midspan, it is always possible to obtain approximate – but sufficiently reliable – results for deflection (as from Equation (a)) and internal forces suitable for preliminary design purposes.

It is obvious that all the boundary conditions of the problem that are expressed through the specific requirements for the rotation angle, θ, the bimoment, M_ω, or the torsional moment, M_T, may thus be attributed to the corresponding terms of the beam in bending (see Figure 4.17). The result is a supervisory and naturally more direct perception of the torsional problem. Clearly, all the results of Stavridis and Georgiadis (2025, Section 7.4) can be appropriately applied according to the above analogy.

4.9.   Worked examples

The aim of this example is to determine the torsional stiffness of a cantilever of length 10.0 m – that is, to determine the torsional moment required to be applied at its free end in order to cause rotation $\theta = 1$ rad. The behaviour of an open I-section as well as a closed box section with the same external dimensions and the same cross-sectional area are examined and compared by imposing a deflection $w = 1$ m at the relevant support, considering in addition in each case the possibility of preventing the warping of the free end. It is assumed that the profile of the cross-sections remains undeformable.

The details of approaching the torsional behaviour through the established analogy with the beam in bending based on second-order theory, according to the previous section, are shown in Figure 4.18.

Appropriate computer software must be used, allowing for geometrically non-linear effects.

The tensile axial force is produced by an equivalent reduction in temperature. The torsional stiffness is determined as the developed shear force due to an imposed support settlement in the beam equal to the assumed twisting angle at the free end.

Figure 4.18 Determination of torsional stiffness of a thin-walled beam

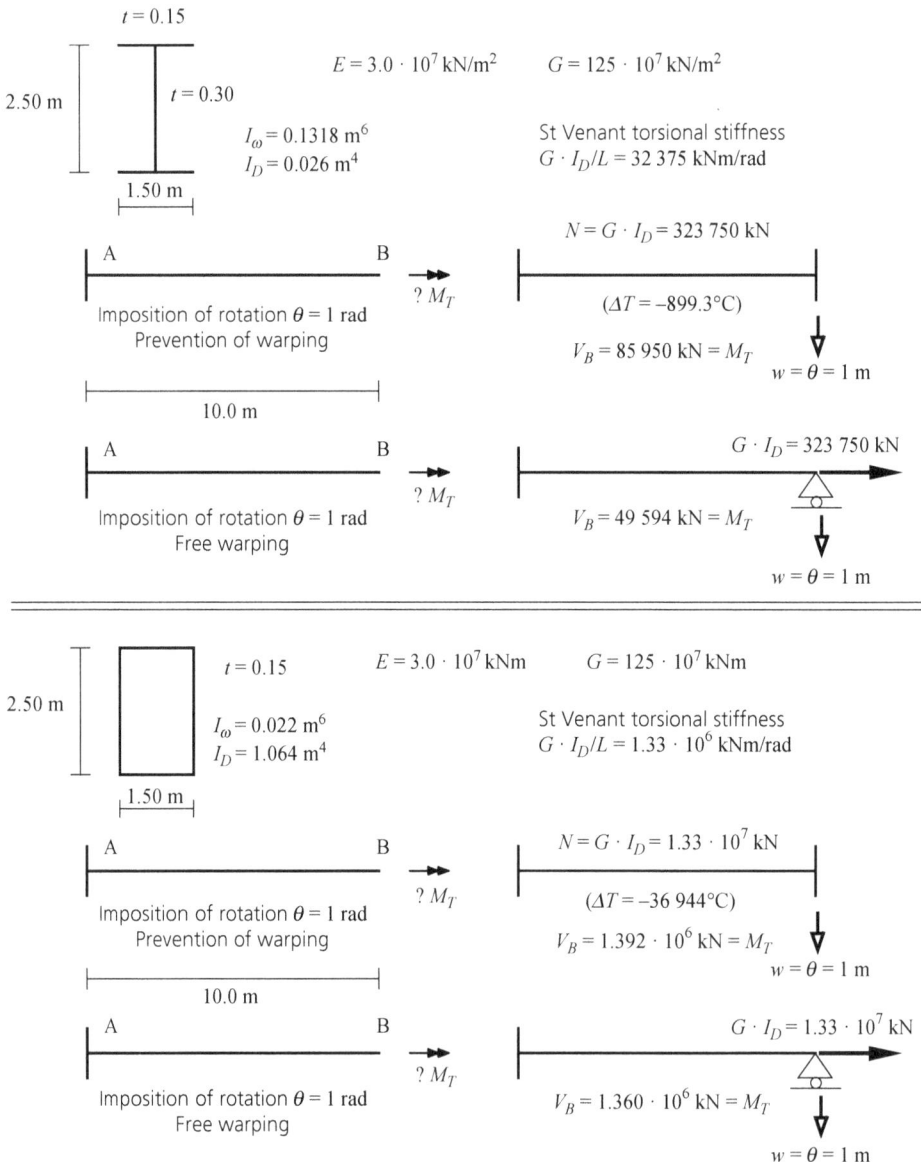

By analysing this case, it is seen that in the open section, warping prevention at the free end increases the torsional rigidity considerably, which, in any case, is significantly higher than the predicted one, according to the classic expression after St Venant $(G{\cdot}I_D/L)$ (see Figure 4.18). In a closed cross-section, the difference in torsional rigidity between prevented and unprevented warping is much smaller and of much less practical importance.

Example 2

The aim of this example is to determine the stress state of a simply supported concrete bridge, having a span of 40 m with an open cross-section, which is loaded at the midspan with an eccentric load 1000 kN, applied directly on the top of one web. It is supposed that the profile of the cross-section remains undeformable thanks to the insertion of transverse diaphragms (see Figure 4.19).

Figure 4.19 Determination of stress state in a thin-walled girder due to eccentric load

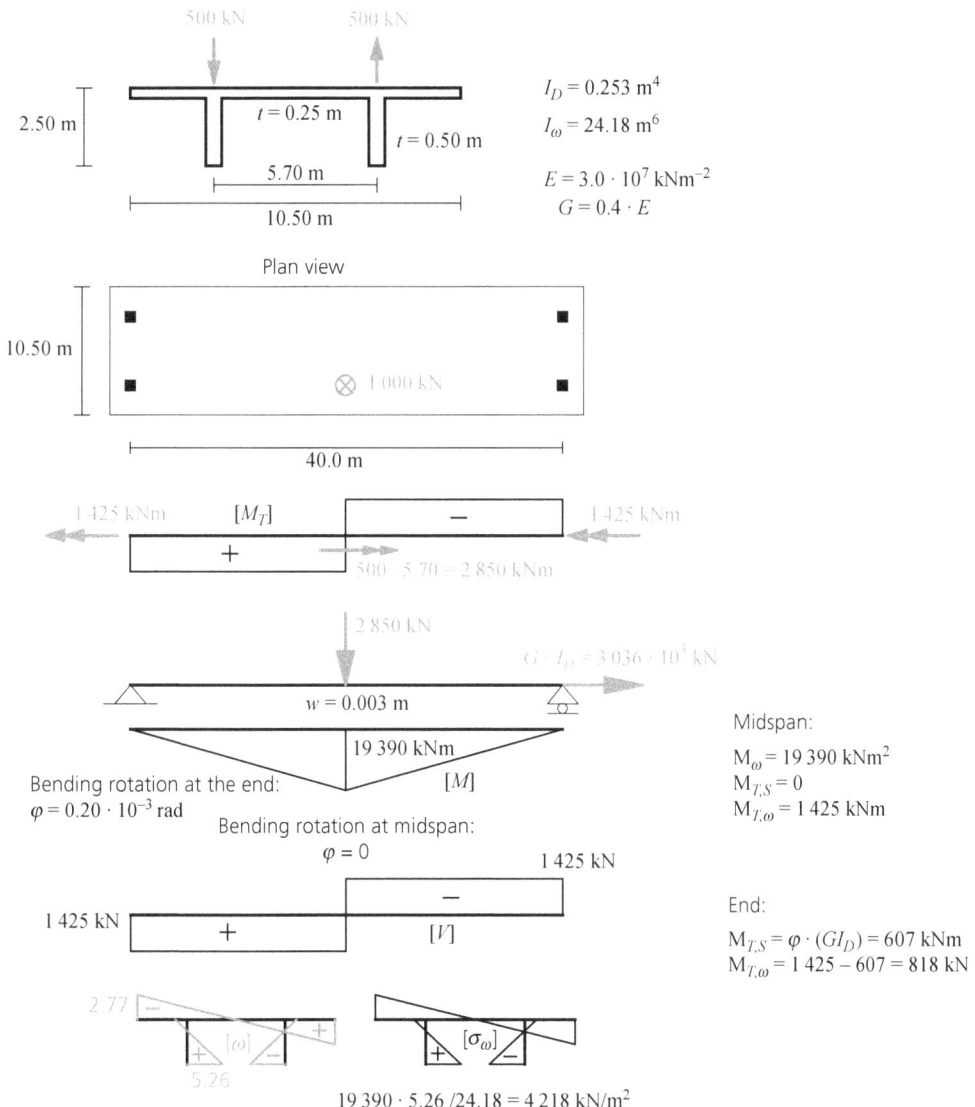

$I_D = 0.253$ m^4

$I_\omega = 24.18$ m^6

$E = 3.0 \cdot 10^7$ kNm^{-2}

$G = 0.4 \cdot E$

Midspan:
$M_\omega = 19\,390$ kNm^2
$M_{T,S} = 0$
$M_{T,\omega} = 1\,425$ kNm

End:
$M_{T,S} = \varphi \cdot (GI_D) = 607$ kNm
$M_{T,\omega} = 1\,425 - 607 = 818$ kN

$19\,390 \cdot 5.26 / 24.18 = 4\,218$ kN/m^2

The load is analysed exactly as in Section 4.1 (see Figure 4.5), in a symmetric and an anti-symmetric part. For the sake of later comparison, it is noted that the symmetric part consisting of the two loads of 500 kN causes at midspan a maximum tensile stress of 6.3 MPa. In this respect, the interest is, of course, focused on the anti-symmetric loads of 500 kN, as shown in Figure 4.19, which causes a corresponding torsion diagram. It is supposed that the bearings layout can offer to the beam the required reactions for the development of the torsional moment 1 425 kNm at its ends.

The analysis is done according to the established analogy in Section 4.8 under a concentrated load of 2850 kN at the midspan and is shown in Figure 4.19. Following the procedure presented in Stavridis and Georgiadis, 2025, Section 7.4), on the basis of a first-order deflection equal to $2850.0 \cdot 40.0^3/$ $(48 \cdot 3 \cdot 10^7 \cdot 24.18) = 0.0052$ m and a buckling load equal to $\pi^2 \cdot 3 \cdot 10^7 \cdot 24.18/40.0^2 = 4\,474\,621$ kN, a second-order deflection of $w = 0.0052/(1 + 3036 \cdot 10^3/4\,474\,621) = 0.003$ m is determined. Thus, the estimated bending moment at the midspan is $M = 2850.0 \cdot 40.0/4 - 3036 \cdot 10^7 \cdot 0.003 = 19390$ kNm. This value differs by just 2.3% compared with the one obtained from appropriate software.

The above moment represents the bimoment on the basis of which the longitudinal warping stresses are determined, according to Section 4.4.4. Based on the corresponding diagram $[\omega]$, the biggest tensile stress is equal to 4.12 MPa – that is, 65% of that caused by bending. Regarding the contribution of $M_{T,S}$ – that is, the St Venant shear flow – in taking up the torsional moment, M_T, this is determined according to Section 4.8. As shown in Figure 4.19, it is zero at the midspan, while at the ends, by considering the corresponding slope, φ, of the deflection line $\varphi = [2850 \cdot 40.0^2/(16 \cdot EI)]/$ $(1 + 3036 \cdot 10^3/4\,474\,621) = 0.0002$, it is equal to $M_{T,S} = \varphi \cdot G \cdot I_D = 607$ kNm. This value differs by less than 13% from the corresponding value obtained from appropriate software. Nevertheless, this difference is deemed acceptable for preliminary design purposes.

REFERENCES

Roik K (1983) *Vorlesungen über Stahlbau. Grundlagen.* Wilhelm Ernst, Berlin, Germany. (In German.)

Stavridis L and Georgiadis K (2025) *Understanding and Designing Structures without a Computer: Plane structural systems.* Emerald Publishing, Leeds, UK.

Vlasov VZ (1961) *Thin Walled Elastic Beams.* Oldbourne Press, London, UK.

Zbirohowski-Koscia K (1967) *Thin Walled Beams – From Theory to Practice.* Crosby Lockwood, London, UK.

emerald
PUBLISHING

ice
Publishing

Leonidas Stavridis and Konstantinos Georgiadis
ISBN 978-1-83662-945-0
https://doi.org/10.1108/978-1-83662-942-920251005
Emerald Publishing Limited: All rights reserved

Chapter 5
Box girders

5.1. Introduction

Box girders, used mainly for bridges, are thin-walled beams having a closed section and, as such, all that has been presented in Chapter 4 is relevant, but on one condition: in order for the basic equation as well as for all the relative concepts (bimoment etc.) concerning torsional response to be valid, the in-plane undeformability of the box section must be ensured. For relatively small section dimensions, this condition may be guaranteed by a certain wall thickness, but for larger dimensions stiff diaphragms are required over each support, as well as at a few places along the span. If these diaphragms are not wanted for constructional reasons, torsional loads will deform the section and, consequently, an additional bending response in the longitudinal as well as in the transverse direction will be developed. This will be examined in the present chapter by following a new approach taking into account the absence of transverse diaphragms, considering both rectilinear and curved girders.

5.2. Rectilinear girders
5.2.1 General loading case

In the typical box section shown in Figure 5.1, the eccentric layout of a traffic load is indicatively considered. In reality, this is specified by the relevant design standards.

External actions are considered to be applied to the nodes A and B of the upper slab, which eliminate both their displacement and rotation. These *fixing actions* (see Stavridis and Georgiadis, 2025, Section 3.2.1) consist of longitudinally distributed vertical loads q_A and q_B, as well as longitudinally distributed moments m_A and m_B, which are determined directly from the fixed-end actions on the assumed clamped nodes A and B.

Thus, following the same logic as in Stavridis and Georgiadis (2025, Section 3.2.1), it may be considered that the state of stress of the girder results from the superposition of fixed state (I) containing the loads with the appropriate nodal actions and state (II) containing only these nodal actions q_A and m_A, together with q_B and m_B, acting, however, in the opposite sense. The above nodal actions q and m exhibit a certain longitudinal distribution along the two upper edges of the hollow girder and are expressed in kN/m and kNm/m, respectively.

Fixed state (I) can be determined directly (similar to a frame). Thus, state (II) is predominantly discussed below. The loading at state (II) – as long as the section is symmetric – may be split to a symmetric and an antisymmetric part. The symmetric part consists of the symmetrically distributed loads $q_R/2$ and moments $m_R/2$ applied on the edges A and B, respectively. The antisymmetric part consists of the mutually opposite loads $q_S/2$ as well as the moments of the same sense $m_S/2$ (see Figure 5.1).

The symmetric loading activates the longitudinal bending stiffness of the girder, resulting in the well-known sectional force diagrams of bending moments and shear along the beam. Moreover, it causes a transverse response. If the equilibrium of a segmental part of unit length is considered, it can be seen that this girder strip is under the resultant of the shear flows applied on each of its two faces (see Figure 5.2). Given that the shear flow, v, is expressed as $v = V \cdot S/I$, it is $(dv/dx) = (S/I) \cdot (dV/dx) = (S/I) \cdot q_R$ so that the above resultant is exactly the differential shear flow, dv, acting on the strip, being equal to $dv = (S/I) \cdot q_R$. It is noted that S represents the static moment of inertia of the considered cut-out part of the section with respect to its centroidal axis.

Figure 5.1 Derivation of nodal actions on the beam profile from the acting eccentric loading

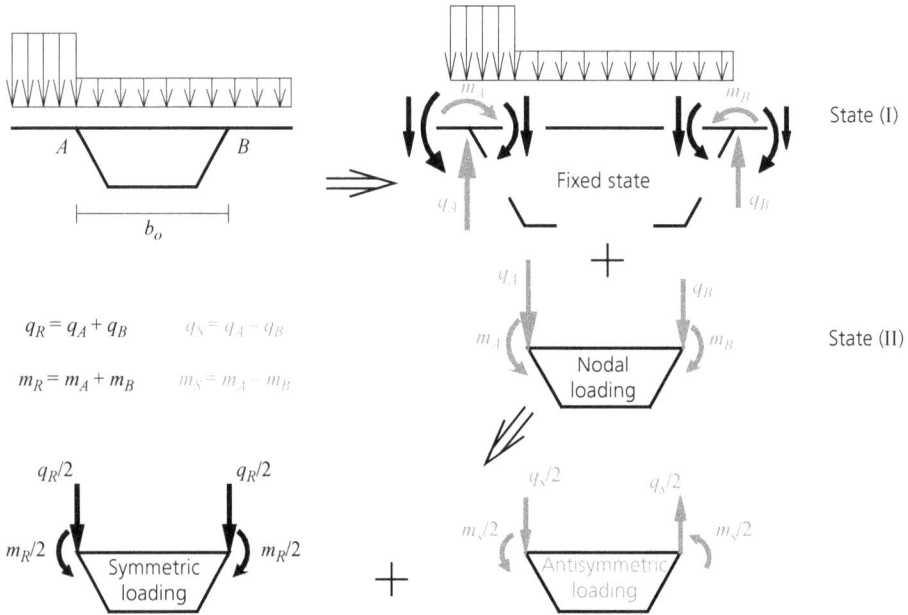

Figure 5.2 Transverse response under symmetric loading

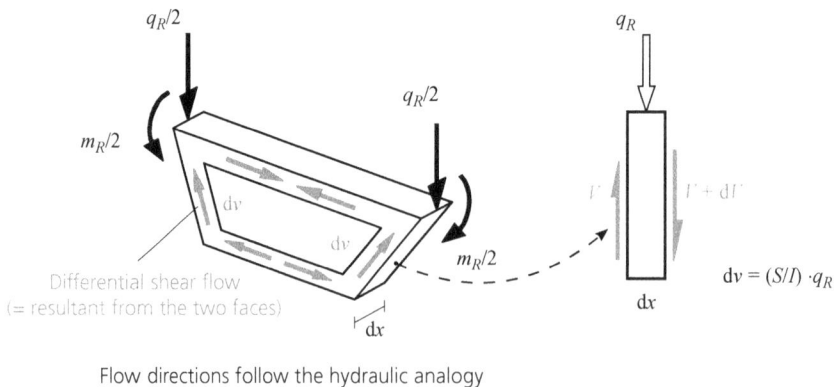

Flow directions follow the hydraulic analogy

Thus, the girder strip is in equilibrium under the symmetric loads $q_R/2$, $m_R/2$, as well as under the differential shear flow, dv. The loads $q_R/2$ and dv do not induce any bending at all in the plane frame, but only axial stresses. The loads $m_R/2$ cause both bending and axial stresses. This state of stress regarding the three-times statically indeterminate closed frame may be determined directly by using appropriate computer software for the analysis of plane frames. ■

Two key points are added here regarding haunched bridge girders under symmetric loading, either in a cantilevered system or, analogously, in a continuous one over its internal supports – in addition to what has been mentioned in Stavridis and Georgiadis (2025, Sections 4.6 and 5.5.2) – mainly with respect to the free cantilevered method of construction.

The first and most important point concerns the variation of the tensile axial force, Z, in the top slab of a cantilever, being interrelated with the longitudinal shearing flow, v, as shown in Figure 5.3. As explained in Stavridis and Georgiadis (2025, Section 4.6), the inclined compressive force of the bottom slab helps with its vertical component in offering to the cross-section the required vertical shear force, thus achieving, for a parabolic curving of the bottom slab, an approximately constant shear response throughout the cantilever length. By cutting out from the web an elementary segment of the top slab subjected to the axial tensile forces, as well as to the acting shear flow, and considering its horizontal equilibrium, it is concluded that:

$$v = \tau \cdot b^w = \frac{\Delta Z}{\Delta x}$$

where

$$\tau \cdot b^w = \frac{1}{z}\left(V - \frac{M}{z} \cdot \frac{\Delta z}{\Delta x}\right) \text{ (see Stavridis and Georgiadis, 2025, Section 4.6)}$$

The above relations are generally valid. In the case of a constant depth ($\Delta z/\Delta x = 0$), the tensile force, Z, exhibits a parabolic variation along the girder ($\Delta Z/\Delta x = V/z$). However, in the case of a variable depth having a constant flow, v, the tensile force, Z, is linearly varied that, regarding the balanced cantilevered construction for continuous bridges as mentioned in Stavridis and Georgiadis (2025, Section 5.5.2) and suggested in Figure 5.3, means a practically constant amount of prestressing 'dosage' to be added at each consecutive construction stage, thus having a constructional advantage. It is pointed out that the prestressing steel for the balanced cantilevered construction method (see also Stavridis and Georgiadis, 2025, Sections 5.4 and 4.6) is principally placed at the top slab of the girder section.

The second point regards the fact that the ever-changing inclination of the compressive force of the bottom slab due to bending creates an inward deviation pressure $p_u = D/r$ (see Figure 5.3), which stresses the bottom slab in the transverse direction, obviously counteracting its self-weight, but necessarily affecting the whole closed transverse segmental strip in bending, as shown in Figure 5.3. ■

As the symmetric loading does not present any other peculiarity, the whole problem is shifted to the antisymmetric loading part. This acts as a distributed torsional load $m_D = (q_S/2) \cdot b_o + m_S$ (see Figure 5.1), causing a certain torsional moment diagram (see Figure 5.4). At each position of the beam, the developed torsional moment is taken up by the respective Bredt shear flow. Thus, a shear force diagram may be attributed to each wall of the section considered in its longitudinal direction.

Figure 5.3 Structural behaviour of cantilevered haunched girder

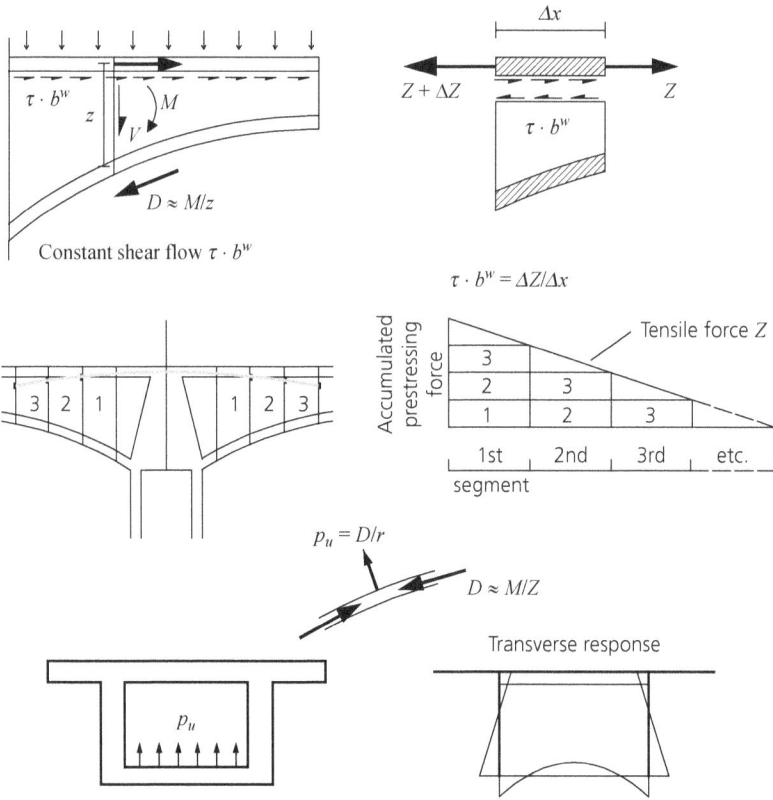

Figure 5.4 Influence of support layout on the distribution of the torsional response

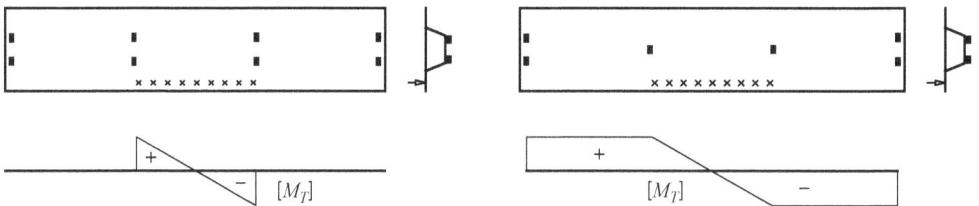

If now the non-deformability of sections is ensured through the layout of transverse diaphragms, as mentioned previously, an additional longitudinal response consisting of axial and shearing stresses will take place, according to Chapter 4. The fact that the section is closed certainly leads to limited stresses arising from the restraint of the already small warping of the section.

However, the presence of transverse diaphragms is generally not desirable for constructional reasons. Thus, the girder section must be considered as deformable. The development of torsional response is not governed any more by the basic equation given in Chapter 4, Section 4.8 and, consequently, the structure has to be accordingly considered as a folded system.

In this case, as will be examined in the following sections, an additional response to the Bredt shear flow is developed. This consists of a longitudinal bending for each wall, as well as a transverse response for the entire section profile.

5.2.2   Response due to the deformability of the profile section

The imposed torsional loading $[m_D = (q_S/2) \cdot b_o + m_S]$ all along the girder, due to the antisymmetric couple, causes deformation of the closed section, resulting in longitudinal bending of the section walls (see Figure 5.5), which is coupled with the resulting transverse bending of the section profile itself (Schlaich and Scheef, 1982). It is again pointed out that this response is in addition to the initially existing Bredt shear flow in the section walls.

In order now to understand the influence of the deformability of the section on the response, the equilibrium of a cut-out girder strip of unit length is at first considered with the antisymmetric loads $q_S/2$ and $m_S/2$ acting at the section edges A and B (see Figure 5.5). It is clear that the segment is in equilibrium under the above external forces and the *differential shear flow*, Δv, which is obtained as the resultant of the Bredt shear flows on the two faces of the considered strip.

Given that:

$$v = \frac{M_T}{2 \cdot F_k}$$

Figure 5.5 Analysis of antisymmetric loading

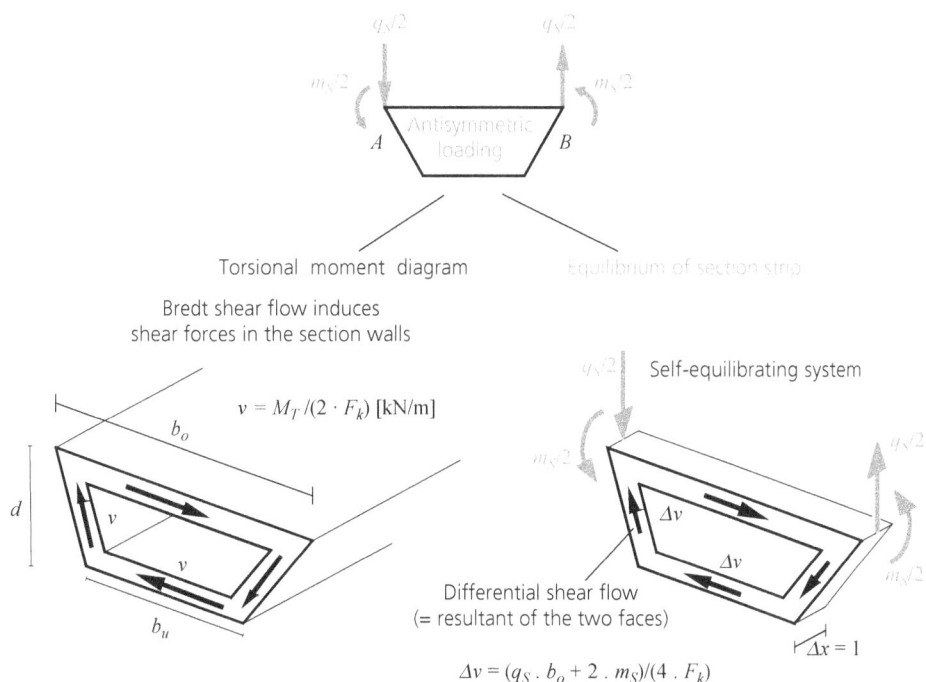

Torsional moment diagram

Equilibrium of section strip

Bredt shear flow induces
shear forces in the section walls

$v = M_T/(2 \cdot F_k)$ [kN/m]

Self-equilibrating system

Δv

Δv

Differential shear flow
(= resultant of the two faces)

$\Delta x = 1$

$\Delta v = (q_S \cdot b_o + 2 \cdot m_S)/(4 \cdot F_k)$

it is:

$$\frac{\Delta v}{\Delta x} = \frac{\Delta M_T}{\Delta x} \cdot \frac{1}{2 \cdot F_k} = \frac{m_D}{2 \cdot F_k}$$

and, consequently:

$$\Delta v = \frac{q_S \cdot b_o + 2 \cdot m_S}{4 \cdot F_k}$$

According to the section dimensions shown in Figure 5.5, it is:

$$F_k = d \cdot (b_o + b_u)/2$$

The differential flow, Δv, in question is essentially the differential shear flow Δv (kN/m) required to be applied on all four sides of the closed transversal girder strip in order to keep it in equilibrium under the twisting action $[(q_s/2) \cdot b_o + 2 \cdot (m_s/2)]$. For that purpose, the same Bredt formula can be directly applied giving exactly the previously derived result $\Delta v = M_T/(2 \cdot F_k) = [(q_s/2) \cdot b_o + m_s]/(2 \cdot F_k)$. This means that the Bredt formula can be applied to determine either the shear action, v, on a certain section subjected to a torsional moment, M_t (which is statically equivalent to the latter), or the resultant differential shear, Δv, acting on a cut strip, being itself in equilibrium as a free body under a certain applied torsional load.

Obviously, the examined strip tends to deform under the equilibrating forces $q_s/2$, $m_s/2$ and Δv. This deformation consists essentially in a change of length of its diagonals and induces an additional state of stress for each section wall beyond the shearing force caused by the torsional moment itself according to Bredt's formula.

The situation can be treated, in an analogous way as in Section 5.2.1, by inserting a hinged strut of an infinite axial stiffness in the augmenting diagonal (i.e. $A = \infty$) – that is, a non-deformable one (see Figure 5.6). With the insertion of this element throughout the length of the girder, the strip deformability is eliminated, whereas the strip itself develops a bending and axial state of stress, the former being induced by the acting moments $m_s/2$. It is clear that this diagonal element develops a tensile axial force, R (kN/m), distributed along the length of the girder. Considering now the opposite of the acting forces, R, on the corresponding longitudinal edges of the girder, as the diagonal strut does not really exist, it is obvious that the superposition of the thus resulting state of stress with that of the blocked strip will give the finally developed response (see Figure 5.6). ∎

For the evaluation of the above diagonal loading it may be first considered that the force, R, acts on a girder having hinged connections at the section edges instead of monolithic ones (see Figure 5.7). The force, R, may be resolved equivalently at each edge into the two concurring walls; therefore, each of them can be considered as a longitudinal beam loaded by the relevant component, developing bending moments, M_0, and corresponding normal stresses, σ, according to the classical technical theory of bending. It can be seen, however, that the resulting strain, σ/E, at the common edges of the walls are not equal – as they ought to be – and for that reason some distributed longitudinal forces have to be additionally introduced along the edges of each wall in order to establish the

Figure 5.6 Additional girder response due to antisymmetric loading

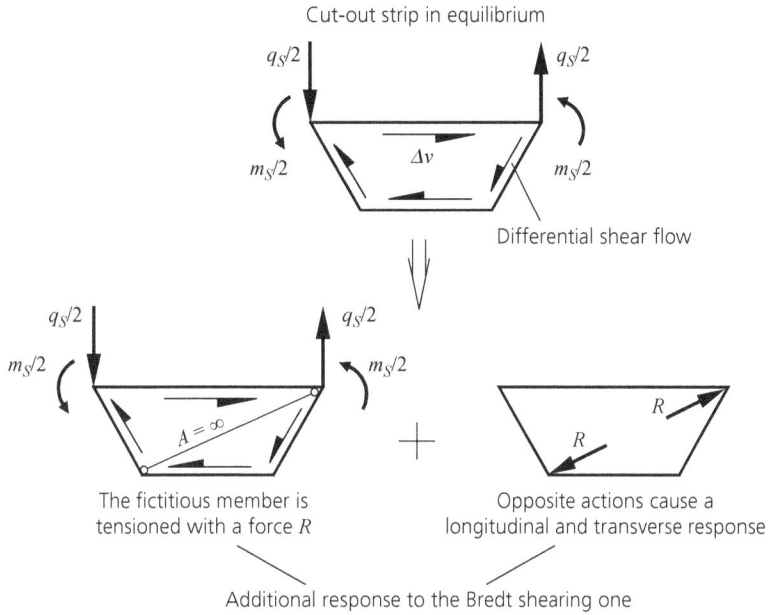

Cut-out strip in equilibrium

$q_S/2$ $q_S/2$

$m_S/2$ Δv $m_S/2$

Differential shear flow

$q_S/2$ $q_S/2$

$m_S/2$ $m_S/2$

$A = \infty$

$+$

R

R

The fictitious member is
tensioned with a force R

Opposite actions cause a
longitudinal and transverse response

Additional response to the Bredt shearing one

Figure 5.7 Taking up of the diagonal loads through the longitudinal wall-beams

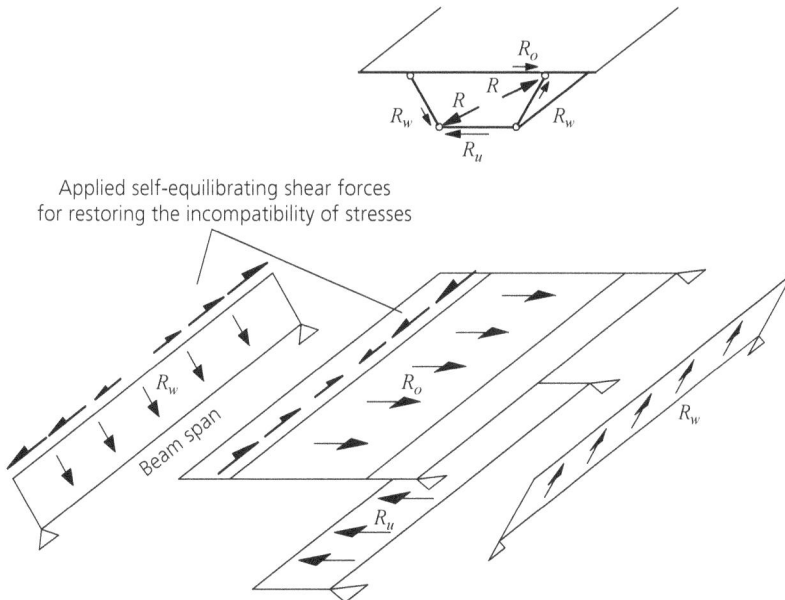

R_o
R
R
R_w R_w
R_u

Applied self-equilibrating shear forces
for restoring the incompatibility of stresses

R_w
Beam span
R_o
R_w
R_u

155

strain compatibility at each edge (see Figure 5.7). It is clear that in this way the initially determined normal stress, σ, will be changed.

By following the above analysis, the longitudinal axial stresses can be determined – for example, for the left web of the *hinged* box section as shown in Figure 5.8, through the classical bending formula on the basis of the aforementioned moment, M_0. In this case, the moment of inertia, I^*, is equal to the normal value, I_w, for the web multiplied by a factor, k^w. Thus, the value of I^* is a bit greater than the normal value of I_w for the web. In this case, the new 'neutral axis' lies not at the midheight of the section, but at a distance, y_0, from the top fibre. This distance, y_0, is somewhat less than its 'normal value' – that is, half of the web height (Schlaich and Scheef, 1982).

The moment, M_0, results from the loading of the left web with the corresponding component, R_w, of the *diagonal force, R* (see Figure 5.7). It is found that $R_w = R \cdot b_w/s$. Thus, it is:

$$\sigma_o = -\frac{M_0}{I^*} y_o \quad \text{and} \quad \sigma_u = \frac{M_0}{I^*} \cdot (b_w - y_o)$$

where

$$I^* = I_w \cdot k^w$$

Figure 5.8 Bending response of the longitudinal section walls in the hinged system

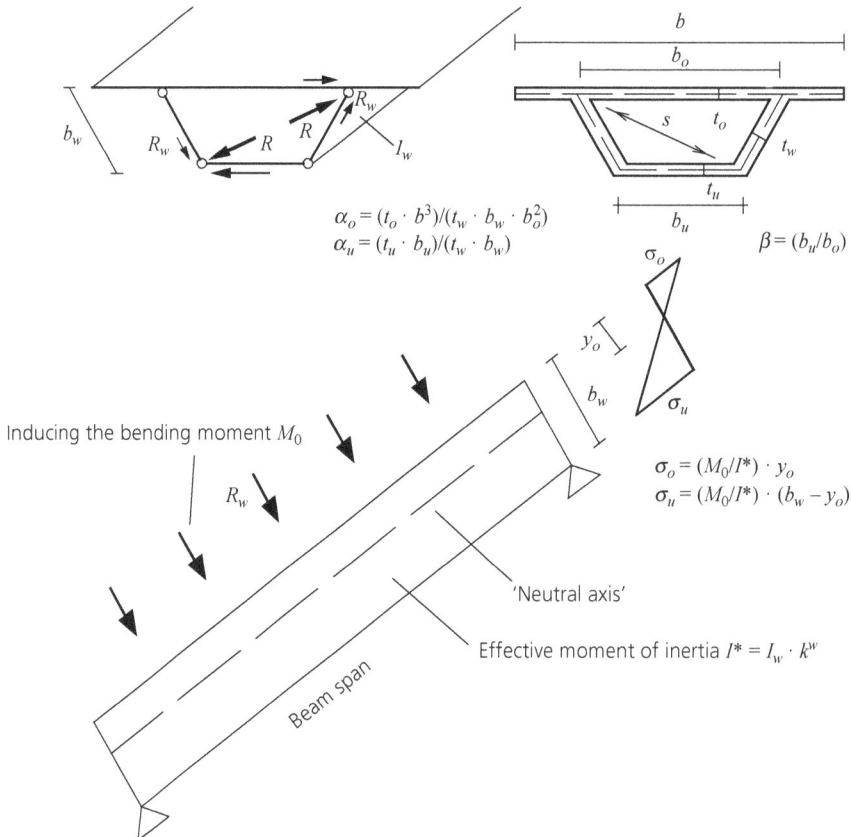

$\alpha_o = (t_o \cdot b^3)/(t_w \cdot b_w \cdot b_o^2)$
$\alpha_u = (t_u \cdot b_u)/(t_w \cdot b_w)$

$\beta = (b_u/b_o)$

$\sigma_o = (M_0/I^*) \cdot y_o$
$\sigma_u = (M_0/I^*) \cdot (b_w - y_o)$

Inducing the bending moment M_0

'Neutral axis'

Effective moment of inertia $I^* = I_w \cdot k^w$

Beam span

Furthermore, it is found (see Figure 5.8) that:

$$k^w = \frac{2 \cdot \beta \cdot \left[(\alpha_o + 2) \cdot (\alpha_u + 2) - 1\right]}{(1 + \beta) \cdot (3 + 3 \cdot \beta + \alpha_o + \alpha_u \cdot \beta)} \quad \text{and} \quad y_o = \frac{1 + 2 \cdot \beta + \alpha_u \cdot \beta}{3 + 3 \cdot \beta + \alpha_o + \alpha_u \cdot \beta} \cdot b_w$$

Thus, the longitudinal web beam satisfies the following typical differential equation (see Stavridis and Georgiadis, 2025, Section 2.3.6):

$$EI^* \frac{\mathrm{d}^4 w}{\mathrm{d}x^4} = R_w$$

where R_w and I^* are the initial distributed web loading and the equivalent moment of inertia, respectively, as explained above.

It should be noted that w, which is unrelated to the overall girder deflection, represents the in-plane deflection of the web and arises solely from the assumed deformability of the section (see Figure 5.8).

Under the action of diagonal forces, R, both webs undergo a certain bending deflection, w, (downwards and upwards, respectively) leading to an increment of the diagonal dimension of the section profile by δ (see Figure 5.9) according to the geometrical relation:

$$\delta = w \cdot D$$

where

$$D = \frac{2 \cdot b_w}{\beta \cdot s} \cdot \frac{(1 + \beta) \cdot (2 + 2 \cdot \beta + 2 \cdot \beta^2 + \alpha_o + \alpha_u \cdot \beta^2)}{3 + 3 \cdot \beta + \alpha_o + \alpha_u \cdot \beta}$$

In the case in which the box section is rectangular, the factor k^w is practically equal to 1.0 (meaning that $I^* = I_w$) and the geometric parameter D is equal to $(4 \cdot h/s)$.

While this diagonal deformation in the case of the hinged frame examined above can take place freely, in a real situation the monolithic connection cannot be developed without resistance. This means that the corresponding transverse stiffness, C, of the closed monolithic frame will be automatically mobilised according to the relation:

$$r = \delta \cdot C$$

The stiffness $C = r/\delta$ can be found by determining the diagonal deformation, δ, of the closed frame under the action of an arbitrary diagonal pair of forces, r, using appropriate computer software for plane frames. However, in Appendix B of this chapter, an analytical expression of C for an orthogonal box girder profile is given.

Thus, the intention of the web to be deformed by w will be counteracted by the respective component of the resistance, r, of the monolithic closed frame and in this way the web will be subjected, apart from its 'initial' loading, R_w, also to the loading of the component r_w of the force r, obviously with the opposite sense. It is:

$$r_w = r \cdot (b_w/s) \qquad \text{(see Figure 5.9)}$$

The component, r_w, can also be expressed through w by the relation:

$$r_w = w \cdot D \cdot C \cdot (b_w/s) = w \cdot K$$

where

$$K = D \cdot C \cdot (b_w/s)$$

The distributed load, r_w, acts on the web beam in addition to the load, R_w, obviously in the opposite sense; thus, the above beam equation can be written as:

$$EI^* \frac{\mathrm{d}^4 w}{\mathrm{d}x^4} = (R_w - r_w) = R_w - w \cdot K$$

This equation is recognised as the typical equation of a beam on an elastic foundation with modulus of subgrade reaction, K (see Chapter 8, Section 8.3.3.1), meaning indeed that the web is carried by the elastic support offered by the profile resistance when deforming under the distributed load, r_w, by $w = r_w/K$ (see Figure 5.9).

In this sense, the beam can also be considered as being supported on elastic springs with an equidistance of s and having an axial stiffness, k_s. By considering the upward distributed load $(K \cdot w)$ as being represented by the spring force $(w \cdot k_s)$ distributed over the distance, s, according to the

Figure 5.9 Structural action of section wall in the monolithic system

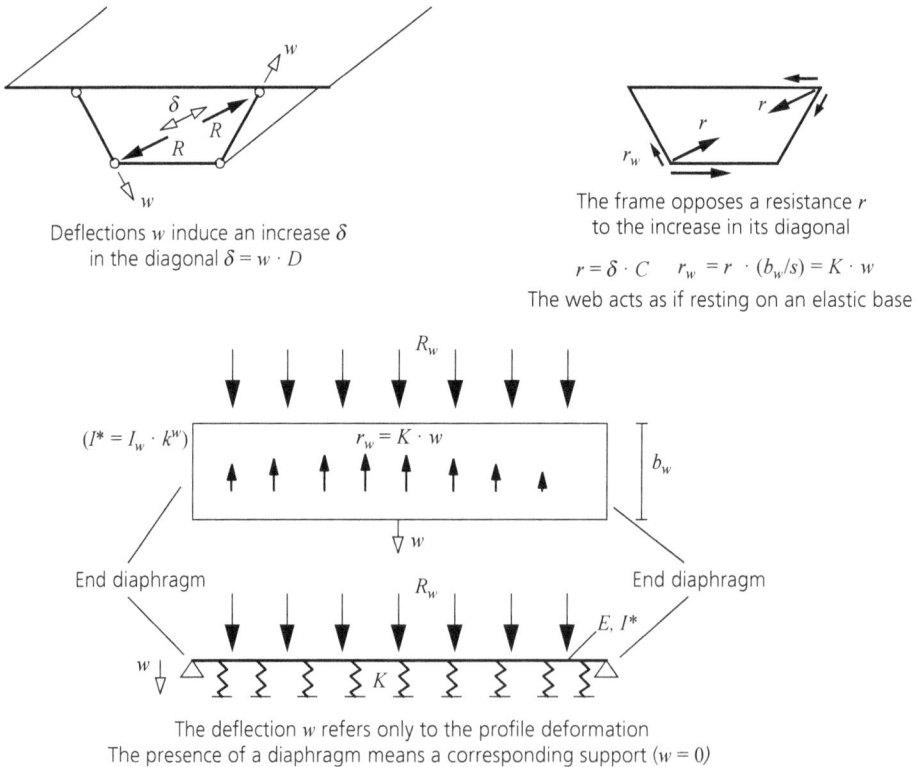

Deflections w induce an increase δ
in the diagonal $\delta = w \cdot D$

The frame opposes a resistance r
to the increase in its diagonal

$$r = \delta \cdot C \qquad r_w = r \cdot (b_w/s) = K \cdot w$$

The web acts as if resting on an elastic base

$(I^* = I_w \cdot k^w)$ $r_w = K \cdot w$

The deflection w refers only to the profile deformation
The presence of a diaphragm means a corresponding support $(w = 0)$

relation $K \cdot w = w \cdot k_s/s$, the spring stiffness may be readily determined as $k_s = K \cdot s$ (see Figure 5.9). In Appendix A of this chapter some analytical results concerning the deformation and bending response of beams resting on an elastic foundation are presented.

It is clear that the resulting longitudinal bending response of the web will also lead, through the developed axial stresses, to the development of edge stresses of the remaining section walls, as at each edge the strain, ε, and, consequently, the stress ($\sigma = \varepsilon \cdot E$) are common.

It should be particularly again emphasised that the deflection, w, concerns only that deformation which arises from the deformability of the section profile and has nothing to do with the overall deflection of the girder or the web itself. Thus, at each position, where the profile remains undeformed, due to the presence of a transverse diaphragm, the condition $w = 0$ must be satisfied. This means that at the corresponding place of the elastically supported beam model, an unmovable support must be inserted (see Figure 5.10). In such a case, the diaphragm must be appropriately

Figure 5.10 Behaviour of box girder under an eccentric concentrated load

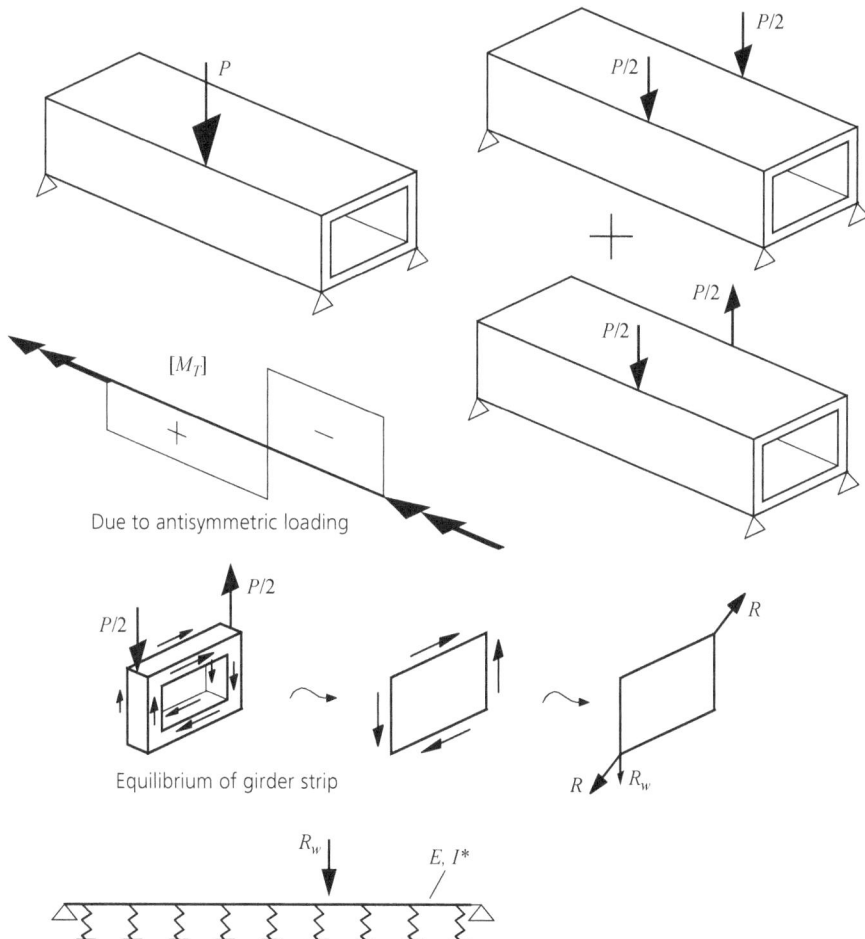

designed to undertake the corresponding diagonal forces, R, determined as shown above, at the particular position of the box girder. ■

The presence of a single concentrated load on the web will lead to an applied concentrated torsional load and a respective shearing flow (Bredt) and, finally, to a concentrated vertical load applied on the model of the elastically supported beam (see Figure 5.10). If this eccentric load is the only one considered acting on the girder, then in the above equation the distributed load, R_w, on the right side must be replaced by a single force representing the vertical component of the diagonal pair of equal and opposite forces, R. These forces, in the case in which there is a transverse diaphragm at this place, are taken up directly by it, without causing any bending stresses in the girder as this can be easily understood by considering the appropriately inserted support in the elastically supported beam model. If such a transverse diaphragm does not exist, then the diagonal forces are taken up by the deformable section of the girder itself, as previously examined in detail (see Figure 5.10).

Finally, the treatment of an internal support of a continuous girder, having no transverse dia-phragm, under the developed support reactions that arise from torsional response is examined (see Figure 5.11).

In this case, the developed vertical reactions that offer the required torsional reaction should be considered as antisymmetric loads exerting a concentrated torsional moment and, according to the above, they lead to a concentrated force on the elastically supported web model. ■

Of course, the simultaneous transverse response of the section profile should not be overlooked.

Figure 5.11 Behaviour at a support without a transverse diaphragm

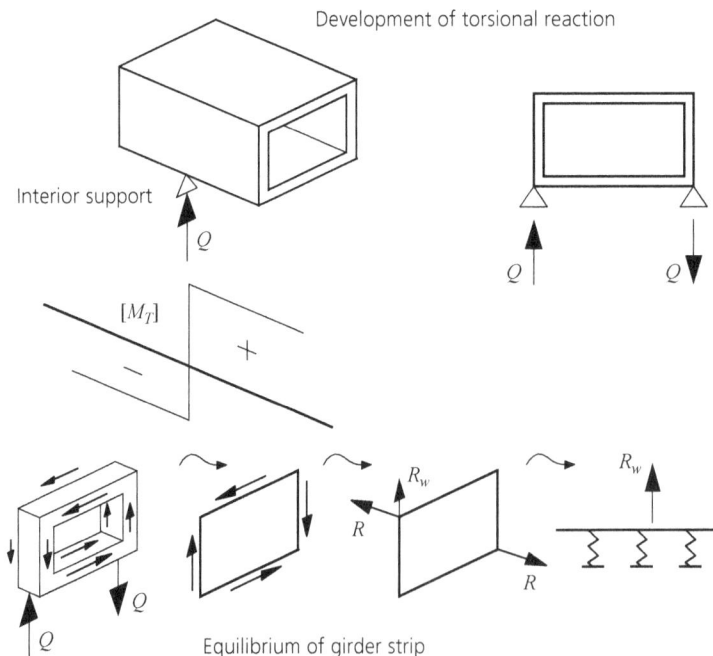

Figure 5.12 Transverse response due to profile deformation

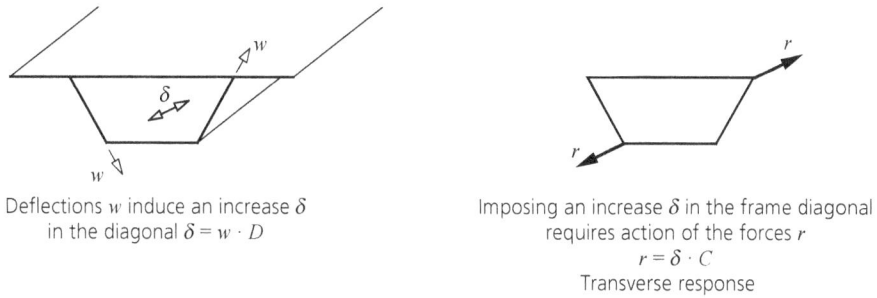

Deflections w induce an increase δ
in the diagonal $\delta = w \cdot D$

Imposing an increase δ in the frame diagonal
requires action of the forces r
$$r = \delta \cdot C$$
Transverse response

The implied deformation, w, of the web induces a transverse response that results from applying the self-equilibrating diagonal loading, r, on the closed rigid profile, as shown in Figure 5.12, which can be readily determined according to the above. Thus, it is:

$$r = w \cdot K \cdot \frac{s}{b_w}$$

Worked example

A simply supported orthogonal concrete box girder with a 40 m span, as shown in Figure 5.13, is examined. At the midspan and over the left web a concentrated load 1000 kN is considered, which, according to Figure 5.10, induces torsion through the antisymmetric couple of forces of 500 kN each. On the basis of the foregoing expressions in the Section 5.2.2, the following cross-section data are determined:

$$\beta = 1, \quad \alpha_o = 1.24, \quad \alpha_u = 0.79, \quad k^w = 1.002, \quad y_o = 1.18 \text{ m}, \quad D = 1.495, s = 6.69 \text{ m}$$

$$I_w = 0.504 \text{ m}^4$$

Thus, it is: $I^* = 0.504 \cdot 1.002 = 0.505 \text{ m}^4$

Moreover, the stiffness, C (kN/m^2), of the unit length's girder strip with respect to a unit increase (1.0 m) of the diagonal distance of its edges is determined as $C = 21\,524.0$ kN/m^2 (see Appendix B).

As it comes out from the equilibrium of the girder strip at midspan, the web is under a concentrated force 250 kN. More specifically, the strip is in equilibrium under the antisymmetric couple of 500 kN as well as the peripheral differential shear flow, v, which according to Section 5.2.2 is equal to:

$$v = \frac{500 \cdot 6.20}{2 \cdot 2.50 \cdot 6.20} = 100 \text{ kN/m}$$

The force, R_w, acting on the left web is:

$$R_w = 500.0 - v \cdot 2.50 = 250.0 \text{ kN} \qquad \text{(see Figure 5.13)}$$

The web, as explained previously, acts as a beam on elastic foundation with a modulus of subgrade reaction equal to $K = 1.495 \cdot 21\,524 \cdot 2.50/6.69 = 12\,025$ kN/m².

The corresponding bending moment diagram is obtained, as can be seen in Figure 5.13, and can be determined according to Appendix A whereby the elastic length:

$$L_s = \left(\frac{4 \cdot EI}{K} \right)^{\frac{1}{4}} = 8.42 \text{ m}$$

Figure 5.13 Antisymmetric loading of the girder caused by an eccentric load

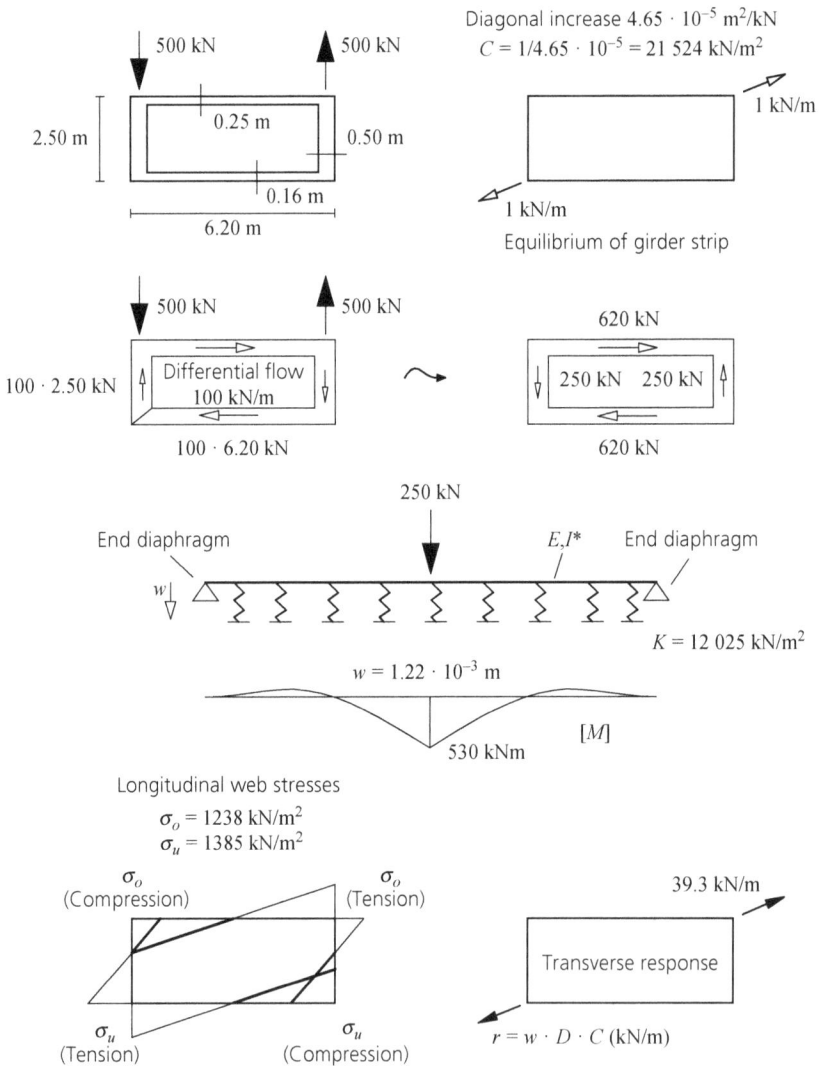

and,

$$\lambda = 1/L_s = 0.118/m$$

Thus, the maximum moment at midspan is:

$$M = \frac{F}{4 \cdot \lambda} = \frac{250}{4 \cdot 0.118} = 530 \text{ kNm}$$

whereas the corresponding deflection is:

$$w = \frac{F \cdot \lambda}{2 \cdot K} = \frac{250 \cdot 0.118}{2 \cdot 12025} = 1.22 \cdot 10^{-3} \text{ m}$$

Thus, the normal stresses at the top and bottom fibres are obtained according to the relations of the previous section as:

$$\sigma_o = 530 \cdot 1.18/0.505 = 1238 \text{ kN/m}^2 \qquad \text{(compression)}$$

$$\sigma_u = 530 \cdot (2.50 - 1.18)/0.505 = 1385 \text{ kN/m}^2 \qquad \text{(tension)}$$

The above longitudinal stresses are exclusively due to the torsional response of the girder, by taking into account the deformability of its cross-section. The complement of the stress image for the whole section is shown in Figure 5.13.

The development of deflection, w, causes an increase of the diagonal length of the profile, which requires the action of the corresponding diagonal forces, r. These are, according to the above, equal to:

$$r = 1.22 \cdot 10^{-3} \cdot 1.495 \cdot 21524 = 39.3 \text{ kN/m}$$

On the basis of these forces, the respective bending, axial and shearing response of the closed girder section may be readily deduced (see Figure 5.13).

5.2.3   Summary

According to previous analysis, the whole examination of the behaviour of rectilinear box girders under an eccentric loading consists in the following.

▨ Evaluation of the fixation state of stress $[m^0]$ (with $[v^0]$), as well as the (non-symmetric) node loading.
▨ The non-symmetric loading is split into a symmetric and an antisymmetric part.
▨ The symmetric part causes a longitudinal bending and shearing response determined by the corresponding sectional force diagrams $[M]$ and $[V]$, as well as a transverse response concerning a transverse girder strip of unit length arising from a self-equilibrated loading consisting of the acting loads and the differential shear flow. The transverse response is represented by the corresponding bending moment, shear force and axial force diagrams $[m_{sym}]$, $[v_{sym}]$ and $[n_{sym}]$, respectively, plotted along the section profile.

- The antisymmetric part causes a torsional response along the girder represented by a torsional diagram $[M_T]$, which enables the determination of the Bredt shear flow at each position. This shear flow causes, in turn, on the one hand shear of the section walls depicted by a diagram $[V_T]$ for each wall and, on the other, a longitudinal 'diagonal loading' $\{R\}$, arising from the resulting differential shear flow of a strip. To this last response, the transverse one arising from fixation of the section edges has to be added, represented by the diagrams $[m_f]$, $[v_f]$ and $[n_f]$.
- The diagonal loading $\{R\}$ causes a longitudinal bending and shear response represented by the sectional force diagrams $[M^R]$ and $[V^R]$, respectively, arising from the consideration of a characteristic section wall (usually the web) as a beam resting on elastic foundation. This beam is supported only at places where the deformation of the section profile is prevented.
- The above response is always accompanied by the transverse response of the section profile, determined on the basis of its mobilised diagonal stiffness and is represented by the respective sectional diagrams $[m^R]$, $[v^R]$ and $[n^R]$ along the profile perimeter.

Thus, for the response of each wall, the following diagrams must be taken into account.

Longitudinal bending: $[M] + [M^R]$

Longitudinal shear: $[V] + [V_T] + [V^R]$

Transverse bending: $[m^0] + [m_{sym}] + [m_f] + [m^R]$

Transverse axial response: $[n_{sym}] + [n_f] + [n^R]$

Transverse shear: $[v^0] + [v_{sym}] + [v_f] + [v^R]$

The transverse shear is usually of minor importance.

5.3. Curved girders
5.3.1 Overview
Box girders are particularly suitable for bridges curved in plan, as these are subjected to permanent torsional response caused even by non-eccentric loads, such as, for example, the self-weight of the girder.

The skeletal model used for a curved girder is of the grid type (see Chapter 1) – that is, a plane structure loaded perpendicular to its plane – and as such it develops three sectional forces at each position, namely a vertical shear force, V, a bending moment, M_B, and a torsional moment, M_T (see Figure 5.14).

As will be shown below, an equilibrium interrelation exists between bending and torsional moments and thus any redistribution of the bending response of the girder, for reasons concerning its plastic design (see Stavridis and Georgiadis, 2025, Section 6.6.2), has to be necessarily accompanied by a corresponding adjustment of the torsional moments.

5.3.2 Determination of bending and torsional response
5.3.2.1 Evaluation of the equilibrium equations
A curved girder is assumed to have a constant radius of curvature equal to R. The applied load consists of a vertical distributed load, q, passing through the shear centre of the cross-section, as well as a distributed torsional load, m_D (see Figure 5.14).

Figure 5.14 Curved beam model with acting loads and the developed sectional forces

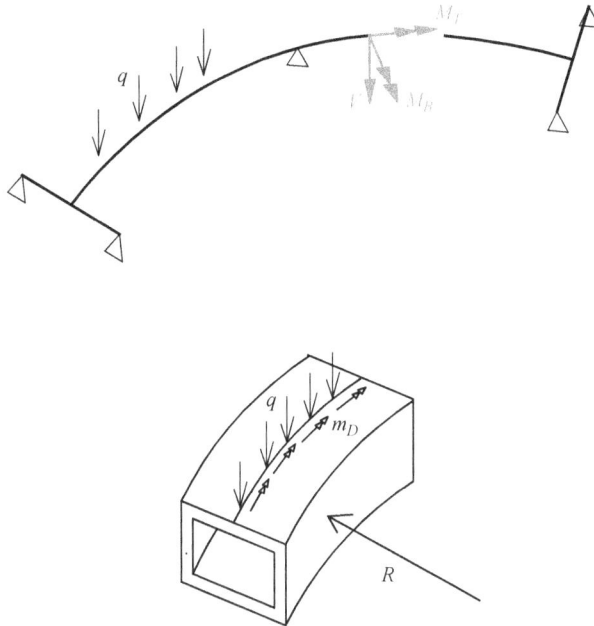

The main reason that a curved box girder is subjected to torsion even under the load, q, only is the presence of bending along a curved axis, which implies the imposition of a distributed torsional load, m_q, all over the length of the girder (see Figure 5.15).

The acting compressive forces, D, and tensile forces, Z, in the curved top and bottom flanges, respectively, cause distributed deviation forces, q, according to the well-known relation:

$$\text{Distributed deviation force} = \text{Axial force}/\text{Radius of curvature}$$

It is clear from Figure 5.15 that the transversely distributed equal and opposite forces q_D and q_Z, which the top and bottom slab, respectively, are obliged to take up, create a torsional load, m_q, per unit of curved length leading directly to a torsional response. It can be written as:

$$q_D = \frac{D}{R} = \frac{M_B}{h \cdot R} \qquad q_Z = \frac{Z}{R} = \frac{M_B}{h \cdot R}$$

and, given that $D = Z$, it results in:

$$m_q = q_D \cdot h = \frac{M_B}{R}$$

Thus, it is clear that, even without the action of an externally applied torsional moment, m_D, merely the presence of bending along a curved axis implies the imposition of a distributed torsional load (M_B/R).

Figure 5.15 How the self-weight produces torsional load on the girder

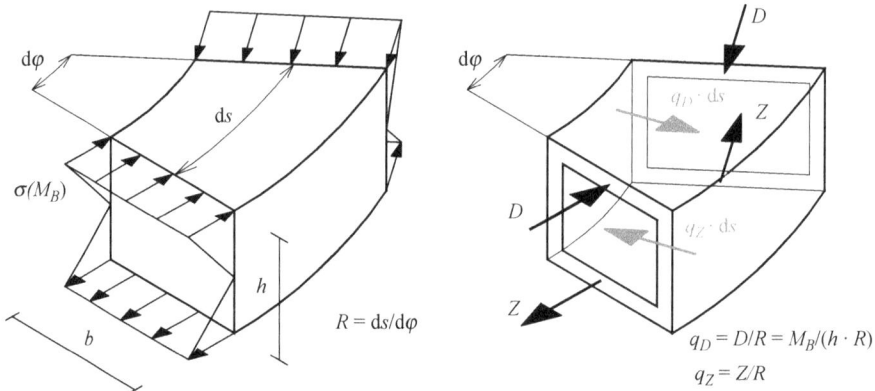

Apart from the above consideration, it should be remembered here that, exactly as in the case of grillages, any part of the beam must satisfy three conditions of equilibrium, namely with respect to vertical forces, as well as with respect to the projections of moment vectors on two arbitrary horizontal axes.

The equilibrium equations of an elementary segment of length, ds, forming an angle dφ ($1/R = \mathrm{d}\varphi/\mathrm{d}s$), as shown in Figure 5.16, may be set up as follows.

Regarding first the equilibrium of moment vectors with respect to the tangential axis, it can be written as:

$$\frac{\mathrm{d}M_T}{\mathrm{d}s} = -m_D - m_q = m_D - \frac{M_B}{R} \tag{a}$$

following in principle the relation between torsional moment and torsional load exhibited in Section 1.1.6. The acting torsional load on the right side consists of the externally applied load, m_D, and the implied torsional load $m_q = M_B/R$ by the existing bending moments, as explained above. The term (M_B/R) may also be understood as the vectorial sum of the two bending moment vectors acting on the curved segment's ends, as shown also in Figure 5.16.

The vectorial equilibrium of moments about the perpendicular axis to the beam centre line can be written as:

$$\mathrm{d}M_B - (M_T/R)\,\mathrm{d}s = V \cdot \mathrm{d}s$$

The term (M_T/R)ds renders simply the fact that the existence of torsion along a curved axis is correlated to bending through its *deviation term* in an analogous way, as discussed previously regarding the bending vectors acting on the segment's edges. The last relation can be written as:

$$\frac{\mathrm{d}M_B}{\mathrm{d}s} = V + \frac{M_T}{R}$$

Box girders

Figure 5.16 Evaluation of equations of equilibrium

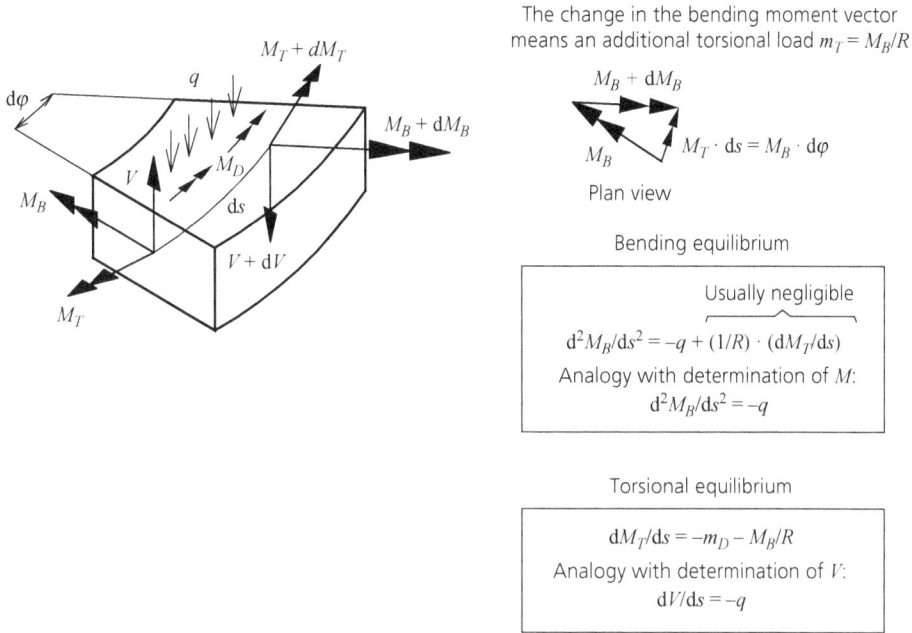

The change in the bending moment vector means an additional torsional load $m_T = M_B/R$

$M_T \cdot ds = M_B \cdot d\varphi$

Plan view

Bending equilibrium

Usually negligible

$$d^2 M_B/ds^2 = -q + (1/R) \cdot (dM_T/ds)$$

Analogy with determination of M:
$$d^2 M_B/ds^2 = -q$$

Torsional equilibrium

$$dM_T/ds = -m_D - M_B/R$$

Analogy with determination of V:
$$dV/ds = -q$$

and by differentiating again with respect to s and taking also into account that for the vertical equilibrium it must hold that $\dfrac{dV}{ds} = -q$, the above equation takes the form:

$$\frac{d^2 M_B}{ds^2} = -\left(q - \frac{1}{R} \frac{dM_T}{ds} \right) \tag{b}$$

If the arc span length, L, is much smaller than the radius of curvature, R, (i.e. $L/R < 0.3$) and, moreover, the distributed torsional load, m_D, is much smaller than $(q \cdot R)$, then, by expressing M_B as $q \cdot L^2/c$ (where c is in the order of magnitude of about 10) from Equation (a), it may be concluded that:

$$\left(-\frac{1}{R} \frac{dM_T}{ds} \right) \Big/ q = \frac{L^2}{R^2 \cdot c} + \frac{m_D}{R \cdot q}$$

As the right-hand term is very small, it can be concluded that on the right side of Equation (b) the term $-(1/R) \cdot (dM_T/ds)$ is much smaller than q and thus can be neglected. Thus, the above Equation (b) may be written much more simply as:

$$\frac{d^2 M_B}{ds^2} = -q$$

that is, like the equilibrium relation of a rectilinear beam between bending moment and external load (see Stavridis and Georgiadis, 2025, Section 2.2.3). This means that, under the above conditions, the bending moments of a curved girder may be approximated by the bending moments, M_B, of a straight beam of span, L, equal to the arc length of the girder (Menn, 1990).

Once the moments, M_B, are determined in this way, then the right side of Equation (a) takes a concrete value and the equation itself refers just directly to the equilibrium relation of a beam $dV/ds = -p$ between shear forces and distributed load, as discussed in Stavridis and Georgiadis (2025, Section 2.2.3). This means that the torsional moments, M_T, may result as the shear forces of a straight beam of length, L, carrying the load $(M_B/R + m_D)$ (see Figure 5.16).

If the above conditions regarding the ratios L/R and $m_D/q \cdot R$ are not met and a better approximation is needed, then on the basis of the above initial values of M_B, one may return to Equation (b) and evaluate the term:

$$ q - \frac{1}{R}\frac{dM_T}{ds} = q - \frac{1}{R}\cdot\left(\frac{M_B}{R} + m_D\right) $$

as a new loading of the beam in order to obtain 'improved' values of bending moments, M_B, and afterwards through Equation (a) to determine anew the torsional moments, M_T, as shear forces for the loading $\left(\dfrac{M_B}{R} + m_D\right)$. The repetition of the above procedure converges very fast.

5.3.2.2   Summary
▨   First, the 'stretched' girder is considered.
▨   On the basis of vertical loads, the bending moment diagram M_B is considered.
▨   The 'straight' girder is loaded with the torsional load M_B/R, according to the direction dictated by the deviation actions in compression and tension slabs, respectively, or by the composition of vectors M_B as applied on the corresponding elementary curved girder segment considered in plan view. The distribution of this torsional loading is obviously identical to the corresponding bending diagram $[M_B]$. Any existing load m_D must be added.
▨   The torsional loads M_B/R are considered as distributed vertical downward loads as long as the bending moments M_B cause tension at the bottom fibres; otherwise, they are directed upwards. The possibly present m_D must also be considered as a distributed vertical load with the same direction as that of M_B/R if its direction coincides with that of the torsional load M_B/R; otherwise, it has the opposite direction.
▨   The torsional moment diagram $[M_T]$ is identical to the diagram of shear forces for the above fictitious loading; however, not with respect to its sign. ∎

It is clear that the above equations concern only the equilibrium and, as they do not take into account the compatibility of deformations in case of redundant supports, they do not lead to a strictly accurate solution, unless the ratio $\kappa = EI/GI_T$ is considered equal to zero. Nevertheless, their use for purposes of preliminary design is quite appropriate. It is noted that for a simply supported curved girder and for a ratio L/R less than 0.3, which means a central angle not greater than $20°$, the discrepancies from the 'exact' solution are less than 1%. Moreover, for a fixed-ended girder and thus for the approximately equal internal spans of a continuous girder with $L/R \leq 0.3$ and $\kappa \leq 5$, the discrepancies are less than 1.5%. Furthermore, as mentioned initially, the above equations may have a direct use when, for the purposes of plastic design, the bending moments obtained from a computer solution have to be altered, in which case the redistribution of torsional moments must necessarily satisfy the equilibrium criterion.

Worked examples
The above procedure is applied to the following examples.

Example 1

In the first example (see Figure 5.17), a simply supported girder allowing the uptake of torsional moments at its supports, as well as a fixed-end girder, are considered. Figure 5.17 shows that the fixed-end girder develops a clearly lower torsional response, while its torsional reactions are zero. This may be understood on the basis of the remark made in Stavridis and Georgiadis (2025, Section 2.3.6), as the beam is loaded by its bending moment diagram, divided of course by the constant radius of curvature and thus its reactions (i.e. the torsional ones) must correspond to the developed end rotation angles, which are obviously zero. In both cases examined, attention must be paid to the correct determination of the direction of the torsional loading (M_B/R), as shown in Figure 5.17. In this respect, either the resultant of the two adjacent bending moment vectors is considered or, alternatively (to gain a direct physical insight), in a separate sketch, the developed

Figure 5.17 Determination procedure for the bending and torsional response of a curved girder

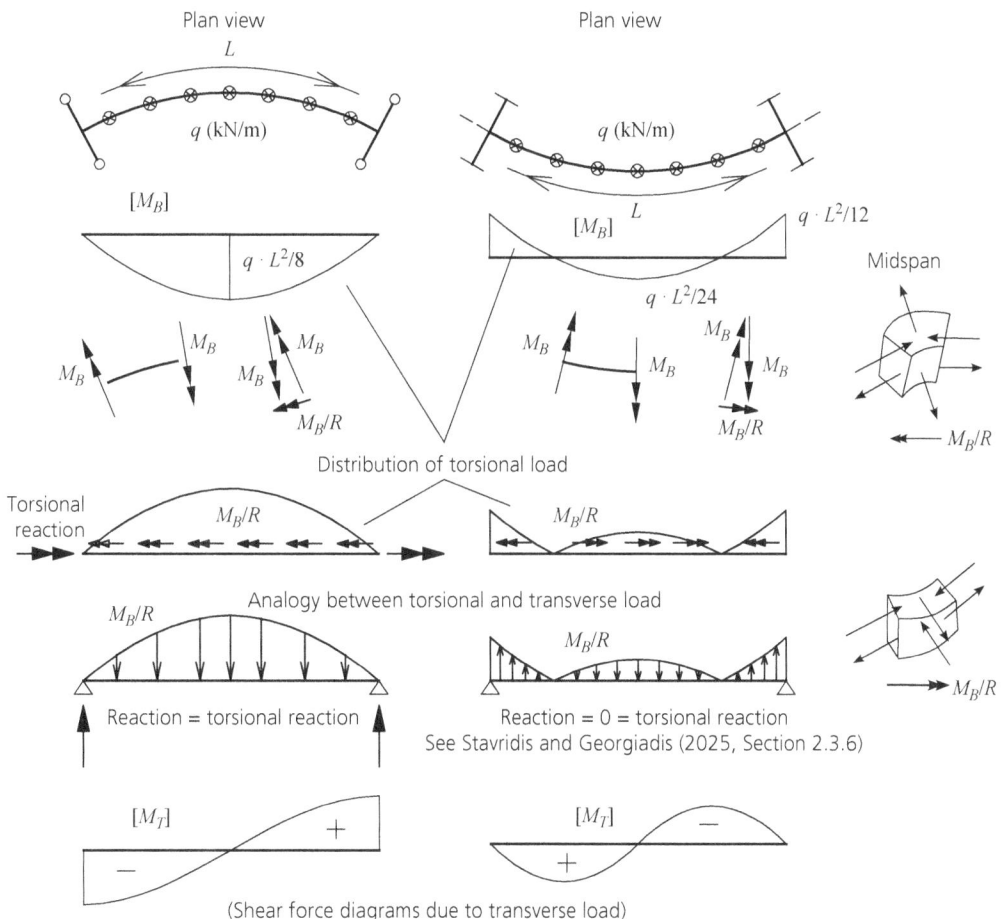

deviation force on the top and bottom slabs and, consequently, the resulting torsional load are deduced (see Figure 5.17).

Concerning the internal span of a continuous beam, the consideration of a fixed-end beam instead does not essentially alter the torsional response.

Example 2

In the second example (see Figure 5.18), a beam is examined that can develop a torsional reaction only at its left support, being in the first case supported on an intermediate point, whereas in the second case it acts as a cantilever.

Despite the fact that the intermediate support cannot develop a torsional reaction, it nonetheless relieves the beam significantly with respect to torsional response, just because it restricts the

Figure 5.18 Evaluation of bending and torsional response with cantilever action

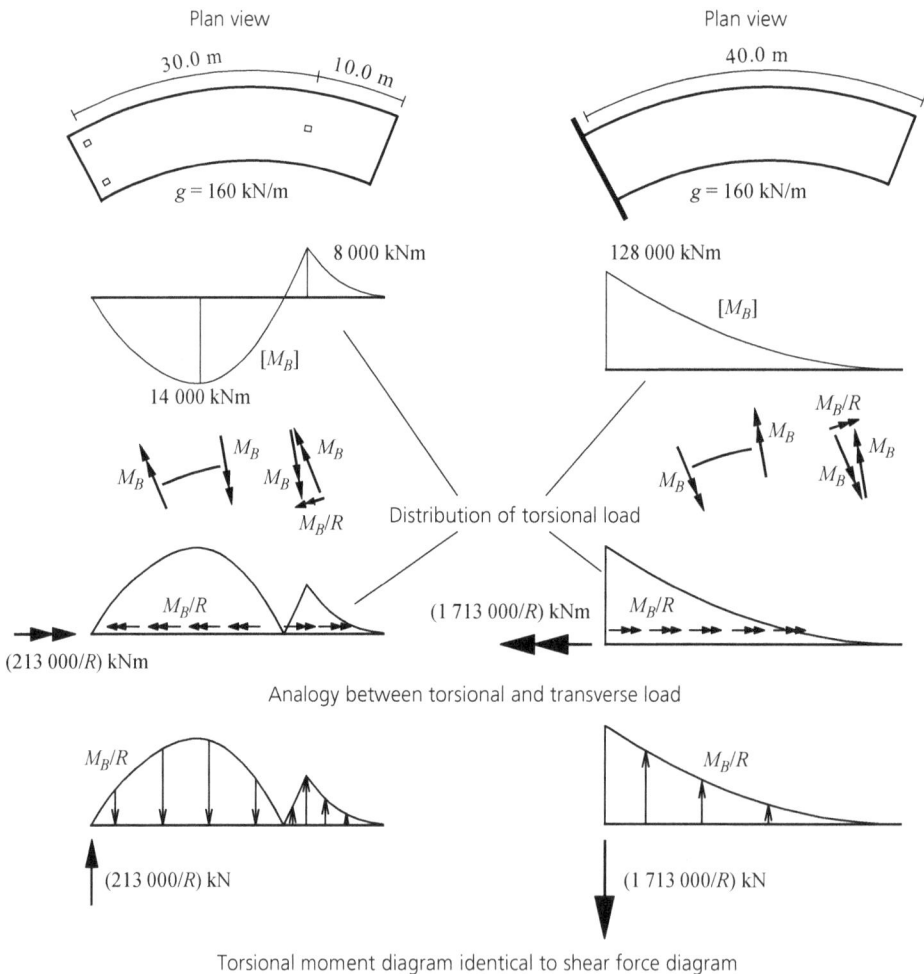

bending response, leading to a reduced torsional loading in comparison with the case of a cantilever. The torsional reaction in the case of a cantilever comes out about eight times bigger. Again, due attention is paid to the direction of the torsional loading (M_B/R), according to Figure 5.18. ∎

5.3.2.3   The case of a uniformly distributed torsional load

It is interesting now to examine the bending and torsional response of a girder under a distributed torsional moment, m_D, acting along its whole length. At first, it is obvious that a straight girder, where the ratio L/R is equal to zero, develops only torsional response and no bending at all.

Regarding a curved girder, from Equations (b) and (a) in Section 5.3.2.1, it can be deduced that:

$$\frac{d^2 M_B}{ds^2} = \left(\frac{1}{R} \frac{dM_T}{ds} \right) = \frac{m_D}{R} - \frac{M_B}{R^2}$$

Considering the term m_D/R as a gravity load and neglecting the term M_B/R^2, for a span, L, it comes out approximately that $M_B = (m_D/R) \cdot L^2/8$. This expression referring to a semicircle with radius, R, results in $M_B = (m_D/R) \cdot (\pi \cdot R)^2/8 \approx 1.2 \cdot m_D \cdot R$.

However, in this respect, the behaviour of a circular ring with radius, R, subjected to a uniformly distributed torsional loading moment, m_D, seems quite particular as it develops no torsion at all but only a constant bending, something that has been already ascertained in Section 3.4.1.2 (see Figure 3.29). Such a ring also develops a constant twisting angle all along its perimeter.

The above property of a ring makes possible the bridging of a span length ($2R$) through a girder in the form of a semicircle of radius, R, suspended along its internal edge developing exclusively a bending response without any torsion. This goal, however, is achieved under the condition that the girder is being offered an appropriate bending moment at both its supported ends and this is, in turn, directly accomplished if the girder is connected to two rectilinear beams of appropriate length, as shown in Figure 5.19.

Figure 5.19 Layout of semicircular cable-stayed bridge

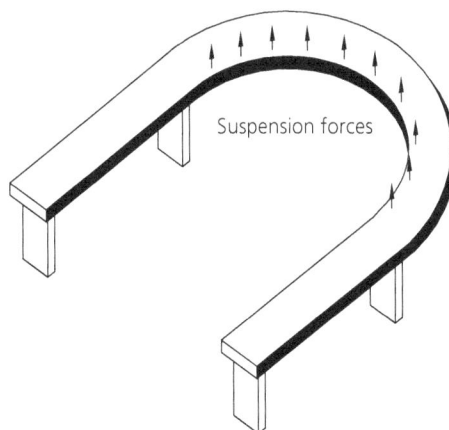

Suspension forces

Figure 5.20 Structural action of the system

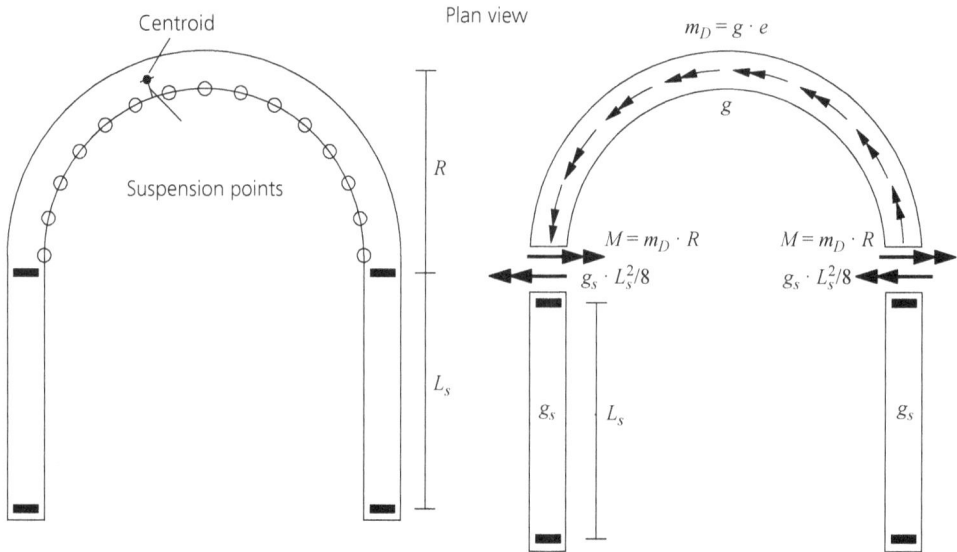

More specifically, the eccentrically suspended semicircular girder under the uniform vertical load, g_S, is subjected to a constant torsional load $m_D = g_S \cdot e$, where e is the distance of the resultant vertical load from the suspension line (see Figure 5.20). According to Chapter 3, Section 3.4.1.2 (see Figure 3.29), no torsional moment is developed while the ring undergoes a constant bending moment $M_B = m_D \cdot R$ all over its length. This value of M_B is comparable in magnitude to the one deduced above.

Thus, the equilibrium condition of the curved girder, considered as part of the whole ring, in order to allow the development of bending response exclusively, requires the above bending moment to be applied at both its ends.

This specific bending moment can be effectively provided by a straight beam rigidly connected to both ends of the semicircular girder, acting as the support moment developed in a simply supported beam of appropriate length, L_s, subjected to a downward vertical load, g_S. Both last values must comply with the obvious requirement $m_D \cdot R = g_s \cdot L_s^2/8$ (see Figure 5.20).

5.3.3   Response of the cross-section walls

The torsional response of the girder, as examined in Section 5.3.2.1, implies, beyond the Bredt peripheral flow acting on the cross-section itself, an additional response for the box section walls. This response results from the way the gravity loads are introduced to the girder.

At first, it is recognised that under gravity loads, the acting compressive forces, D, and tensile forces, Z, in the curved top and bottom flanges, respectively, cause distributed deviation forces, q, as explained in Section 5.3.2.1. It is clear from Figure 5.21 that the transversely distributed equal and opposite forces q_D and q_Z, which are introduced to the top and bottom slab, respectively, create

Figure 5.21 Introduction of torsional response in the box girder walls

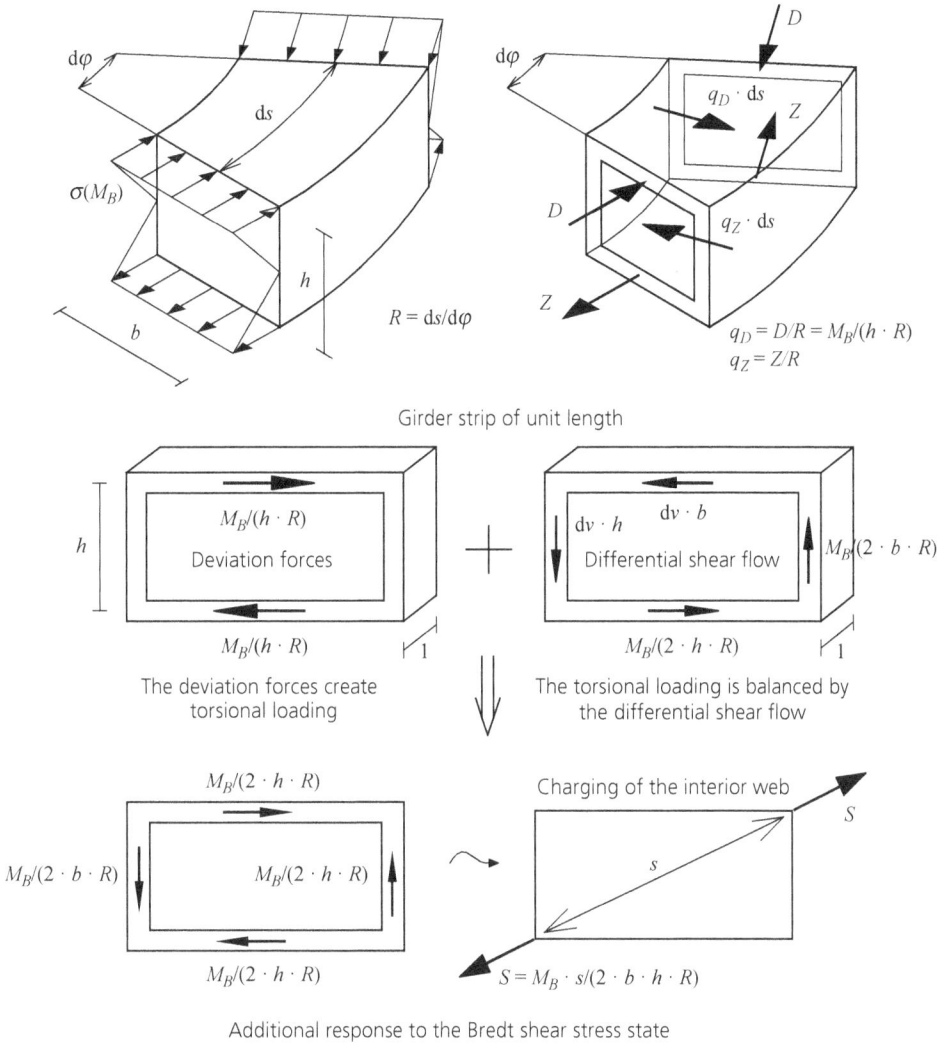

Girder strip of unit length

The deviation forces create torsional loading

The torsional loading is balanced by the differential shear flow

Charging of the interior web

Additional response to the Bredt shear stress state

a torsional load, m_q, per unit of curved length, being nothing else than that resulting from the vectorial variation of bending moments, as discussed in Section 5.3.2.1. It is:

$$q_D = \frac{D}{R} = \frac{M_B}{h \cdot R}$$

$$q_Z = \frac{Z}{R} = \frac{M_B}{h \cdot R}$$

and, as $D = Z$, it yields:

$$m_q = q_D \cdot h = \frac{M_B}{R}$$

Referring now to a cut-out transverse strip of unit arc length it can be seen that the strip is in equilibrium under the above deviation loads ($q_D \cdot 1.0$) and ($q_Z \cdot 1.0$) acting on the top and bottom slab respectively, as well as the developed Bredt shear flow at both of its faces (see Figure 5.21).

The resultant of these two flows is the so-called *differential shear*, Δv (kN/m), acting on all four sides of the strip. As discussed in Section 5.2.2 and following Bredt's formula, it can be written as:

$$\Delta v = \frac{q_D \cdot 1.0 \cdot h}{2 \cdot b \cdot h} = \frac{M_B}{R \cdot 2 \cdot b \cdot h}$$

It is clear that the differential shear, Δv, results in applied single forces that are equal to ($\Delta v \cdot b$) for the top and bottom slab and to ($\Delta v \cdot h$) for both webs. The resulting forces acting on each side of the strip are shown in Figure 5.21. Thus, the strip being in equilibrium as a plane structure, under the above forces q_D, q_Z and Δv, gives rise to the self-equilibrated diagonal loading of the profile:

$$S = \frac{M_B}{2 \cdot R} \sqrt{\frac{1}{b^2} + \frac{1}{h^2}}$$

as shown in Figure 5.21, acting all along the length of the girder, causing longitudinal bending of the walls as well as transverse bending of the section profile, as examined in detail for a rectilinear girder.

Although the analogy is not quite accurate, for the preliminary design and for limited curvatures, it may be safely considered that the left web wall takes the downward uniform load $S_w = M_B/(2 \cdot R \cdot b)$, acting, like a beam resting on an elastic foundation with a *subgrade modulus, K*, on the basis of stiffness of the section profile against a diagonal length change, exactly as in the case of a rectilinear beam. Of course, if the resulting response is high, transverse diaphragms should be provided. Then, the deviation forces of the top and bottom slab that trigger the whole response will be transferred through corresponding horizontal bending to each diaphragm. This, in turn, will transfer a concentrated torque to the section by eliminating at the same time its deformation that would otherwise take place.

The above diagonal loading of the girder cross-section, when transverse diaphragms are not present, makes it clear – according to Figure 5.21 concerning the longitudinal response – that the internal web is burdened whereas the external one is relieved. However, it is interesting to note that in the case of an open section, as shown in Figure 5.22, quite the opposite happens, as explained below.

More specifically, the consideration of an elementary curved segment, as shown in Figure 5.22, shows that the action of bending moments, M_B, is equivalent – as explained previously – to the application of a distributed torsional load (M_B/R). It is clear that this torsional load may be taken up by the *opposite bending* of the two webs, through the resolution in the continuously distributed vertical loads $q^* = M_B/(R \cdot b)$, which obviously give an additional load to the external web while decreasing the load on the internal one. If it is assumed that the existing vertical load, q, is equally distributed between the two webs, then their bending response will be obtained as:

$$M_{B,ex} = (1/2) \cdot M_B(q) + M_B(q^*)$$

Figure 5.22 Uptake of the torsional load by the webs of the open section

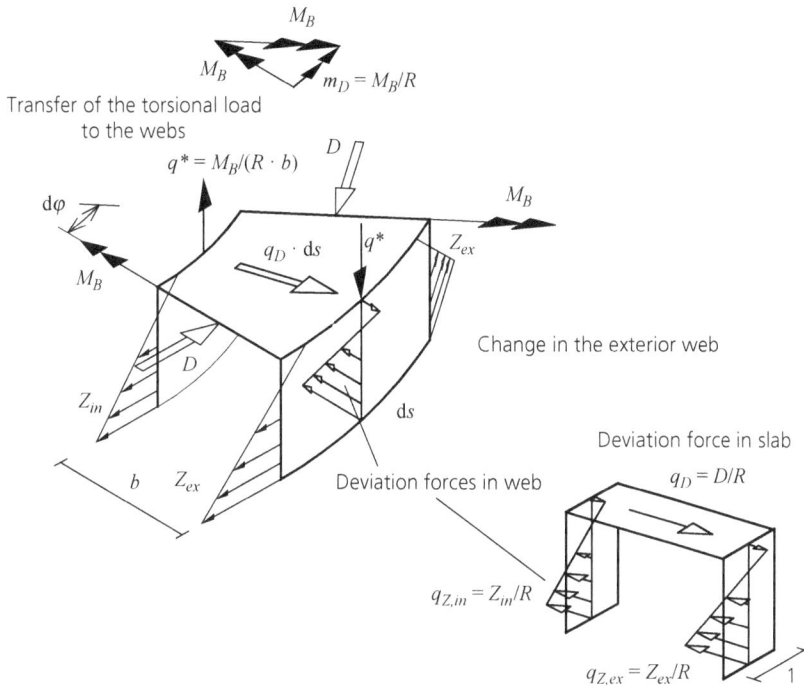

and,

$$M_{B,in} = (1/2) \cdot M_B(q) - M_B(q^*)$$

for the external and the internal web, respectively.

Certainly, the compressive forces in the slab due to the bending moments, M_B, compel it to take up outward deviation forces, whereas the existence of compressive and – especially – tensile stresses in the webs, create transverse distributed deviation forces that equilibrate those of the slab, causing at the same time transverse bending of the open frame of the section, as shown in Figure 5.22 (Menn, 1990).

Worked example

A simply supported concrete curved girder is considered with a span of 50 m and a radius of curvature $R = 200$ m having the cross-section shown in Figure 5.23. The thickness of horizontal and vertical section walls is 0.20 m and 0.50 m respectively. Besides its self-weight, which is equal to 120 kN/m, the girder is subjected to an eccentric linear load 40 kN/m acting along the right web (2), as shown below. In order to estimate the bending and torsional as well as the transverse response of the girder, the loading is considered as consisting of a symmetric and an antisymmetric part, which will be evaluated separately as explained below.

SYMMETRIC LOADING

The symmetric loading of the girder consists of its self-weight plus two linear loads equal to 20.0 kN/m acting separately over the two webs. Thus

The bending moment is: $M_B = (120.0 + 40.0) \cdot 50.0^2/8 = 50\,000$ kNm

The torsional load parabolically distributed is: $m_D = 50\,000.0/200 = 250$ kNm/m

The torsional moment at each support is: $M_T = (1/2) \cdot (2/3) \cdot 250.0 \cdot 50.0 = 4167$ kNm

The torsional moment diagram is identical to the shear force diagram of the rectilinear simply supported beam under the above parabolic distributed load of 250 kN/m at midspan. The peripheral

Figure 5.23 Simply supported concrete curved girder under eccentric loading (continued on next page)

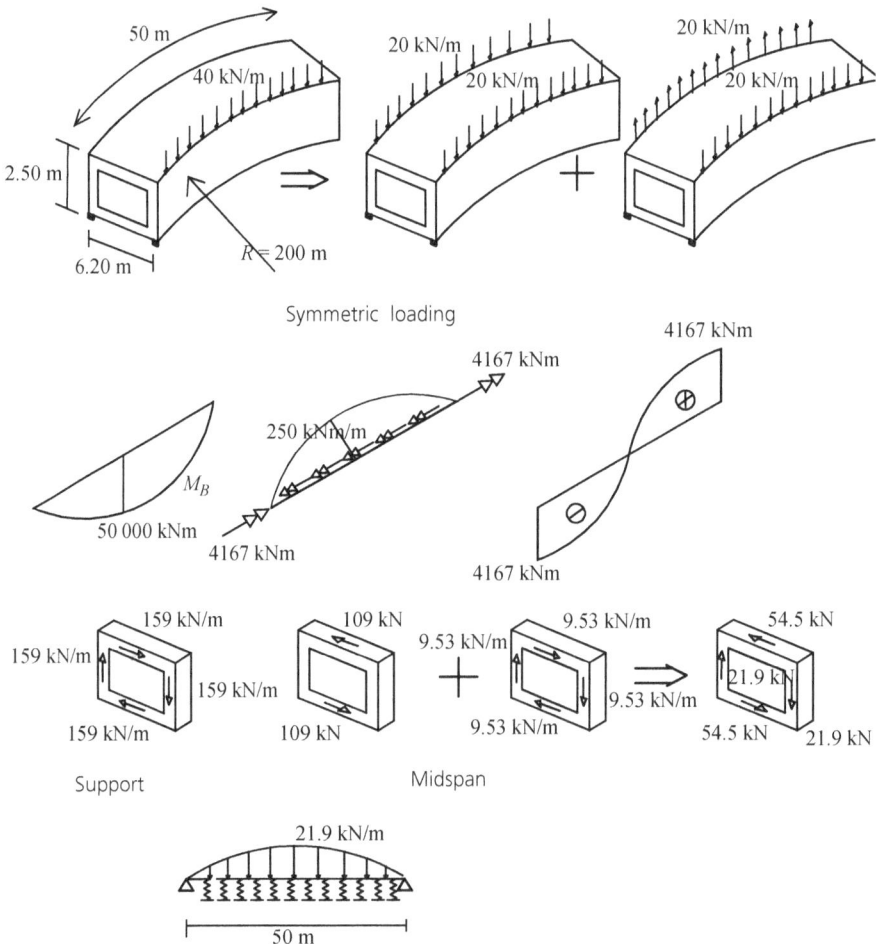

Figure 5.23 continued

Antisymmetric loading

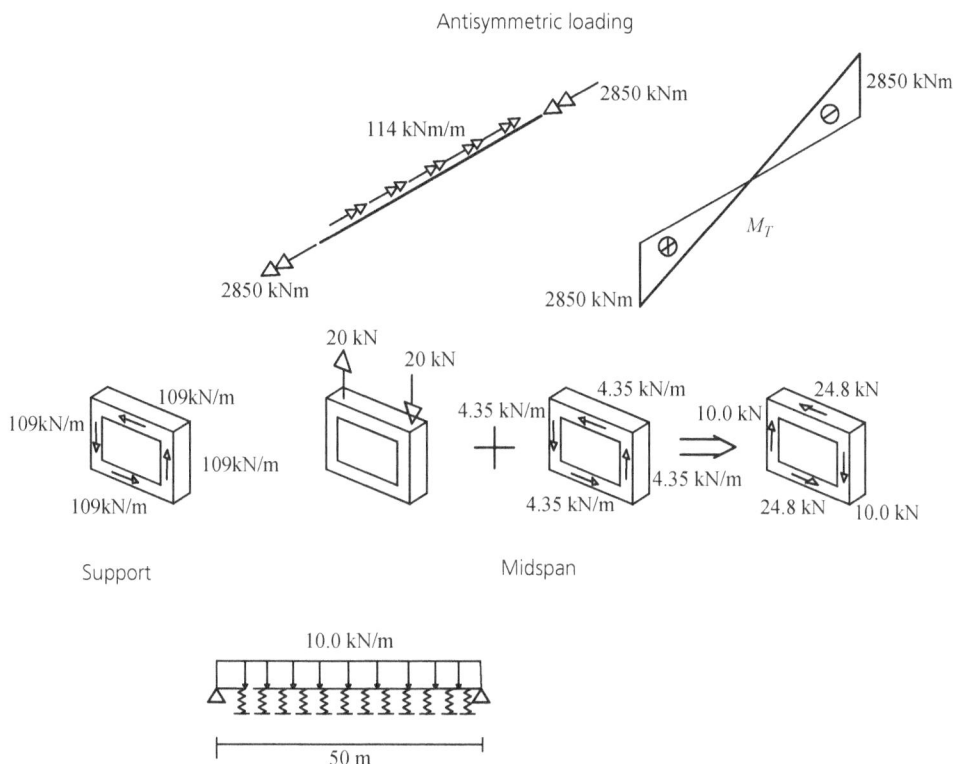

Support   Midspan

shear flow at the supports according to Bredt's formula is equal to $4167/(2 \cdot 2.3 \cdot 5.7) = 159.0$ kN/m. This gives shear forces equal to $159.0 \cdot 5.7 = 906$ kN and $159.0 \cdot 2.3 = 366$ kN to each horizontal slab and to each web, respectively. Therefore, the shear stresses are equal to $159.0/0.20 = 795$ kN/m^2 and $159.0/0.50 = 318$ kN/m^2 for the slabs and the webs, respectively.

TRANSVERSAL RESPONSE

Both top and bottom slabs are subjected to a transverse horizontal deviation load having at midspan the value of $q = (M_B/2.30)/200.0 = 109$ kN/m, which is distributed parabolically according to M_B in opposite directions. For the equilibrium of a transverse strip of the cross-section with a length of 1 m, which is subjected to a torsional moment $109.0 \cdot 2.30 = 250.0$ kNm, the action of a differential shear Δv with respect to both faces of the strip is required that, according to the appropriate interpretation of the Bredt formula, results in $\Delta v = 250.0/(2 \cdot 2.30 \cdot 5.70) = 9.53$ kN/m.

Superposing the two previous actions, the resulting horizontal loading parabolically distributed for both the top and bottom (curved) strip slabs is equal to $(109.0 - 9.53 \cdot 5.70) = 54.5$ kN/m in opposite directions, while each web of the strip takes up a vertical load with the same parabolic distribution equal to $\Delta v \cdot 2.50 = 9.53 \cdot 2.30 = 21.9$ kN/m downwards for the internal web and upwards for the external web.

It must be noted that in the case considered, the maximum transverse response occurs at a place where the torsional moment vanishes. However, each slab receives its respective loading working as resting on an elastic base with modulus of subgrade reaction, K, being approximately the same (as discussed above) as that considered in Section 5.2.2. Obviously, the internal web is considered to be more critical because it is adversely affected in comparison with the external one.

ANTISYMMETRIC LOADING

The antisymmetric loading of the girder consists of two linear loads equal to 20.0 kN/m acting in opposite directions over the two webs (see Figure 5.23). The girder is subjected only to a uniformly distributed torsional load equal to $20.0 \cdot 5.70 = 15$ kNm/m. The resulting torsional moment diagram has a linear distribution giving at both supports a torsional moment equal to $M_T = 15.0 \cdot 50.0/2 = 2850$ kNm (see Figure 5.23).

The peripheral shear flow at the supports, according to Bredt's formula, is equal to $2850/(2 \cdot 2.3 \cdot 5.7) = 109$ kN/m, inducing a shear force $109.0 \cdot 2.30 = 251$ kN to each web and $109.0 \cdot 5.70 = 621$ kN to each horizontal slab. The shear stresses are equal to $109.0/0.20 = 545$ kN/m^2 and $109.0/0.50 = 218$ kN/m^2 for the slabs and each web, respectively.

TRANSVERSAL RESPONSE

To establish the equilibrium of a transverse strip with a length of 1 m that is subjected to an antisymmetric loading of 20 kN/m producing a clockwise torsional moment of $20.0 \cdot 5.70 = 114.0$ kNm, the action of a differential shear, Δv – with respect to both faces of the strip – is required (see Figure 5.23). According to the appropriate interpretation of the Bredt formula, it is equal to $\Delta v = 114.0/(2 \cdot 2.30 \cdot 5.70) = 4.35$ kN/m.

Superposing the two previous actions, the horizontal uniformly distributed loading for both the top and bottom (curved) strip slabs is equal to $\Delta v \cdot 5.70 = 4.35 \cdot 5.7 = 24.8$ kN/m in opposite directions, whereas each web of the strip takes up a vertical load with the same uniform distribution equal to $20.0 - 4.35 \cdot 2.30 = 10.0$ kN/m, which is downwards for the internal web and upwards for the external web. However, each section wall receives its corresponding loading, working as being supported on an elastic base with a modulus of subgrade reaction, K, being approximately the same as that considered in Section 5.2.2 (as discussed in Section 5.3.3). The internal web is again considered as finally more loaded than the external web.

5.3.4   Influence of prestressing on the curved girders

As was made clear in Stavridis and Georgiadis (2025, Chapter 4) concerning beams, the layout of prestressing tendons aims to provide such deviation forces that counteract the gravity loads for which the structure has to be designed. In the case of curved girders, it has to be recognised that gravity loads create indirectly torsional loads through the existing curvature, being eventually increased by their eccentric layout on the bridge deck. Thus, it is useful to examine how the presence of prestressing in a curved box girder influences its torsional response and, moreover, to consider the possibility of torsional relief through an appropriate tendon layout.

Given that the tendons are arranged within the thickness of each web by using all its available depth, it should first be pointed out that a cable inclined with respect to the girder axis implies torsion (see Figure 5.24). This is explained by the fact that the inclined compressive force on the section has a component in its own plane that causes a torsional moment equal to the product of this component by its distance to the shear centre of the section. Of course, in the usual prestressing layouts where the tendons are arranged symmetrically to the girder axis, the torsional moments induced from each side on the concrete cross-section cancel each other.

Figure 5.24 Torsional response due to tendon inclination

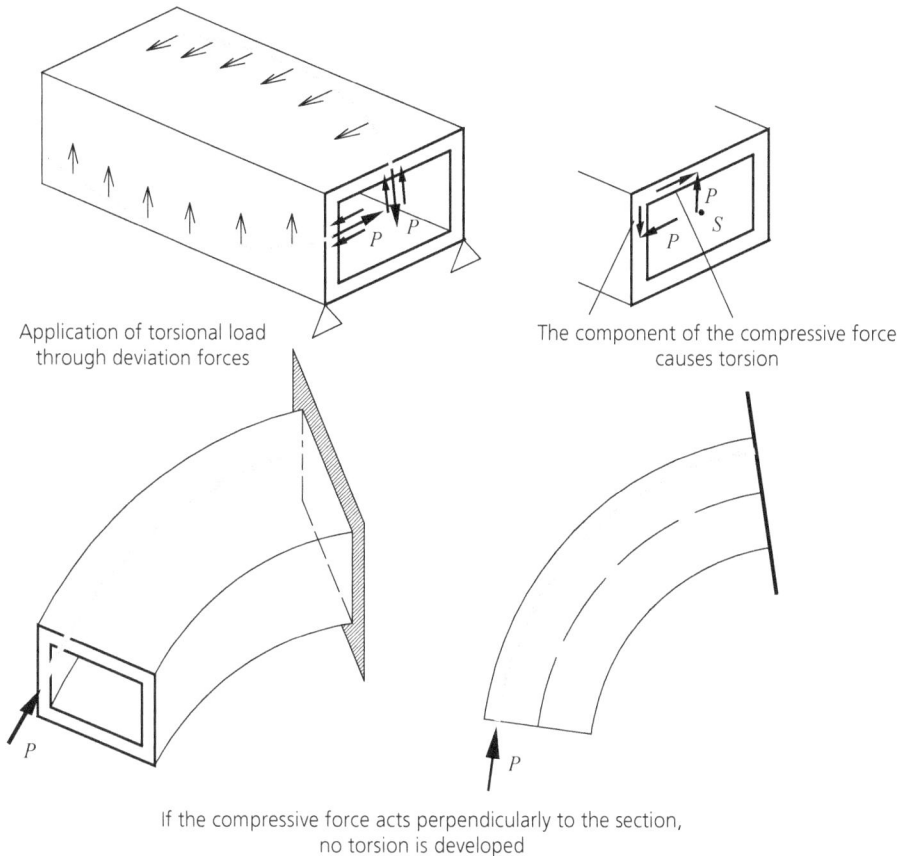

Application of torsional load
through deviation forces

The component of the compressive force
causes torsion

If the compressive force acts perpendicularly to the section,
no torsion is developed

It is also clear that a cable lying in the top or bottom slab with an inclined position to the girder axis may also cause a torsional moment with respect to the shear centre through the component of its compressive action on the section (see Figure 5.24).

Moreover, it is understood that in a curved beam, where the prestressing cable runs parallel to the shear centre axis and is anchored at its two ends, no torsional loads are imposed as the prestressing force does not offer anywhere a component on the respective section plane (see Figure 5.24).

On the basis of the foregoing, it has to be understood that in a curved statically determinate girder, an arbitrary tendon layout with a given prestressing force causes exactly the same response (i.e. bending and torsional moments) as that developed in the respective rectilinear (i.e. 'stretched out') girder. The reason for this is explained below.

A statically determinate structure containing a prestressed tendon appropriately anchored, if subjected only to the prestressing forces (anchor and deviation forces), does not develop any reaction at the points of support. Hence, any bending or torsional moment, or even shear force of a section, results from the acting prestressing force on it. The point of application and the direction of this

Figure 5.25 External loads and prestressing in a curved statically determinate beam

Prestressing of a
statically determinate beam

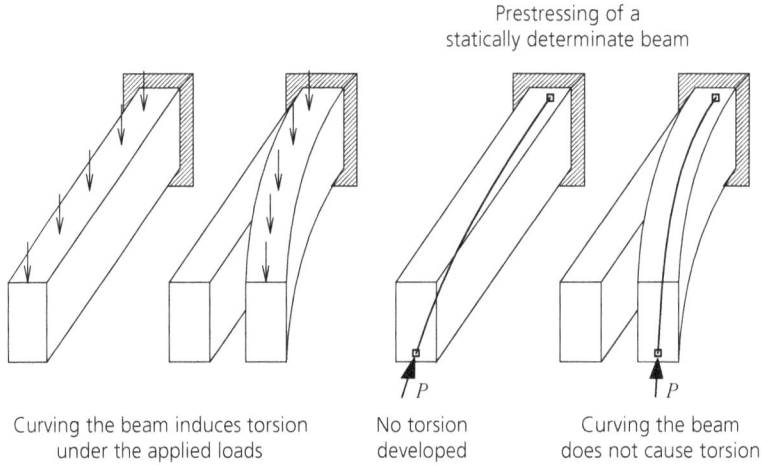

Curving the beam induces torsion
under the applied loads

No torsion
developed

Curving the beam
does not cause torsion

force correspond always, as explained in Stavridis and Georgiadis (2025, Section 4.3.1.1), on the respective trace and local tangent of the tendon. If the component of this action on the section plane does not pass from the shear centre, then a torsional moment is caused, as previously mentioned. Otherwise, no torsion is imposed at all. This situation remains unchanged if the initially considered straight girder becomes curved (see Figure 5.25).

Thus, while a rectilinear statically determinate girder – for example, under its self-weight – will develop torsion if it becomes curved, a prestressing tendon not producing torsion in the rectilinear girder will continue to do so, even if the girder becomes curved. Accordingly, it may be concluded that in a curved statically determinate girder, if the prestressing deviation forces counterbalance the external loads completely (e.g. self-weight), the bending shall be eliminated but the torsion shall not.

It should, however, be understood at this point that the bending moment, M_p, due to prestressing should not be perceived as creating – according to Section 5.3.1 – a torsional load M_p/R, just because the bending moment, M_p, is not 'genuine' – that is, it is not directly required for the equilibrium, as shown in Figure 5.26.

Of course, all the above is changed once the girder becomes statically indeterminate. A curved continuous girder, for example, with a prestressing tendon that in the assumed 'stretched' structure does not induce torsional moments, certainly develops a torsional response, as explained below, according to Figure 5.27.

The continuous girder considered in Figure 5.27 has simple supports at both ends capable of developing torsional reactions, whereas it is simply supported in the middle. The primary – statically determinate – structure obtained through removal of the intermediate support under the existing deviation and anchor forces does not develop any torsion, as already explained. However, the redundant vertical force required to be applied in order to eliminate the occurring deflection at the midpoint of the primary structure does certainly induce a torsional response in it. This externally

Figure 5.26 Prestressing in a statically determinate beam does imply an internal moment

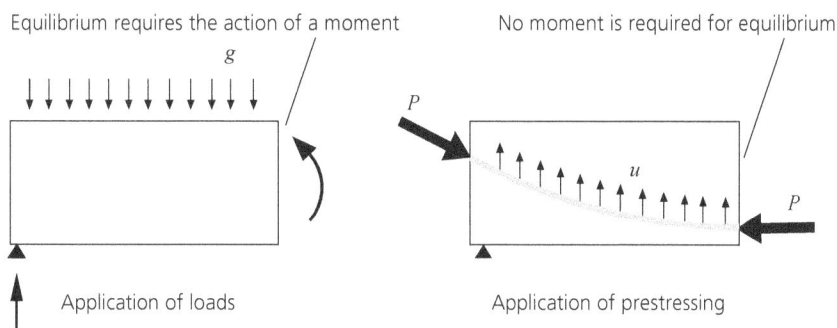

Figure 5.26 Prestressing in a statically determinate beam does imply an internal moment

imposed redundant force – essentially equivalent to the corresponding reaction of the rectilinear continuous beam, provided that the ratios L/R and κ are sufficiently small (see Section 5.3.2.1) – will induce a triangular bending moment diagram in the primary structure.

This represents the statically indeterminate prestressing moments $[M_{SP}]$, as explained in Stavridis and Georgiadis (2025, Section 5.4) (see also Figure 5.27). Certainly, the bending diagram $[M_{SP}]$ applied to the curved statically determinate girder will automatically produce the torsional loading $[M_{SP}/R]$ because M_{SP} is due to non self-equilibrated actions that do create reactions. The implied torsional response will then be obtained as the shear force diagram of the respective rectilinear beam due to the transverse triangular loading $[M_{SP}/R]$ (see Figure 5.27). It is clear, then, that a prestressing tendon in a curved continuous girder causes a torsional response due exclusively to the statically indeterminate prestressing moments, M_{SP}, of the corresponding rectilinear beam, in contrast to the statically determinate bending moments which, by corresponding to a self-equilibrated loading do not produce any support reactions.

It is clear, then, that a prestressing tendon in a curved continuous girder induces a torsional response solely due to the statically indeterminate prestressing moments, M_{SP}, of the corresponding rectilinear beam, since the statically determinate bending moments – associated with self-equilibrated loading – produce no support reactions and, therefore, no torsional effects.

Thus, according to the above and referring to Figure 5.28, the torsional moment diagram $[M_T]$ of a curved statically indeterminate prestressed girder may be obtained on the basis of the bending moment diagram $[M_P]$ due to deviation and anchor forces (see Stavridis and Georgiadis, 2025, Section 5.4) as follows. Once the statically indeterminate moments ($M_{SP} = M_P - P \cdot e$) have been determined, where P and e are the prestressing force and the tendon eccentricity at each position, respectively, the torsional load $[M_{SP}/R]$ is applied as a distributed vertical load on the statically determinate rectilinear girder. The resulting shear force diagram represents, then, the torsional moment diagram $[M_T]$ of the statically indeterminate curved girder.

Worked example

A prestressed concrete continuous curved girder is considered to have two spans of 50 m, with a central simple support and a radius of curvature $R = 200$ m (see Figure 5.29). At all three supports, a transverse diaphragm is provided. For the resultant tendon geometry representing the total of the

Figure 5.27 Influence of prestressing on the curved redundant beam

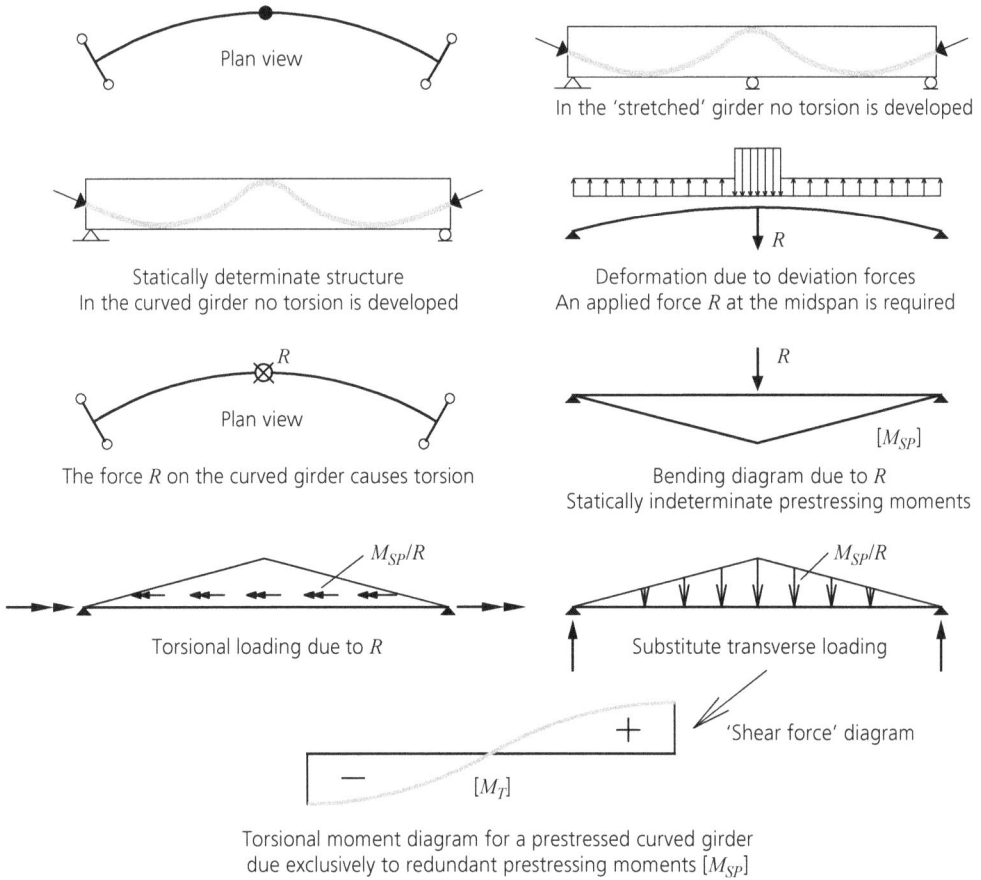

Plan view

In the 'stretched' girder no torsion is developed

Statically determinate structure
In the curved girder no torsion is developed

Deformation due to deviation forces
An applied force R at the midspan is required

R

Plan view

The force R on the curved girder causes torsion

R

Bending diagram due to R
Statically indeterminate prestressing moments

$[M_{SP}]$

M_{SP}/R

Torsional loading due to R

M_{SP}/R

Substitute transverse loading

'Shear force' diagram

$[M_T]$

Torsional moment diagram for a prestressed curved girder
due exclusively to redundant prestressing moments $[M_{SP}]$

embedded tendons, an estimation is to be carried out of the overall bending, torsional and transverse response of the girder due to prestressing only.

For the tendon, a prestressing force of 24 000 kN is applied. The radius of curvature is 169 m at span and 24 m over the internal support and, thus, the deviation forces are 24 000/169 = 142 kN/m upward at span and 24 000/24 = 1000 kN/m downward at middle support, respectively (see Figure 5.30). The downward reaction at middle support is 1354 kN and represents the redundant force which, through the produced bending moment $M_{SP} = 1354 \cdot (2 \cdot 50.0)/4 = 33\ 850$ kNm, induces on the girder the torsional load $M_{SP}/R = 33\ 850/200.0 = 169.2$ kNm/m with a triangular distribution, as shown in Figure 5.30. The induced torsional reaction at the two ends is $M_A = (1/2) \cdot 169.2 \cdot 50.0 = 4230$ kNm, while at the support the torsional moment vanishes.

The peripheral shear flow at the supports according to Bredt's formula is equal to $4830/(2 \cdot 2.3 \cdot 5.7) = 184$ kN/m, inducing a shear force $184.0 \cdot 2.30 = 423$ kN to each web and $184.0 \cdot 5.70 = 1049$ kN to each horizontal slab. The shear stresses are equal to $184.0/0.20 = 920$ kN/m^2 and $184.0/0.50 = 368$ kN/m^2 for the slabs and each web, respectively.

Figure 5.28 Assessment of the influence of prestressing on the statically indeterminate curved girder

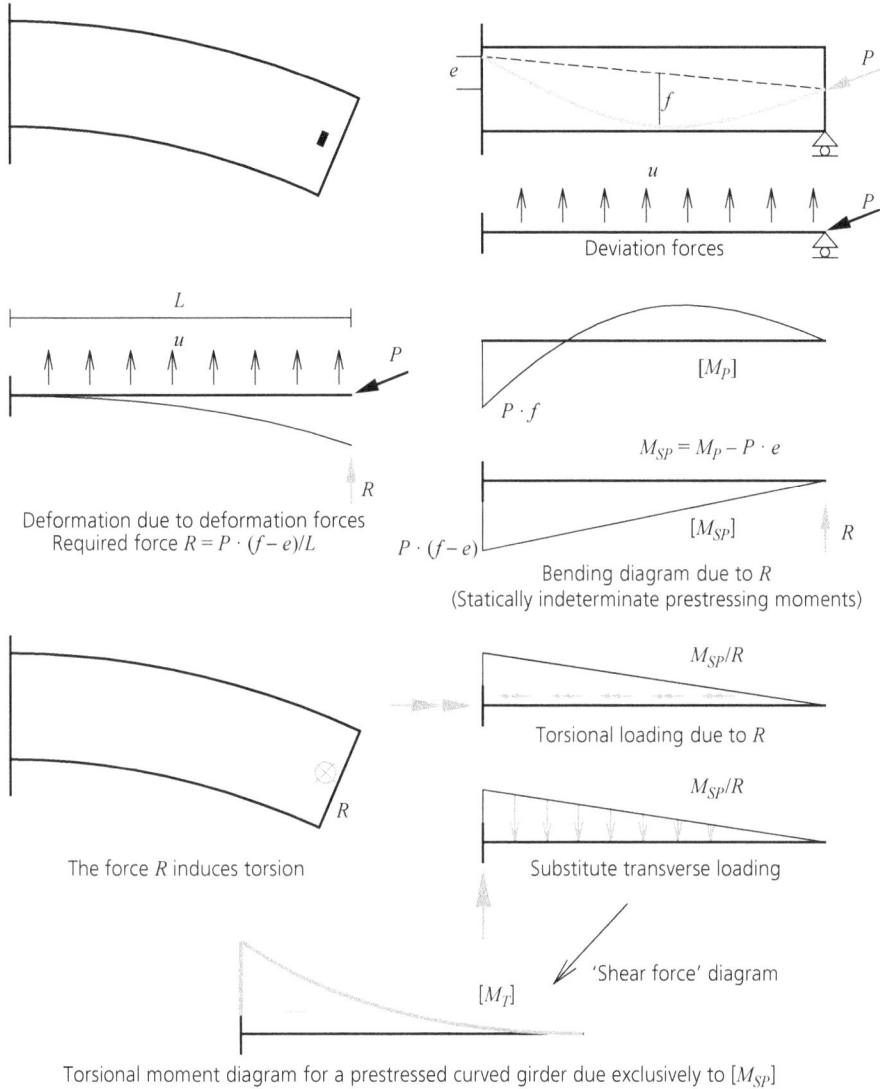

Deviation forces

L

u

P

R

Deformation due to deformation forces
Required force $R = P \cdot (f - e)/L$

$P \cdot f$

$[M_P]$

$P \cdot (f - e)$

$M_{SP} = M_P - P \cdot e$

$[M_{SP}]$

R

Bending diagram due to R
(Statically indeterminate prestressing moments)

M_{SP}/R

Torsional loading due to R

M_{SP}/R

The force R induces torsion

R

Substitute transverse loading

'Shear force' diagram

$[M_T]$

Torsional moment diagram for a prestressed curved girder due exclusively to $[M_{SP}]$

TRANSVERSAL RESPONSE

Both top and bottom slabs are subjected to a transverse horizontal deviation load that is distributed linearly according to M_{SP} having at midspan the value of $(33\,850/2.30)/200 = 73.60$ kN/m in opposite directions (see Figure 5.31). To establish the equilibrium of a transverse strip with a length of 1 m, which is subjected to an anticlockwise torsional moment $73.6 \cdot 2.30 = 169.3$ kNm, the action of a differential shear, Δv, with respect to both faces of the strip is required. According to the appropriate interpretation of Bredt's formula, it results in $\Delta v = 169.3/(2 \cdot 2.30 \cdot 5.70) = 6.46$ kN/m. Superposing the two previous actions, the resulting horizontal loading linearly distributed for both the top and bottom (curved) strip slabs is equal to $(73.6 - 6.46 \cdot 5.70) = 36.8$ kN/m in opposite directions, while

Figure 5.29 Prestressing layout on a continuous curved girder

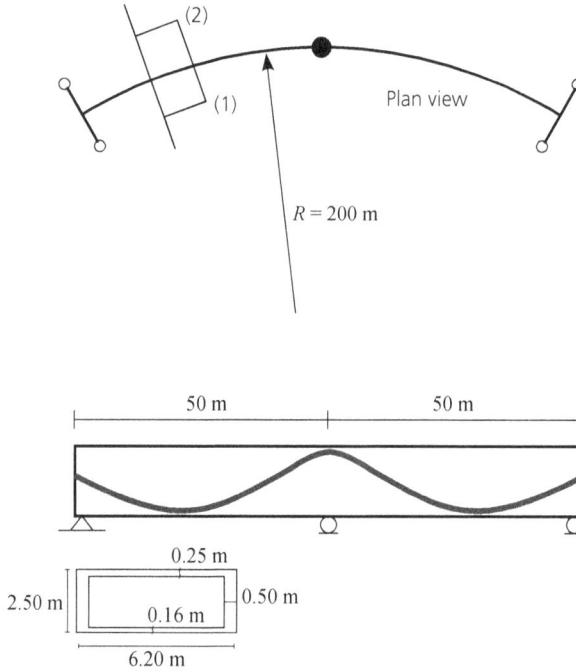

each web of the strip takes up a vertical load with the same triangular distribution equal to $\Delta v \cdot 2.30 = 6.46 \cdot 2.30 = 14.9$ kN/m upwards for the external web and downwards for the internal web.

It must be noticed that in the considered case the maximum transverse response occurs at a place where the torsional moment vanishes. However, at the middle support the presence of the transverse diaphragm prevents the deformation of the profile of the girder and for that reason a support is inserted in the Winkler model, as can be seen in Figure 5.31.

5.3.5   Reducing the torsional response through prestressing
The torsional actions and the consequent development of a torsional moment diagram $[M_T]$ in a box girder may be reduced to some degree through prestressing by providing torsional actions on the girder that counteract the existing ones.

As has been previously asserted, the presence of an inclination at some point of a prestressing cable in a girder web principally implies a torsional action. If the cable is straight, the torsional response developed is constant. If the cable is parabolic, the constant deviation forces, u, induce a constant distributed torsional load $m_{D,P} = u \cdot b/2$ (see Figure 5.32).

The arrangement of two parabolic cables of opposite curvatures in the two webs, as shown in Figure 5.32, will impose, apart from an additional centrally applied compressive force, a constant torsional load equal to $m_{D,P} = u \cdot b$ where u and b represent the deviation load and the width of the section, respectively.

Figure 5.30 Bending and torsional response due to prestressing

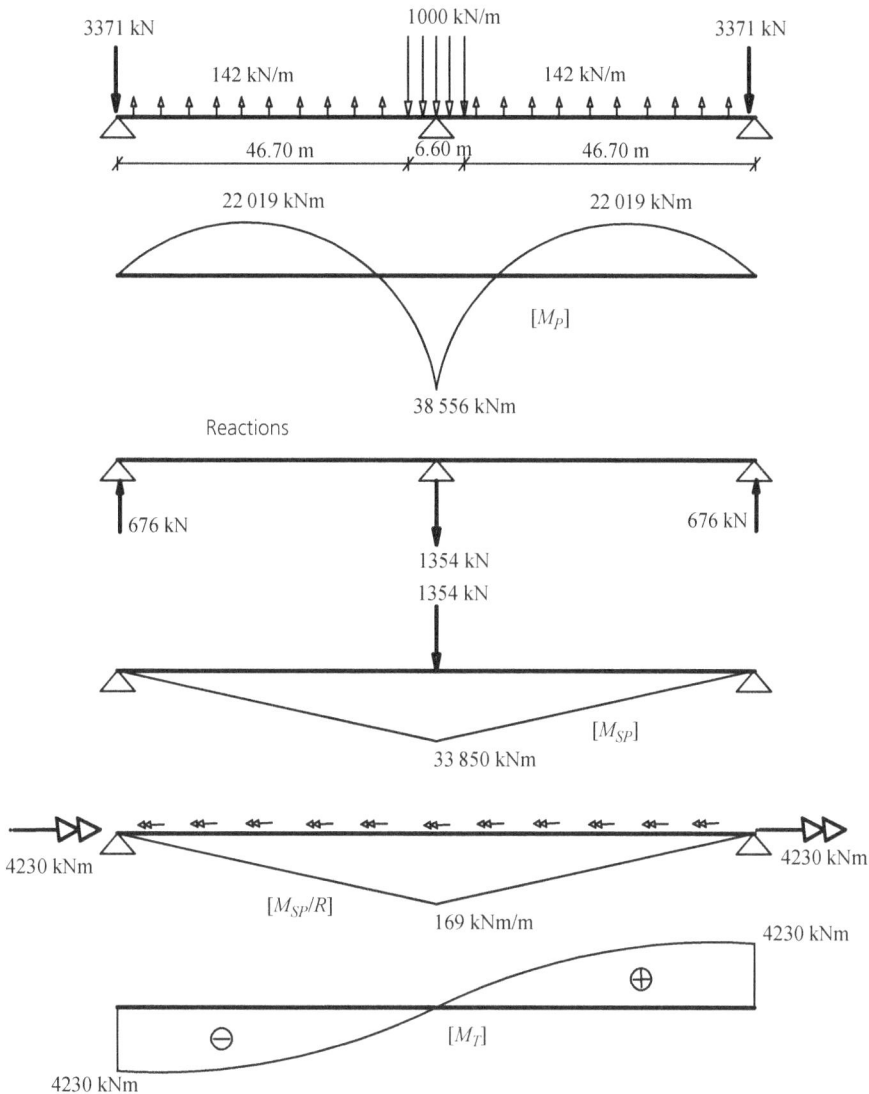

It is clear that this layout, which can counteract the existing torsional load due to gravity, does not affect the bending response as it does not produce any additional deviation load to the girder. Likewise, the arrangement of two parabolic cables of opposite curvatures in the top and bottom slabs (see Figure 5.33) will impose a constant torsional load only, equal to $m_{D,P} = u \cdot h$ where h represents the height of the cross-section. Clearly in both cases, $m_{D,P}$, which represents a constant torsional load, cannot eliminate the parabolically distributed self-weight torsional load. Although the separate implementation of such an arrangement is generally questionable, as it represents an extra cost for the structure, it nevertheless gives an idea of how to proceed with the already available tendons.

As a matter of fact, the basic design of a prestressing tendon is actually decided based on the rectilinear layout of the girder and the influence of prestressing on the torsional response itself

185

Figure 5.31 Transversal response due to prestressing

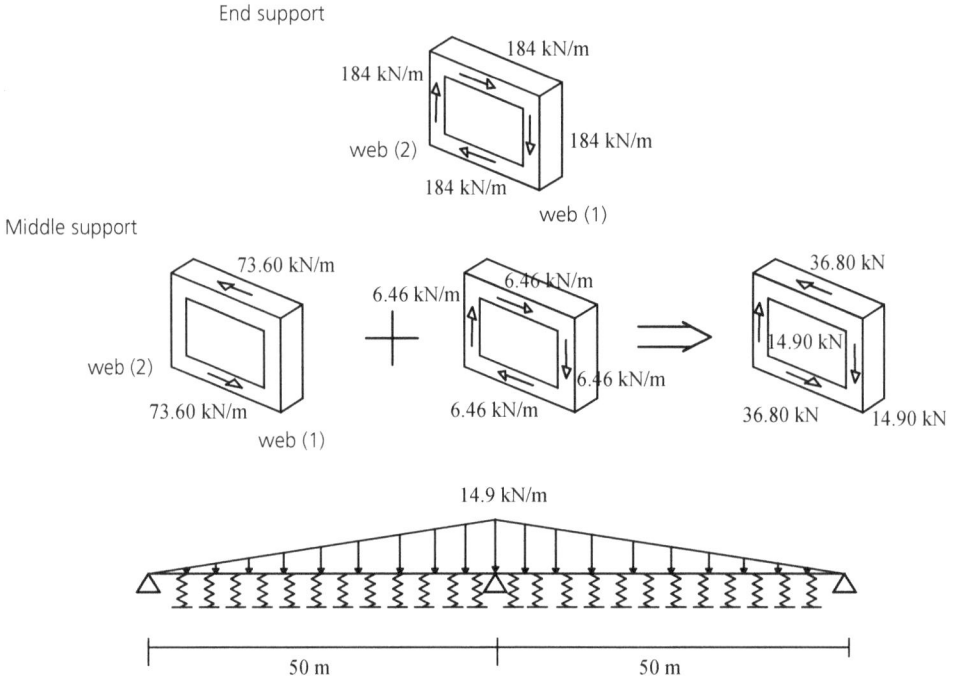

End support

184 kN/m

184 kN/m

web (2)

184 kN/m

184 kN/m

web (1)

Middle support

73.60 kN/m

web (2)

73.60 kN/m

web (1)

6.46 kN/m

6.46 kN/m

6.46 kN/m

6.46 kN/m

6.46 kN/m

36.80 kN

14.90 kN

36.80 kN   14.90 kN

14.9 kN/m

50 m   50 m

Figure 5.32 Creating torsional loading through separate prestressing cables

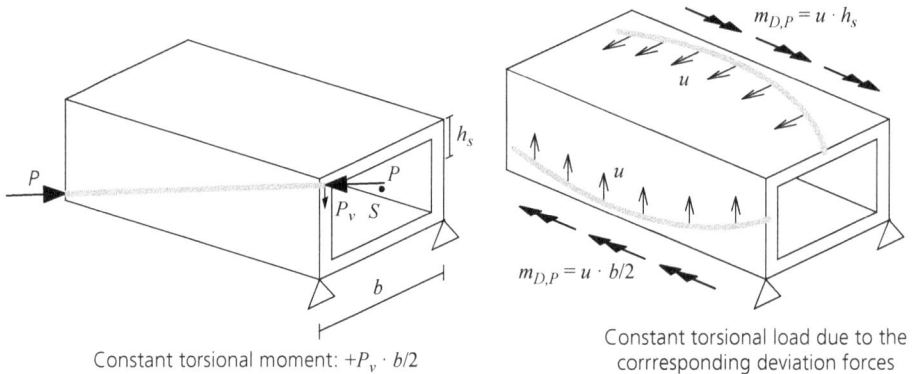

P

P

P_v S

h_s

b

Constant torsional moment: $+P_v \cdot b/2$

$m_{D,P} = u \cdot h_s$

u

u

$m_{D,P} = u \cdot b/2$

Constant torsional load due to the corrresponding deviation forces

is examined after. If the resulting torsional behaviour under gravity loads is found to be unsatis-factory, it is possible to modify the tendon layout – without altering the total number of tendons or the overall prestressing force – in order to generate unequal upward deviation forces in the two webs. This adjustment can maintain the total upward deviation force needed to counteract gravity loads, while introducing a consistent torsional load that offsets the girder's inherent torsional response.

Figure 5.33 **Figure 5.33** Basic layouts of prestressing cables for inducing torsional action

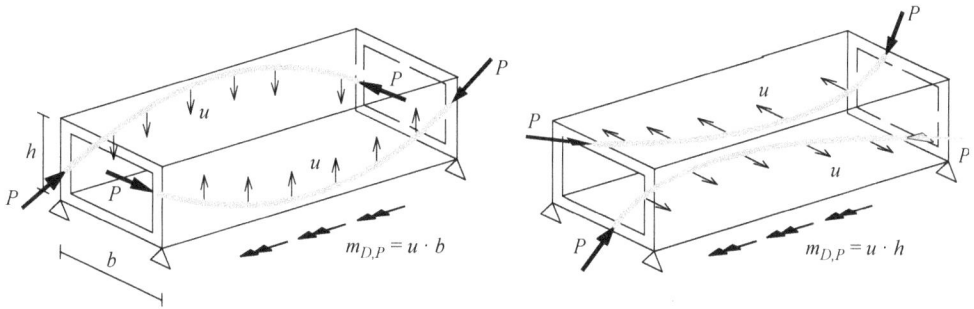

Worked example

The above is illustrated in the example of the simply supported curved girder shown in Figure 5.34, where a certain differentiation of the cable profile is applied for both webs. The final tendon profile for the outer web exhibits a stronger curvature than for the inner web, both of them of course being convex.

Figure 5.34 Relieving influence of tendon layout on the torsional response

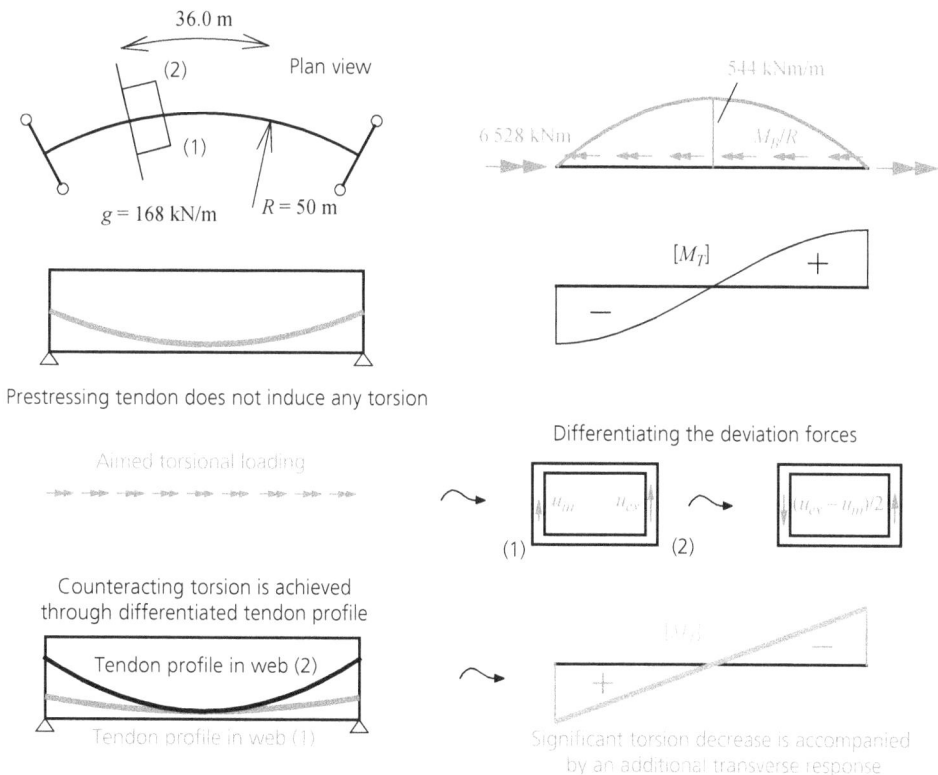

If in the above case the deviation forces of the outer and inner web are u_{ex} and u_{in}, respectively, ($u_{ex} > u_{in}$), it is clear that, while the symmetric total deviation force $(u_{ex} + u_{in})/2$ counteracts the gravity loads without producing any torsional response, the antisymmetric loading of each web with $(u_{ex} - u_{in})/2$ creates a torsional load being equal to $m_{D,P} = (u_{ex} - u_{in}) \cdot b/2$ (see Figure 5.34). This constant torsional load provides a torsional reaction $m_{D,P} \cdot L/2$ counteracting the reaction due to the permanent load.

By selecting the sag f_1 and f_2 for the tendons (1) and (2), respectively, it is $u_1 = u_{in} = P \cdot 8 \cdot f_1/L^2$ and $u_2 = u_{ex} = P \cdot 8 \cdot f_2/L^2$. It is clear that the girder with a permanent load, g, is subjected to an effective transverse load $[g - (u_{in} + u_{ex})]$. On the other hand, the torsional reaction due to self-weight, by taking into account the parabolically distributed torsional load acting on the beam with a maximum value at midspan equal to $g \cdot L^2/(8 \cdot R) = 168 \cdot 36.0^2/(8 \cdot 50) = 544.0$ kN, is determined as the half area of the corresponding parabolic diagram $(1/3) \cdot 544.0 \cdot 36.0 = 6528$ kNm. The resulting torsional reaction of the girder is expressed as:

$$M_{final} = 6528 - \frac{u_2 - u_1}{2} \cdot b \cdot \frac{L}{2} = 6528 - \frac{4Pb}{L}(f_2 - f_1)$$

With the appropriate selection of sags f_1 and f_2, the resulting torsional reaction can be drastically reduced, if not eliminated.

Of course, it should not be overlooked that the introduction of the additional torsional load $m_{D,P}$ by the differentiated tendons creates an additional response of the section walls (see Figure 5.35) through the action of a resulting self-equilibrated diagonal force, in the same way as examined in Section 5.3.3 regarding the similar effect of the gravity loads (see Figure 5.21). However, although the differentiated tendon layout produces a torsional load of opposite direction to that due to gravity loads, the diagonal loading of the considered strip is developed in the same direction as under gravity loads and thus the internal web is further loaded (see Figure 5.35).

Certainly, the task of the restriction of torsional response through prestressing must always be valued for each case as to its techno-economic consequences, without ignoring the construction difficulties associated with a prestressing layout.

Figure 5.35 Influence of prestressing layout on the transverse response of the girder

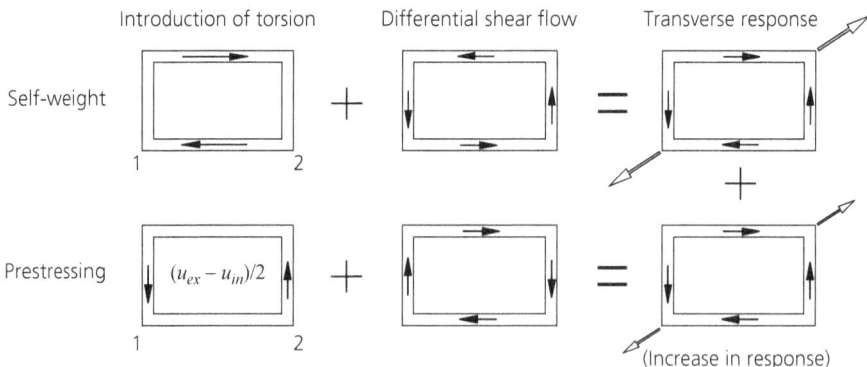

Appendix A

E,I

L

K: modulus of subgrade reactio

Elastic length $L_s = \left(\frac{4EI}{K}\right)^{\frac{1}{4}}$

$\lambda = \dfrac{1}{L_s}$

F

E,I

$\dfrac{L}{2}$

x

F

m

E,I

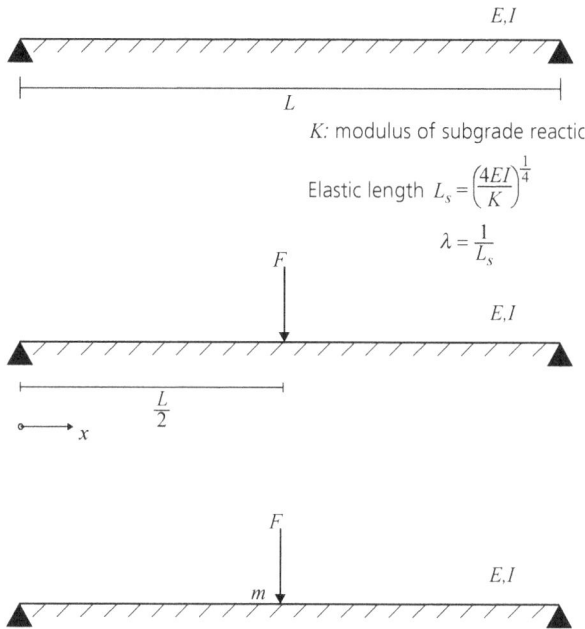

Bending moment: $M(x) = \dfrac{1}{4\lambda} \cdot F \cdot C(|x - L/2|)$

Deflection: $w(x) = \dfrac{\lambda}{2K} \cdot F \cdot A(|x - L/2|)$

where

$$C(x) = e^{-\lambda x}(\cos \lambda x - \sin \lambda x)$$

and

$$A(x) = e^{-\lambda x}(\cos \lambda x + \sin \lambda x)$$

In the case in which $x = L/2$:

$$M_m(x) = \frac{F}{4\lambda}$$

$$w_m(x) = \frac{F\lambda}{2K}$$

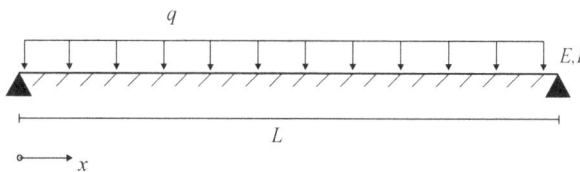

q

E,I

L

x

Bending moment:

$$M(x) = \frac{q}{2.\lambda^2} \frac{\sin h(\lambda x).\sin[\lambda(L-x)] + \sin h[\lambda.(L-x)].\sin(\lambda.x)}{\cos h(\lambda.L) + \cos(\lambda.L)}$$

Deflection:

$$w(x) = \frac{q}{k}\left(1 - \frac{\cos h(\lambda x)\cos[\lambda(L-x)] + \cos h[\lambda(L-x)]\cdot\cos(\lambda x)}{\cos h(\lambda L) + \cos(\lambda L)}\right)$$

(Hetényi, 1946)

Appendix B

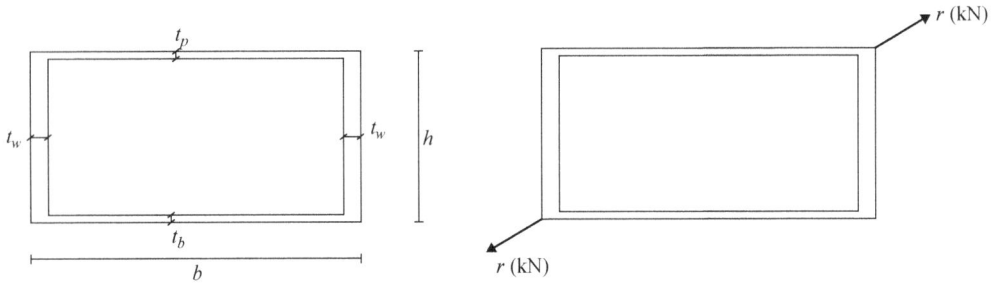

Diagonal elongation:

$$\delta = \frac{r}{C}$$

Diagonal stiffness:

$$C = \frac{2\left(t_p^3 + t_b^3 + \frac{bt_w^3}{2h}\right) + 3\frac{h}{b}\left(\frac{t_p t_b}{t_w}\right)^3}{t_p^3 + t_b^3 + 6\frac{h}{b}\left(\frac{t_p t_b}{t_w}\right)^3}$$

(Menn, 1990)

REFERENCES

Hetényi M (1946) *Beams on Elastic Foundation*. University of Michigan Press, Ann Arbor, MI, USA.

Menn C (1990) *Prestressed Concrete Bridges*. Birkhäuser, Basel, Switzerland.

Schlaich J and Scheef H (1982) *Concrete Box-Girder Bridges*. IABSE Structural Engineering Documents, Zurich, Switzerland.

Stavridis L and Georgiadis K (2025) *Understanding and Designing Structures without a Computer: Plane structural systems*. Emerald Publishing, Leeds, UK.

emerald
PUBLISHING

ice

Leonidas Stavridis and Konstantinos Georgiadis
ISBN 978-1-83662-945-0
https://doi.org/10.1108/978-1-83662-942-920251006
Emerald Publishing Limited: All rights reserved

Chapter 6
Lateral response of multi-storey systems

6.1. Introduction

In this chapter, the behaviour of three-dimensional multi-storey building systems under lateral forces is examined. These forces may result either from an imposed seismic ground motion, from an incoming wind action, or even from a temperature variation that, as will be explained later, activates the lateral stiffness of the structural system. The reason for making the consideration of temperature change an essential design issue not only refers to the obvious environmental temperature influence on the structure, but also to the fact that the shrinkage of the concrete slabs on each storey of the system after the concrete has hardened can be expressed through a temperature fall.

6.2. Formation of the system

Multi-storey systems consist of horizontal diaphragms (slabs) disposing a very high rigidity within their own plane, which are arranged vertically in regular distances of a storey height and connected to each other as well as to the foundation ground through vertical elements of rigidity, according to a certain layout in plan. These vertical elements consist either of frames formed by the columns and beams, single columns or shear walls, or, possibly cores having an open or a closed section (see Figure 6.1).

It is pointed out that the concept of diaphragms is not necessarily restricted to a solid concrete slab but may consist of any structural formation possessing a practically high stiffness in its own plane. The same naturally applies to all the vertical elements, which are not necessarily made of concrete but can be formed from steel elements too.

Each horizontal diaphragm under the action of horizontal forces is moved as a rigid body within its own plane. This movement consists of a horizontal displacement as well as an in-plane rotation and is obviously imposed unchanged to all vertical elements, causing a corresponding response (see Figure 6.2).

In the present examination, which serves in principle the structural design rather than an 'accurate' analysis, it may be assumed that all vertical elements as the supporting skeleton of the diaphragms constitute a set of elements of purely plane stiffness (Stavridis, 1986).

The term *element of purely plane stiffness* or simply *plane element* means that the element is stressed only by that component of the diaphragm displacement which corresponds to its plane. Thus, for example, each frame is assumed to be stressed only by that component of the displacement vector corresponding to the specific frame plane (see Figure 6.3).

Figure 6.1 Layout of horizontal diaphragms and vertical stiffness elements

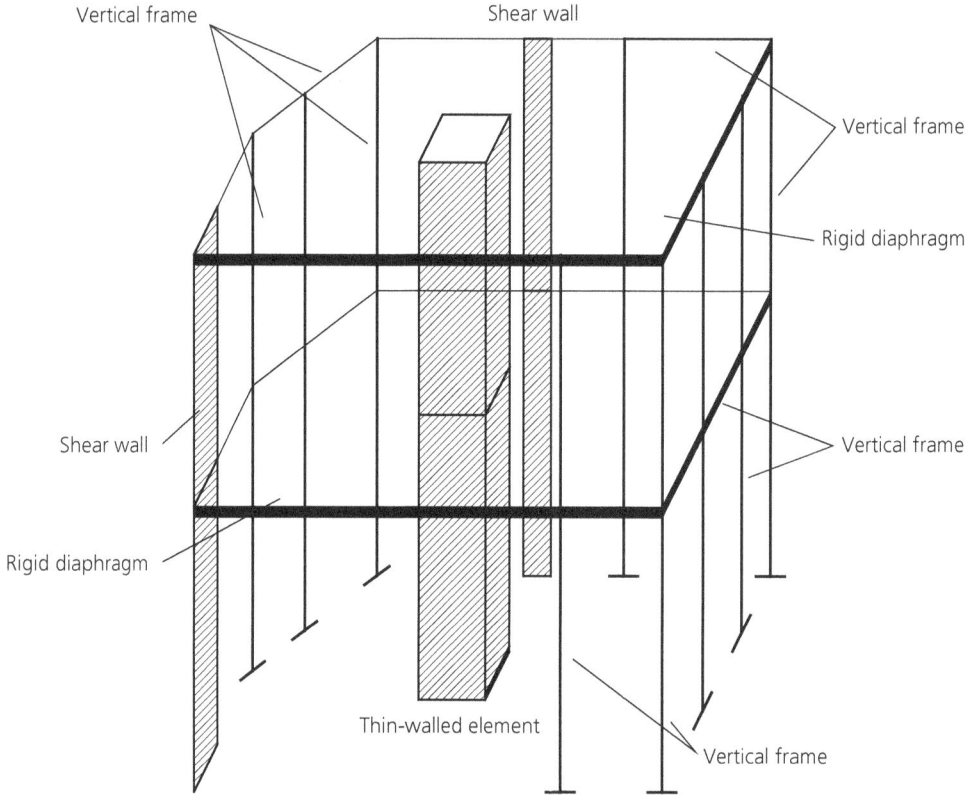

Figure 6.2 Possibility of horizontal movement of the diaphragms

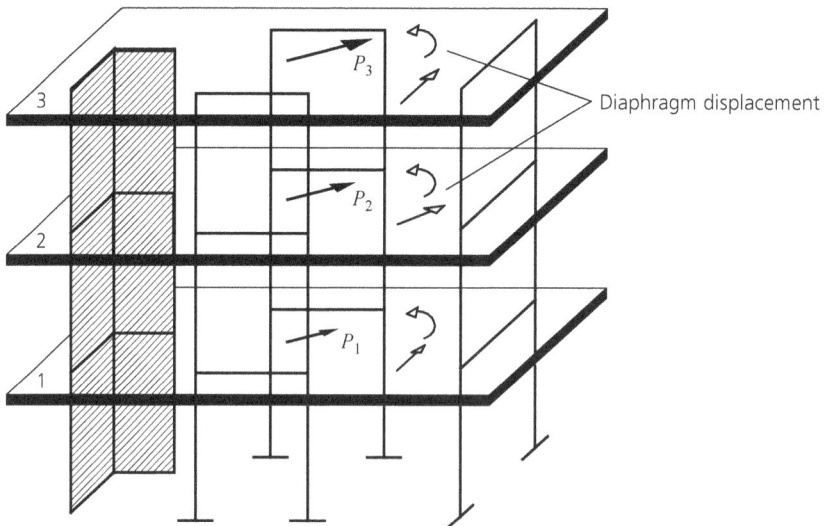

Figure 6.3 Activation of stiffness of the plane elements through the diaphragm

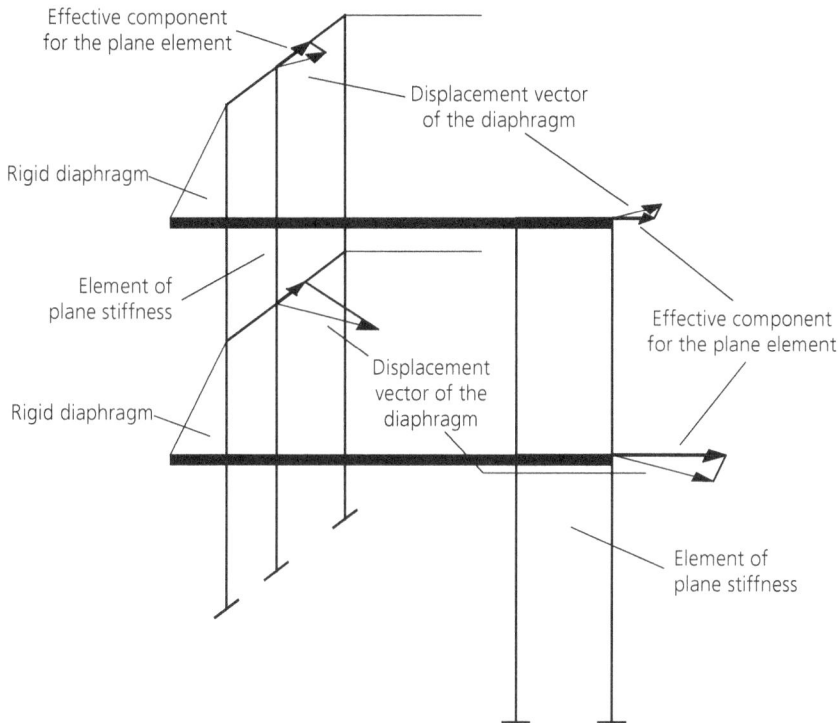

Although the bending stiffness of a frame with respect both to its own plane and transversely to it can be described by two separate plane vertical elements, the preponderant stiffness in its own plane allows the omission of the corresponding transverse plane element. However, in a single column, two plane elements may be considered representative of its stiffness with respect to its two principal directions of stiffness (see Figure 6.4).

Moreover, it may be understood that a vertical core of either open or closed section may be represented by the plane elements constituting the whole section. These plane elements are also able to offer the torsional stiffness of the element as any applied torsional load may be taken up by the plane state of stress of the constituent plane elements themselves (see Figure 6.4).

6.3.   Lateral response
6.3.1   Treatment of load-bearing action
The whole system consists of N diaphragms and n plane elements and is subjected to a certain horizontal force P_i ($i = 1 - N$) acting on a precise point of each diaphragm (e.g. in the case of earthquake loading at the centre of mass). The system may be considered as a superposition of N loading cases, each one comprising the force P_i only. Now, it can be postulated that each loading configuration may be considered as practically identical to that where all the diaphragms above level (i) are omitted, as shown in Figure 6.5.

Figure 6.4 Expressing the stiffness of vertical members through elements of plane stress

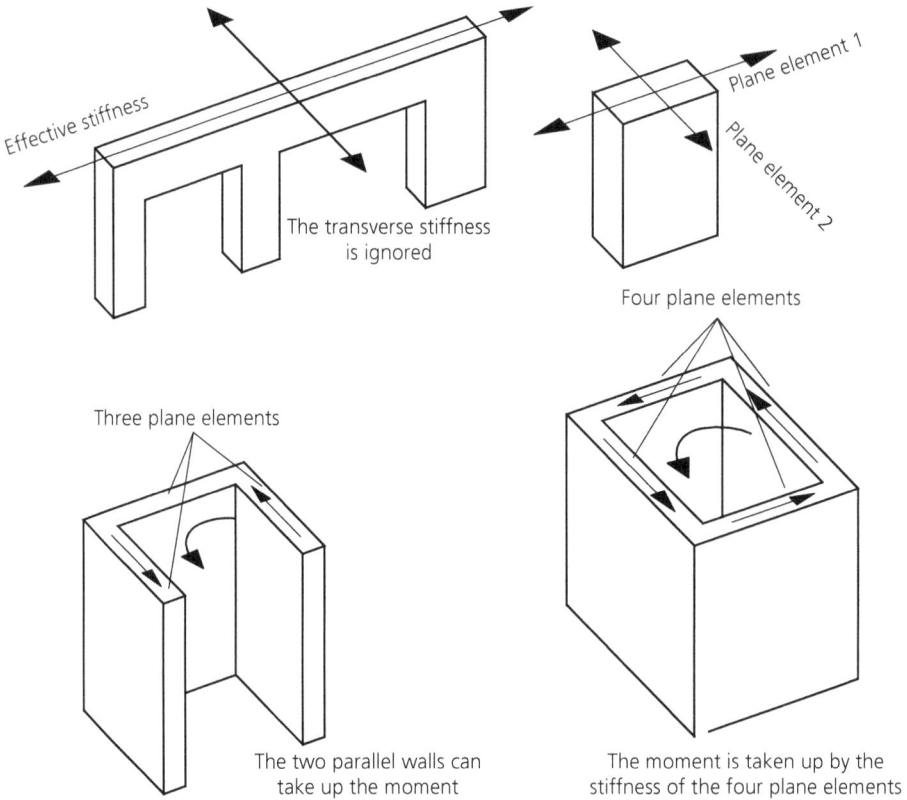

Effective stiffness

The transverse stiffness is ignored

Plane element 1

Plane element 2

Four plane elements

Three plane elements

The two parallel walls can take up the moment

The moment is taken up by the stiffness of the four plane elements

The considered diaphragm at level (i), being referred to fixed orthogonal axes Oxy with their origin O placed at some precise point (the same for all diaphragms), undergoes, through the action of force, P, horizontal movement as a rigid body, represented by the components of displacement Δx and Δy as well as the rotation $\Delta \omega$, taken as positive if anticlockwise. This displacement is obviously resisted by the lateral stiffness of plane elements, as explained above (see Figure 6.5).

The lateral stiffness of each plane element is represented by the required horizontal force, S, acting on the specific level, (i), to induce a unit displacement (1 m) at that level. This force remains practically the same, even if all the upper storey levels are omitted, as shown in Figure 6.6, in accordance with Stavridis and Georgiadis (2025, Section 6.4.3) and in compliance with Figure 6.5.

The resistance offered by the plane elements to the diaphragm movement implies for each of them the action of a certain force at the corresponding level required to produce the imposed displacements on them by the diaphragm, while the same forces are returned with the opposite direction to the diaphragm itself. This must be in equilibrium under the above forces as well as under the directly acting force, P (see Figure 6.7). Given that these resistance forces may be expressed, as will be shown, in terms of the displacement components Δx, Δy and $\Delta \omega$, it is clear that these components can be determined through the three equations of equilibrium of the diaphragm. The above components permit the evaluation of the acting forces on each plane element at the specific level.

Figure 6.5 Horizontal loading of a single diaphragm

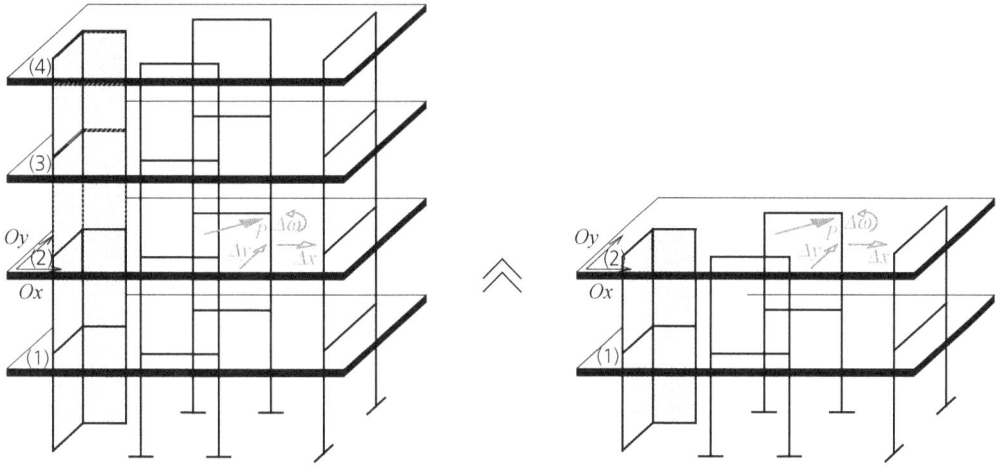

The two systems develop practically the same response

Figure 6.6 Lateral stiffness of a plane element

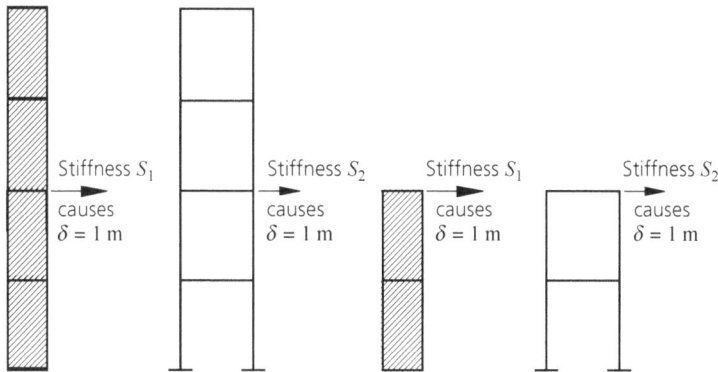

Stiffness S_1 causes $\delta = 1$ m

Stiffness S_2 causes $\delta = 1$ m

Stiffness S_1 causes $\delta = 1$ m

Stiffness S_2 causes $\delta = 1$ m

The lateral stiffness S is practically unaffected
by the upper-lying levels

However, it must be ensured that the diaphragm in question can safely take up these self-equilibrating coplanar forces that it is subjected to.

Of course, the simultaneous loading of all of the diaphragms subjected to corresponding forces, P_i, may be considered by superposition of the loading of each diaphragm separately (ignoring all the above lying levels as previously indicated) and, thus, the final response of each vertical element will be deduced from the total action of the respective forces determined at each level (see Figure 6.8). These forces are used for the design of the plane elements, the assessment of their displacements at the various levels, as well as for the response of the corresponding diaphragm.

195

Figure 6.7 Equilibrium of the loaded horizontal diaphragm

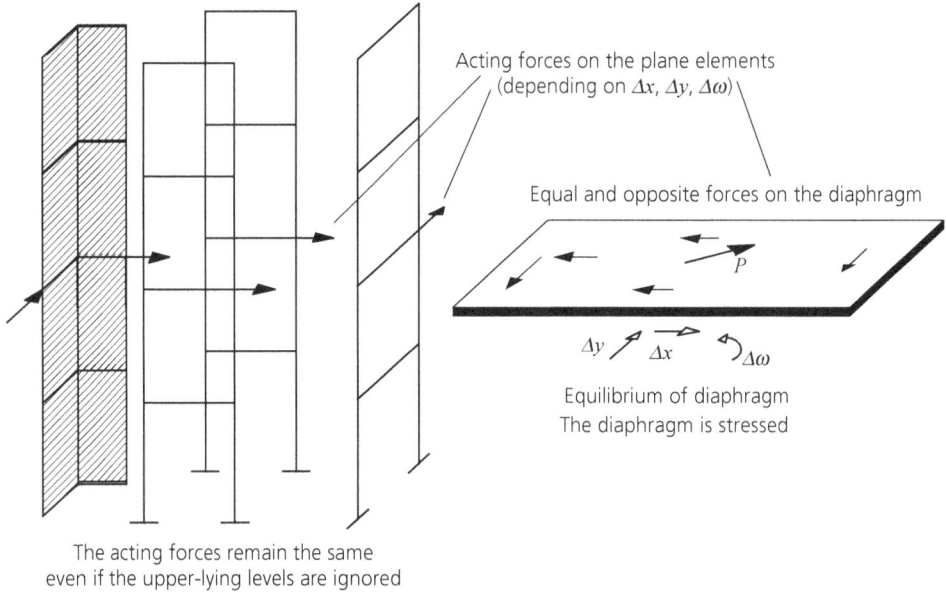

Acting forces on the plane elements
(depending on Δx, Δy, $\Delta \omega$)

Equal and opposite forces on the diaphragm

P

Δy Δx $\Delta \omega$

Equilibrium of diaphragm
The diaphragm is stressed

The acting forces remain the same
even if the upper-lying levels are ignored

Figure 6.8 Treatment of the loading of all diaphragms

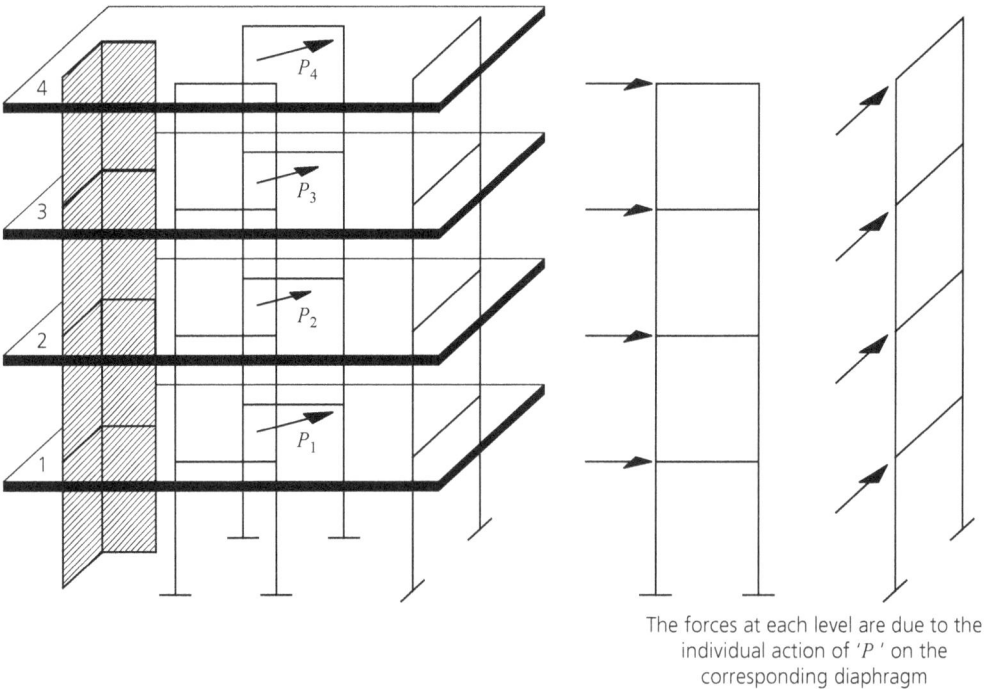

4

P_4

3

P_3

2

P_2

1

P_1

The forces at each level are due to the
individual action of 'P' on the
corresponding diaphragm

6.3.2 Response of a plane element

Each plane element (i) is characterised by the angle, β_i, encompassed in the anticlockwise sense by the axis Ox with a chosen positive direction on the element's trace in ground plan such as to correspond to an anticlockwise rotation about the origin O of the coordinate system. On this trace, an arbitrary point characteristic of the plane element with coordinates x_i and y_i is selected (see Figure 6.9).

By considering a specific level (i) of an element where the displacement components Δx, Δy and $\Delta \omega$ of the corresponding diagram are imposed, the displacement, D_i, of the element, measured on its previously defined positive direction, can be determined by assuming that the diaphragm, as already pointed out, remains undeformable during its movement as a rigid body. It is:

$$D_i = \Delta x \cdot \cos \beta_i + \Delta y \cdot \sin \beta_i + \Delta \omega \cdot e_i$$

where e_i is the (signed) distance of the element's trace from the coordinate axes origin O, being equal to:

$$e_i = x_i \cdot \sin \beta_i - y_i \cdot \cos \beta_i$$

In the considered level, the element (i) may be structurally represented by the corresponding force, S, required to be applied in order to produce a unit displacement (1 m), known also as *lateral stiffness*. This magnitude can be determined by calculating the displacement, δ (m), due to a unit horizontal force (1 kN) applied on that level (see Figure 6.10). Then, it is:

$$S = 1/\delta \, (\text{kN/m})$$

Figure 6.9 Geometric and kinematic characteristics of plane element

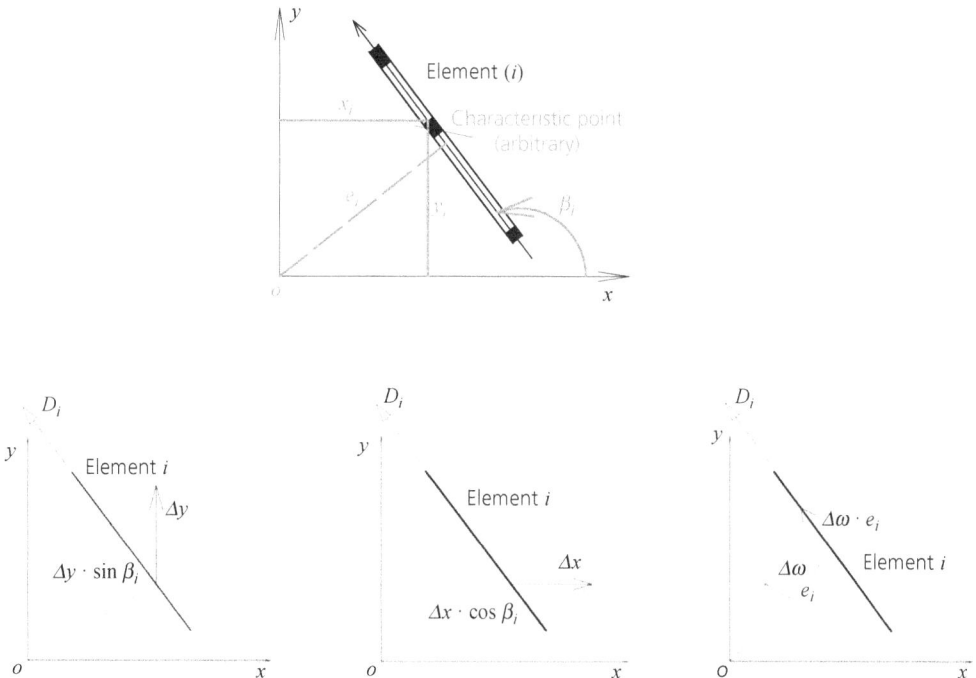

Figure 6.10 Determination and evaluation of the translational stiffness of a plane element

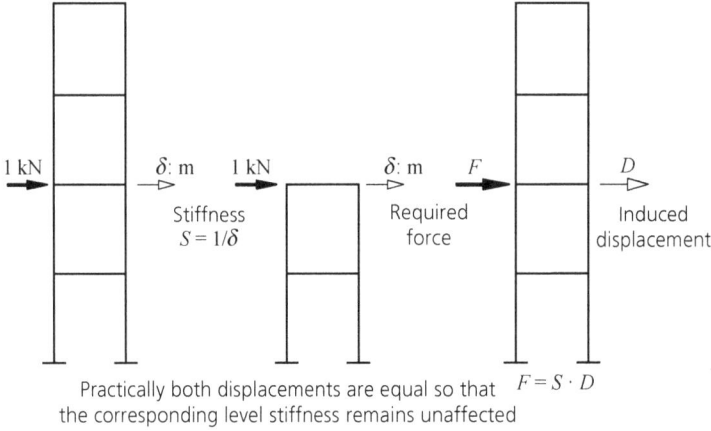

As pointed out in Section 6.3.1, ignoring the overlying levels does not practically affect the above displacement and, consequently, the corresponding stiffness (see Figure 6.10).

Thus, in order for the plane element (i) to join the imposed diaphragm displacement at the considered level, it requires a horizontal force $F = S_i \cdot D_i$ to be imposed on it, that is:

$$F = S_i \cdot (\Delta x \cdot \cos \beta_i + \Delta y \cdot \sin \beta_i + \Delta \omega \cdot e_i)$$

6.3.3   General layout

The diaphragm of a specific level is referred to an arbitrary orthogonal system Oxy and is subjected to a horizontal force, P, applied to a certain point having coordinates x_e and y_e and represented by the components P_x and P_y. As the diaphragm is further subjected to the equal and opposite forces, F, derived in Section 6.3.2 and shown in Figure 6.7, then it may be found from its equilibrium, as previously considered, that the components Δx, Δy and $\Delta \omega$ of its displacement must satisfy the following system of three equations (see Figure 6.11):

$$K_1 \cdot \Delta x + K_2 \cdot \Delta y + K_3 \cdot \Delta \omega = P_x$$

$$K_2 \cdot \Delta x + K_4 \cdot \Delta y + K_5 \cdot \Delta \omega = P_y$$

$$K_3 \cdot \Delta x + K_5 \cdot \Delta y + K_6 \cdot \Delta \omega = P_y \cdot x_e - P_x \cdot y_e$$

where

$$K_1 = \Sigma S_i \cdot \cos^2 \beta_i \quad K_2 = \Sigma S_i \cdot \sin \beta_i \cdot \cos \beta_i \quad K_3 = \Sigma S_i \cdot e_i \cdot \cos \beta_i$$

$$K_4 = \Sigma S_i \cdot \sin^2 \beta_i \quad K_5 = \Sigma S_i \cdot e_i \cdot \sin \beta_i \quad K_6 = \Sigma S_i \cdot e_i^2$$

Figure 6.11 Visualisation of support of the loaded diaphragm through respective springs

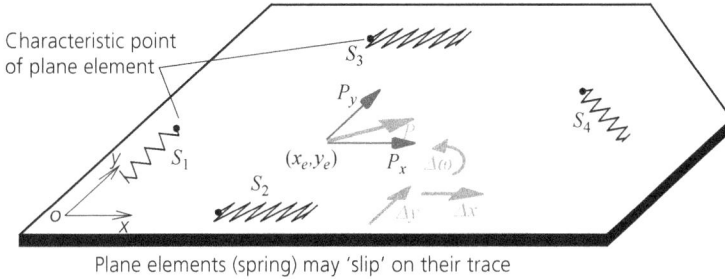

Plane elements (spring) may 'slip' on their trace

Once the components Δx, Δy and $\Delta \omega$ are known, the horizontal forces acting at the corresponding level of all plane elements can be readily determined, according to the previous section (Stavridis, 1986).

The stiffness, S_i, at any level may be thought of as a horizontal spring in the direction of each element, which may be connected to the diaphragm at the position of the corresponding characteristic point of the element (see Figure 6.11). In this way, the diaphragm acted upon by a force, P, maintains its equilibrium thanks to the developed spring forces acting on it, while the same forces are also acting on the corresponding plane elements, obviously in the opposite direction.

As very often the force, P, represents the acting inertia force on the diaphragm due to a specific seismic excitation, its point of application (x_e, y_e) may represent the centre of mass of the considered diaphragm.

From the above relations, it can be deduced that the plane elements may be 'shifted' anywhere along their own trace, since only their respective angle, β_i, and their distance, e_i, from the coordinate axes origin matter. This may also be understood through the concept of spring supports introduced above, which, thanks to the absolute rigidity of the diaphragm, may be connected at any point along their line of action.

However, in this respect, an important question arises regarding the point of application (x_e^*, y_e^*) of the force, P, which would make the ensuing rotation, $\Delta \omega$, of the diaphragm disappear. By considering $\Delta \omega = 0$ in the above equations, it comes out that:

$$x_e^* = \frac{K_2 \cdot K_3 - K_1 \cdot K_5}{K_2^2 - K_1 \cdot K_4}$$

$$y_e^* = \frac{K_3 \cdot K_4 - K_2 \cdot K_5}{K_2^2 - K_1 \cdot K_4}$$

This point is called the *centre of stiffness* and, by referring it to a particular layout consisting of the horizontal diaphragm and its supporting stiffening elements, its structural role is somehow similar to that of the *shear centre* of a thin-walled beam (see Chapter 4, Section 4.3). For any horizontal force, P, acting on this point, the diaphragm exhibits a pure displacement in the direction of the force, P, without any rotation at all. The components Δx and Δy of this displacement can be calculated from the first two equations of the above linear system:

$$K_1 \cdot \Delta x + K_2 \cdot \Delta y = P_x$$

$$K_2 \cdot \Delta x + K_4 \cdot \Delta y = P_y$$

the last equation being satisfied by itself by eliminating $\Delta\omega$ as equal to zero.

It can be seen that under a force, P, not passing through the centre of stiffness, the diaphragm will rotate in its own plane about this very centre, adding some extra stress to the more distant plane elements.

Now, if the centre of stiffness coincides with the centre of mass of the diaphragm where the equivalent inertia force, P, due to an earthquake is always acting, no rotation of the diaphragm will take place and the vertical plane elements will be affected only by the imposition of the displacement components Δx and Δy (see Section 6.3.2).

However, if the acting force, P, on the diaphragm does not pass through the centre of stiffness – no matter if it is applied to the centre of mass or not – it can equivalently be transferred to it together with a pure torsional moment, M_e, acting on the diaphragm. This moment is equal to $P \cdot e$ where e represents the perpendicular distance of the centre of stiffness to the line of action of the acting force. It is clear that the diaphragm will be subjected to a pure translation due to the action of P on the centre of stiffness plus the displacement and rotation due to the torsional moment alone that must satisfy the linear system:

$$K_1 \cdot \Delta x + K_2 \cdot \Delta y + K_3 \cdot \Delta\omega = 0$$

$$K_2 \cdot \Delta x + K_4 \cdot \Delta y + K_5 \cdot \Delta\omega = 0$$

$$K_3 \cdot \Delta x + K_5 \cdot \Delta y + K_6 \cdot \Delta\omega = M_e$$

Figure 6.12 Orthogonal layout of plane elements

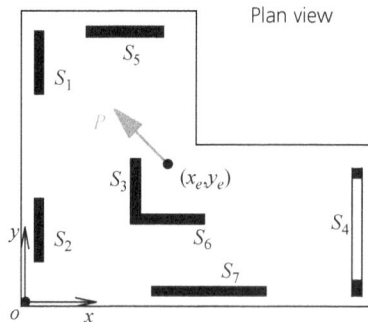

6.3.4   Orthogonal layout

The vertical elements in building structures are usually arranged in mutually orthogonal directions. In this case, the axes Ox and Oy are placed parallel to these two directions (see Figure 6.12). Then, all the elements in the x-direction have $\beta_i = 180°$ and $e_i = y_i$, according to Section 6.3.2, whereas all the elements in the y-direction have $\beta_i = 90°$ and $e_i = x_i$.

Thus (see Section 6.3.3),

$$K_1 = \Sigma S_x \quad K_2 = 0 \quad K_3 = -\Sigma(S_x \cdot y_i)$$

$$K_4 = \Sigma S_y \quad K_5 = \Sigma(S_y \cdot x_i) \quad K_6 = \Sigma(S_x \cdot y_i^2) + \Sigma(S_y \cdot x_i^2)$$

and the linear system of the above section takes the form:

$$K_1 \cdot \Delta x + K_3 \cdot \Delta\omega = P_x$$

$$K_4 \cdot \Delta y + K_5 \cdot \Delta\omega = P_y$$

$$K_3 \cdot \Delta x + K_5 \cdot \Delta y + K_6 \cdot \Delta\omega = P_y \cdot x_e - P_x \cdot y_e$$

Consequently, the displacements Δx, Δy and $\Delta\omega$ can be determined on the basis of the following expressions:

$$\Delta x = \frac{P_x}{K_1} - \frac{K_3}{K_1}\Delta\omega$$

$$\Delta y = \frac{P_y}{K_4} - \frac{K_5}{K_4}\Delta\omega$$

$$\Delta\omega = \frac{P_x\left(\dfrac{K_3}{K_1} + y_e\right) + P_y\left(\dfrac{K_5}{K_4} - x_e\right)}{\dfrac{K_3^2}{K_1} + \dfrac{K_5^2}{K_4} - K_6}$$

For the force acting on the elements in the x-direction at the level under consideration, according to Section 6.3.2 it is:

$$F_x = S_x \cdot (-\Delta x + \Delta\omega \cdot y_i)$$

whereas for the elements in the y-direction:

$$F_y = S_y \cdot \left(\Delta y + \Delta\omega \cdot x_i\right)$$

In the examined case of an orthogonal layout, the coordinates of the centre of stiffness have a simpler expression:

$$x_e^* = \frac{K_5}{K_4}$$

$$y_e^* = -\frac{K_3}{K_1}$$

The above relations lead to the conclusion that if the origin O of the orthogonal axes Ox and Oy is placed on the centre of stiffness itself, the newly evaluated quantities 'K', according to the previous expressions regarding the orthogonal layout, must yield both values of K_3 and K_5 equal to zero. The reason is that the new determination of the stiffness centre for the given layout according to the above relations must obviously provide the coordinate origin itself as that point, which means, of course, $K_3 = K_5 = 0$.

If the horizontal force, P, acts at the centre of stiffness, the diaphragm does not rotate at all and Δx, Δy simply result as:

$$\Delta x = P_x / (\Sigma S_x) \text{ and } \Delta y = P_y / (\Sigma S_y)$$

as can also be deduced from the previous expressions.

If the centre of stiffness coincides with the centre of mass of the diaphragm and the force, P, happens to be applied there, then the plane elements will be stressed only by the above imposed displacements.

The above results may be interpreted and respectively evaluated also according to the spring model of the rigid diaphragm (see Figure 6.13).

Assuming that the diaphragm is subjected to a horizontal force passing through the centre of stiffness with components P_x and P_y, then it undergoes a displacement along the direction of the acting force only and it becomes clear that each component separately will cause a common displacement Δx or Δy on all springs along the x- and y-direction, respectively, while lying on the line of action of the resultant of the corresponding spring forces.

The diaphragm for each direction therefore acts as a rigid beam on spring supports and its equilibrium for each direction assigns the following two conditions:

Direction x: $P_x = \Sigma(S_x \cdot \Delta x) = \Delta x \cdot (\Sigma S_x) \quad \Sigma(S_x \cdot \Delta x) \cdot y = y_e^* \cdot \Sigma(S_x \cdot \Delta x)$

Direction y: $P_y = \Sigma(S_y \cdot \Delta y) = \Delta y \cdot (\Sigma S_y) \quad \Sigma(S_y \cdot \Delta y) \cdot x = x_e^* \cdot \Sigma(S_y \cdot \Delta y)$

These requirements lead directly to the above established relations.

However, a possible application of a load, P, at a point different from the centre of stiffness means the additional action of a certain moment on the diaphragm, which must be balanced by the spring forces – that is, the resistance forces offered by the plane elements. ∎

Figure 6.13 Simulating the diaphragm behaviour through a spring-supported rigid beam

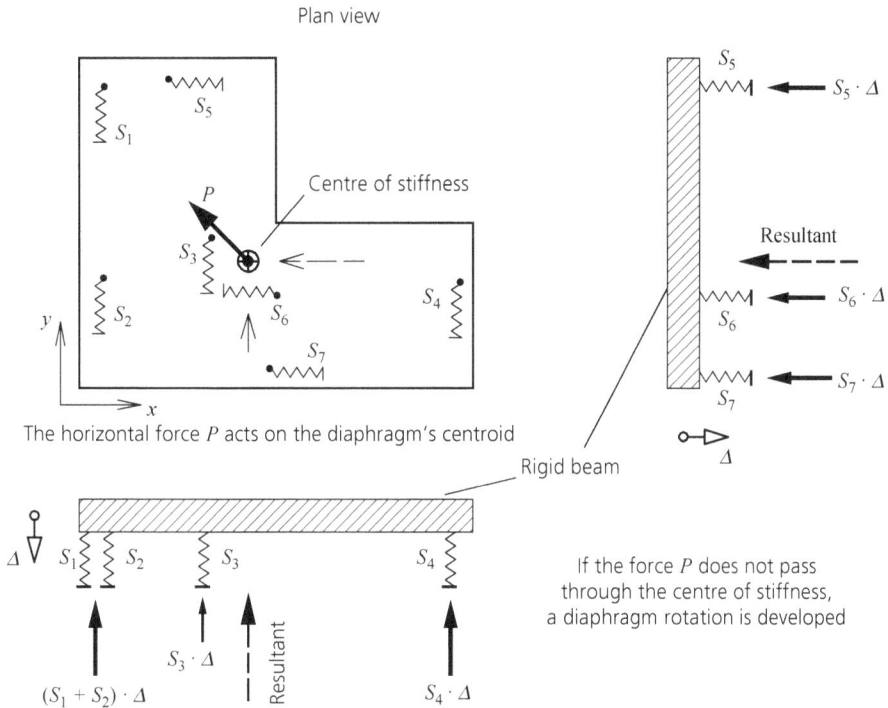

Plan view

Centre of stiffness

The horizontal force P acts on the diaphragm's centroid

Rigid beam

$S_5 \cdot \Delta$

Resultant

$S_6 \cdot \Delta$

$S_7 \cdot \Delta$

If the force P does not pass through the centre of stiffness, a diaphragm rotation is developed

$(S_1 + S_2) \cdot \Delta$

$S_3 \cdot \Delta$

Resultant

$S_4 \cdot \Delta$

Regarding now the spring (plane element) layout, it must ensure generally the uptake of an arbitrary load, P, acting in any direction, as can be seen in Figure 6.14.

In Figure 6.14(a), the x-component of load, P, cannot be balanced by the spring forces and, consequently, the layout is unsafe. In Figure 6.14(b), although springs (i.e. plane elements) are offered to the diaphragm for the uptake of any component of the load, P, these are not able to provide a couple of forces in order to balance the moment of the load with respect to the point of intersection of the spring axes (i.e. the centre of stiffness) and therefore the system may easily collapse. In Figure 6.14(c),

Figure 6.14 Criteria for right or wrong layout of plane stiffness elements

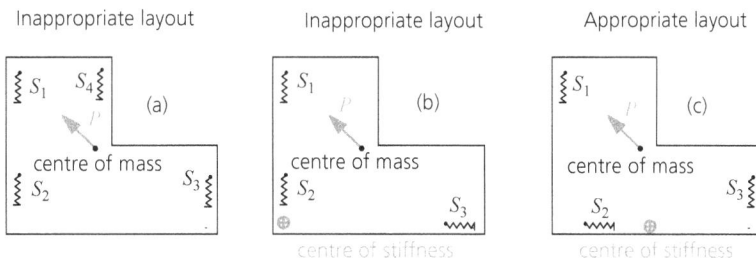

Inappropriate layout

Inappropriate layout

Appropriate layout

centre of mass

centre of mass

centre of mass

centre of stiffness

centre of stiffness

the developed spring forces can balance any component of the acting load, as well as the moment of the load, P, with respect to the centre of stiffness through the offered forces S_1 and S_3.

It is now clear that, while the point of application of the load, P, is generally determined by the specific geometry of the diaphragm itself, the centre of stiffness depends on the selected layout of the plane stiffness elements.

However, since the most important criterion for successfully designing a multi-storey system is to restrict the resulting diaphragm rotation due to seismic forces as much as possible, the designer should aim to minimise the distance between the centre of mass of the diaphragm and the centre of stiffness by arranging the vertical elements of plane stiffness accordingly.

6.3.5   Lateral displacements

In the examined multi-storey system where it is assumed that each diaphragm is subjected to a certain horizontal force, P_i, acting at the respective centre of stiffness in either direction x or y, it is important to estimate the induced deformation on each diaphragm.

Assuming that in each plane frame the stiffness, K, of all storeys remains constant over all its height, it can be concluded that the stiffness K_i at level (i) of the frame is equal to $K \cdot h/h_i$ as the inverse value of the developed displacement caused by a unit force applied at that level, where h_i is the distance of level (i) to the ground (see Stavridis and Georgiadis, 2025, Section 6.4.3). Moreover, the stiffness $S(i)$ at level (i) of a shear wall considered as a cantilever with the fixed end at the foundation is $3EI/h_i^3$. It is assumed, of course, that the existence of storeys above level (i) does not affect the considered stiffness itself (Section 6.3.1, 6.3.2 and Figures 6.6 and 6.10). Thus, the total lateral stiffness, K^i, at level (i) considering each orthogonal direction separately is:

$$K^i = \Sigma k \cdot \frac{h}{h_i} + (3 \cdot E / h_i^3) \cdot \Sigma I$$

whereby Σ extends over the corresponding elements of the same direction (see Figure 6.15). It is clear that K^i that represents the lateral stiffness of level (i) is equal to either K_1 or K_4 (see Section 6.3.4), according to the direction of the acting force, P_i.

Following now essentially the approach described for multi-storey frames in Stavridis and Georgiadis (2025, Section 6.4.3), the shear force diagram corresponding to a fictitious cantilever under the lateral loads, P_i, applied at each level is first considered.

Considering the drift, Δw_1, of the diaphragm at level 1 with respect to the support level 0 (foundation level), by taking into account all the above storeys it can be written as:

$$\Delta w_1 = V_1 / K^1 + w_s$$

where K^1 is the lateral stiffness of storey (1) according to the direction of the acting force P_i, and w_s is, in general, the support drift, which in this case is equal to zero (compare with Figure 6.19).

Considering now the remaining structure above level 1 having fixed supports at level 1, it can be seen that the drift, Δw_2, of the second level with respect to level 1 is:

Figure 6.15 Conceptual layout for the evaluation of lateral drifts

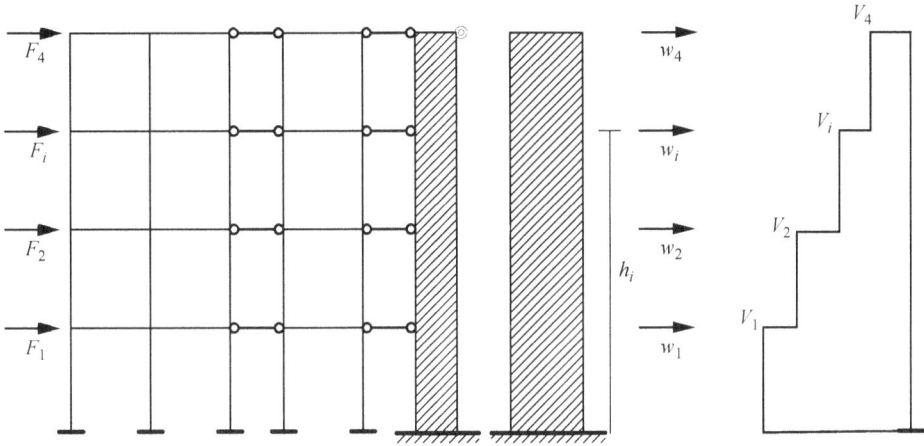

$$\Delta w_2 = V_2 / K^2$$

where K^2 is the lateral stiffness of level 2 being equal to K_1 or K_4 (Section 6.3.4) according to the direction of the acting force, P_i. Taking into account the fact that the fixed supports of this storey undergo a common displacement that is equal to the drift, w_1, it can be concluded that, according to the last relation, the total drift, w_2, at level 2 is equal to:

$$w_2 = \Delta w_2 + w_1 = V_2 / K^2 + V_1 / K^1$$

Following the same logic, it can be concluded that since $w_s = 0$, the total drift, w_i, at level i of a multi-storey frame can be expressed as:

$$w_i = \Sigma \frac{V_i}{K^i}$$

This expression allows the direct calculation of the displacement of a multi-storey system at any level i in the same sense as the corresponding result in Stavridis and Georgiadis (2025, Section 6.4.3).

Worked example

A four-storey system is considered with the layout of its supporting plane vertical elements made of concrete, as shown in Figure 6.16. The structure having a constant storey height of 3.50 m consists of four diaphragms considered rigid in their own plane and is subjected to equal horizontal forces acting on the centre of mass of each diaphragm. The example aims to evaluate the forces that are imposed on the three elements set out in the Oy direction on each level, assuming that each diaphragm receives in this direction a horizontal force 50 kN acting at the centre of each diaphragm. Furthermore, it proceeds to an estimation of the developed diaphragm shifts under the same horizontal force but acting at each level in the parallel direction to the Ox axis.

205

Figure 6.16 Four storey concrete system

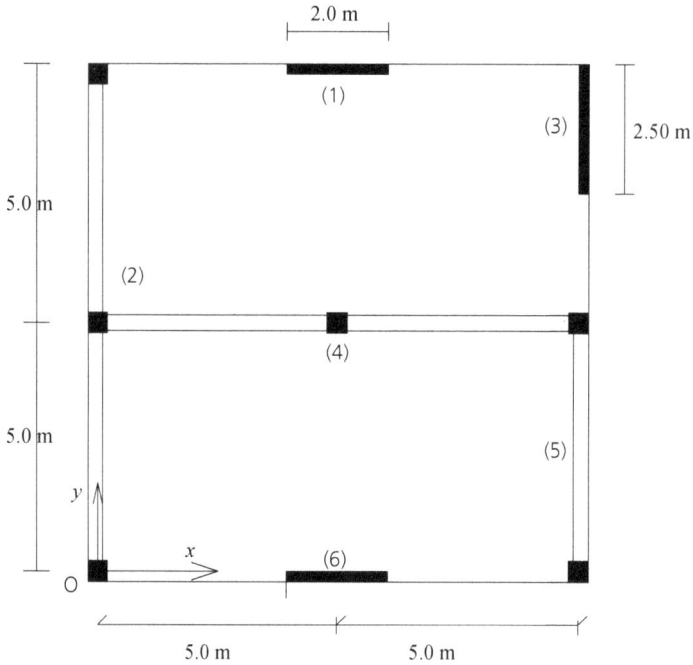

Given that the centre of mass where the external force 50 kN is acting at each level, it can be seen how the increasing discrepancy of this point from the respective centre of stiffness – that is, the increasing eccentricity of the applied force – at each level adversely affects the loading of the more remote elements as numbers 5 and 6.

Considering now the same horizontal force 50 kN acting at the centre of each diaphragm in the Ox direction, one can moreover assess the developing lateral shift at each level according to Stavridis and Georgiadis (2025, Section 6.4.3). The results can be seen in Table 6.3 below.

Table 6.1 Element coordinates and lateral stiffness at each level

Plane element	x	y	Lateral stiffness, S (kN/m)			
			Level 1	Level 2	Level 3	Level 4
1	5.0	10.0	279883	34985	10366	4373
2	0.0	5.0	47160	23580	6720	11790
3	10.0	8.75	546647	68330	20246	8541
4	5.0	5.0	47160	23580	6720	11790
5	10.0	2.50	30672	6336	10224	7668
6	5.0	0.0	279883	34985	10366	4373

Table 6.2 Diaphragm displacements and acting element forces

	Δx	Δy $\cdot 10^{-3}$ m	$\Delta \omega$ $\cdot 10^{-4}$ rad	f_2 (kN)	f_3 (kN)	f_5 (kN)	x_e^* (m)	y_e^* (m)
Level 1	~0	−0.172	0.273	−8.11	55.21	3.09	9.24	5.0
Level 2	~0	0.496	−0.039	11.70	31.22	7.0	7.80	5.0
Level 3	~0	1.42	−0.513	22.32	18.36	9.27	6.60	5.0
Level 4	~0	2.0	−0.438	23.58	13.34	11.98	5.78	5.0

Table 6.3 Lateral shifts due to a horizontal force 50 kN in the *Ox* direction

	Lateral shear (kN)	Level stiffness (kN/m)	Lateral displacement $\cdot 10^{-3}$ m
Level 4	50	20536	7.11
Level 3	100	36452	4.67
Level 2	60	93550	1.93
Level 1	200	606926	0.33

6.4. Temperature effect
6.4.1 Treatment of thermal action

Any temperature variation in a diaphragm always causes a certain stress state on the vertical elements. The reason is that an expansion or contraction of a diaphragm always imposes displacements on the respective level of the plane elements, which are, consequently, stressed (see Figure 6.17).

More specifically, assuming that a diaphragm is subjected to a temperature increase of ΔT by retaining the origin, O, of the coordinate axes absolutely fixed, some radial displacement will be imposed on all vertical plane elements, which will imply a certain movement, δ, along their trace (see Figure 6.18).

Due to this imposed displacement, the plane elements will be subjected to a certain force, F_0, according to their lateral stiffness that will be returned equal and opposite to the diaphragm itself (see Figure 6.18).

It is clear that the diaphragm under the above forces will generally not be in equilibrium as a free body and, therefore, at the point of fixation, O, an appropriate force and a moment have to be externally applied. In this respect, it is purposeful to select the characteristic point for each plane element, as specified in Section 6.3.2, at the middle of its respective trace. The action of these forces ensures the fixation of point O under the temperature variation of the diaphragm by maintaining its equilibrium (see Figure 6.18). However, given that in reality no external actions are applied at point O, equal and opposite actions to these must be additionally applied at this point in order to restore the real situation. Of course, these actions on the 'free' system will cause displacement of

Figure 6.17 Displacements of plane elements due to temperature variation of the diaphragm

Plan view

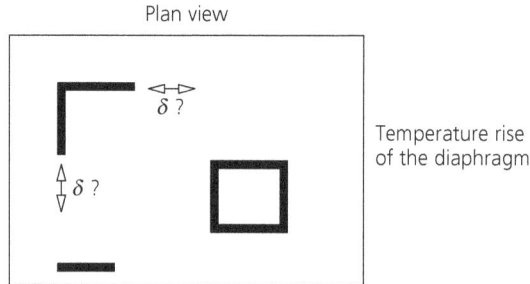

Temperature rise
of the diaphragm

Figure 6.18 Temperature variation under fixation of the diaphragm's origin of coordinates – first stage

Plan view

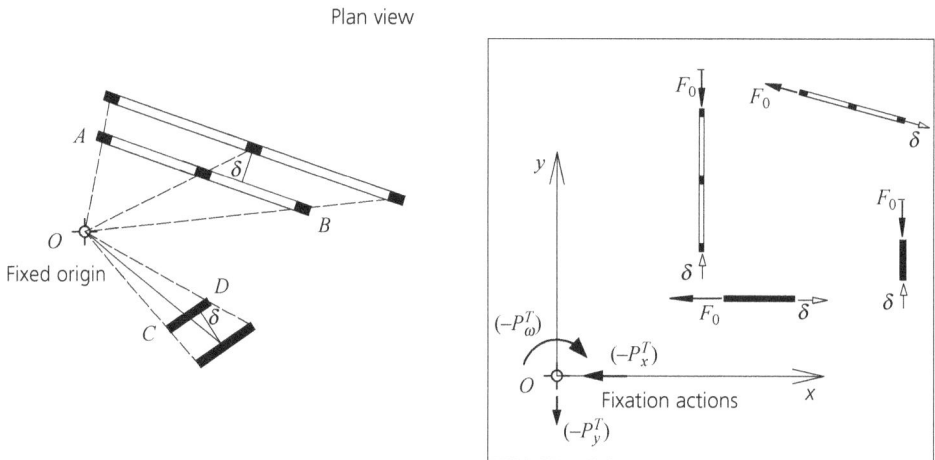

the diaphragm with the components Δx, Δy and $\Delta \omega$, which in turn will cause an additional stress state to the plane elements through the corresponding additional horizontal forces acting on their corresponding level (see Figure 6.19).

The final response of the vertical elements will result in a superposition of the two above stress states (Stavridis, 1994), namely:

- *first stage*: the state of stress resulting from the free dilatation of the diaphragm with its point O fixed, yielding the required applied external actions on this point in order not to be moved or rotated. (see Figure 6.18). This stress state is due to the imposed displacements on the specific level of the plane elements because of the radial dilatation of the diaphragm

- *second stage*: the state of stress and deformation resulting in the free system, through the imposition on the point O of the equal and opposite actions of previous stage (see Figure 6.19).

Figure 6.19 Imposition of the opposite fixation actions to the free system – second stage

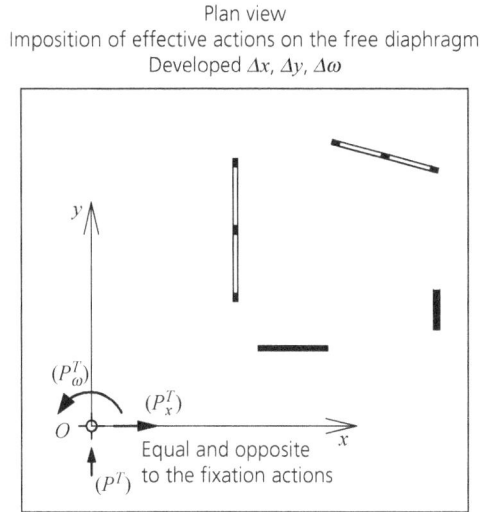

Figure 6.19 Imposition of the opposite fixation actions to the free system – second stage

6.4.2   General layout

According to the foregoing examination, and on the basis of Sections 6.2 and 6.3, it may be deduced that, for the specific level considered, the horizontal displacement, D_i, of the (i) plane element referred to a characteristic point (x_i, y_i) selected at the middle of its trace may be expressed, as follows, by conveying the influence of the two superposed states:

$$D_i = \left[\tau_i \cdot \alpha_T \cdot \Delta T \right] + \left[\Delta x \cdot \cos \beta_i + \Delta y \cdot \sin \beta_i + \Delta \omega \cdot e_i \right]$$

where $\tau_i = x_i \cdot \cos \beta_i + y_i \cdot \sin \beta_i$ and the temperature coefficient $\alpha_T = 10^{-5}/°C$.

The first term on the right side of the above expression gives the projection of the displacement vector of the characteristic point of plane element (i) on the element's trace itself. The second term expresses the displacement on the axis of the element imposed by the diaphragm movement by Δx, Δy and $\Delta \omega$, according to Section 6.3.2. The temperature variation, ΔT, is considered positive if the diaphragm's temperature increases and vice versa.

Thus, the force, F, acting on each plane element at the level examined is:

$$F = S_i \cdot D_i = S_i \cdot \alpha_T \cdot \tau_i \cdot \Delta T + S_i \cdot (\Delta x \cdot \cos \beta_i + \Delta y \cdot \sin \beta_i + \Delta \omega \cdot e_i)$$

The determination of Δx, Δy and $\Delta \omega$ results from the solution of an analogously structured system of equations as in the previous case of lateral loading:

$$K_1 \cdot \Delta x + K_2 \cdot \Delta y + K_3 \cdot \Delta \omega = P_x^T$$

$$K_2 \cdot \Delta x + K_4 \cdot \Delta y + K_5 \cdot \Delta \omega = P_y^T$$

$$K_3 \cdot \Delta x + K_5 \cdot \Delta y + K_6 \cdot \Delta \omega = P_\omega^T$$

209

In the present case of temperature influence, however, the right-hand members represent – with the opposite sign – the required external actions on the point O in order to fix it according to the above examined first stage. It is:

$$P_x^T = -\alpha_T \cdot \Delta T \cdot \Sigma(\cos \beta_i \cdot S_i \cdot \tau_i) \quad P_y^T = -\alpha_T \cdot \Delta T \cdot \Sigma(\sin \beta_i \cdot S_i \cdot \tau_i)$$

$$P_\omega^T = -\alpha_T \cdot \Delta T \cdot \Sigma(S_i \cdot \tau_i \cdot e_i)$$

6.4.3 Orthogonal layout

In the usual case where the plane elements are arranged in an orthogonal layout, the above expressions are simplified as follows. For the plane elements in the x-direction ($\beta_i = 180°$), it is:

$$\tau_i = -x_i \text{ and } e_i = y_i$$

whereas for the elements in the y-direction it is:

$$\tau_i = y_i \text{ and } e_i = x_i \quad \text{(see Sections 6.4.2 and 6.3.2, respectively)}$$

Then, it is:

$$P_x^T = -\alpha_T \cdot \Delta T \cdot \Sigma(S_x \cdot x_i), \quad P_y^T = -\alpha_T \cdot \Delta T \cdot \Sigma(S_y \cdot y_i)$$

and,

$$P_\omega^T = -\alpha_T \cdot \Delta T \cdot \Sigma(S_y \cdot x_i \cdot y_i - S_x \cdot x_i \cdot y_i)$$

while Δx, Δy and $\Delta \omega$ are determined from the linear system:

$$K_1 \cdot \Delta x + K_3 \cdot \Delta \omega = P_x^T$$

$$K_4 \cdot \Delta y + K_5 \cdot \Delta \omega = P_y^T$$

$$K_3 \cdot \Delta x + K_5 \cdot \Delta y + K_6 \cdot \Delta \omega = P_\omega^T$$

According to the expression of D_i in Section 6.3.2, it comes out that the elements in the x- and y-directions are subjected at the examined level to the following forces, respectively:

$$F_x = S_x \cdot D_i = S_x \cdot (x_i \cdot \alpha_T \cdot \Delta T + \Delta x - \Delta \omega \cdot y_i)$$

$$F_y = S_y \cdot D_i = S_y \cdot (y_i \cdot \alpha_T \cdot \Delta T + \Delta y + \Delta \omega \cdot x_i)$$

It is clear, however, that the consideration of an applied temperature difference at more than one diaphragm simultaneously may be treated by superposing the corresponding results for each level, exactly as in the case of lateral loading examined in Section 6.3.1. Thus, each plane element will be subjected to a corresponding force at each specific level where the temperature variation is applied.

However, from the above relations it may be concluded that, contrary to what happens in the case of horizontal loading, where a stiffness element can move along its stiffness direction without affecting the response, for thermal loading a stiffness element cannot move along its direction without changing the response.

As now a diaphragm under a temperature variation undergoes a movement imposing a certain horizontal displacement to every point in its plane, the question arises whether there is a point at all in the diaphragm's level that does not undergo any displacement and remains immovable at the same position as before the imposition of the temperature variation.

If the coordinate origin O is transferred to the centre of stiffness, according to Section 6.3.4 the above linear system is converted to the three simple equations in terms of Δx, Δy and $\Delta \omega$, respectively:

$$K_1 \cdot \Delta x = P_x^T$$

$$K_4 \cdot \Delta y = P_y^T$$

$$K_6 \cdot \Delta \omega = P_\omega^T$$

It can be seen that only when both terms P_x^T and P_x^T vanish – that is, $\sum (S_x \cdot x_i) = 0$ and $\sum (S_y \cdot y_i) = 0$ – then it may be concluded from the previous expressions in this section that the centre of stiffness on the diaphragm does not undergo any displacement and thus represents, indeed, the *point at rest* of the diaphragm under a temperature variation.

The forces F_x and F_y must in any case be taken into account during the design of the arrangement of the vertical elements (frames, shear walls etc.) of structures. They constitute exactly the criteria for the maximum building dimensions regarding a monolithic construction in order to withstand not only a possible temperature variation, but also the ever existing concrete shrinkage of the diaphragms (slabs), being equivalent to a temperature fall of about 25°C. The maximum length usually considered of 30 m, where the expansion joints are normally provided, must clearly always be adjusted for more or for less, depending in each case on the plane stiffness elements themselves, as well as on their specific layout in the ground plan.

It is thus clear that the design of the vertical members of a multi-storey building structure, apart from ensuring the safe uptake of any horizontal action either as seismic or wind loading, must also cover the temperature variation of the diaphragms.

REFERENCES

Stavridis L (1986) Static and dynamic analysis of multi-story systems. *Technika Chronika Scientific Journal of the Technical Chamber of Greece* **6(2)**: 187–219.

Stavridis L (1994) Beanspruchung von mehrstöckigen Bauten infolge Temperaturänderung. *Bauingenieur* **69(3)**: 117–122. (In German.)

Stavridis L and Georgiadis K (2025) *Understanding and Designing Structures without a Computer: Plane structural systems.* Emerald Publishing, Leeds, UK.

emerald
PUBLISHING

ice

Leonidas Stavridis and Konstantinos Georgiadis
ISBN 978-1-83662-945-0
https://doi.org/10.1108/978-1-83662-942-920251007
Emerald Publishing Limited: All rights reserved

Chapter 7
Dynamic behaviour of discrete mass systems

7.1.   Introduction

The static loads that all load-bearing structures have to take up may be considered as completely independent of the structure itself. This uptake of static loads is realised through the stiffness of the structure, depending on its geometric and material characteristics. The closed process [Load action↔Uptake through stiffness] is accomplished through equilibrium – that is, the immobility of the structure. In this static process, the *mass, m*, of the structure obviously remains inactive. It is activated, however, as a factor of generating loads according to Newton's second law of motion ($F = m \cdot \gamma$) from the moment that the structure is somehow set in motion. There are many factors that can set a structure in motion (being itself connected to the ground) such as wind forces, earthquake ground motions, 'sudden' load applications, moving loads (e.g. vehicles or pedestrians on road or footbridges, respectively) and so on.

The problem created through the motion of the structure's mass is that it implies accelerations (γ) leading, according to *d'Alembert's principle*, to inertial forces equal to $-m \cdot \gamma$. The structure is thus stressed under these loads, which may clearly be treated as static loads in the usual manner. However, the problem lies in the fact that the above accelerations, which certainly vary in time, are unknown right from the beginning. As a matter of fact, they depend on the masses and the stiffness of the system and the determination of their variation in every localised mass of the structure and at every instant of time constitutes the actual dynamic problem.

Thus, the whole dynamic problem may be put as follows: the load-bearing system consists of specific masses that are connected to each other as well as to the ground through a massless *structural web*, which, apart from its own geometric layout, exhibits specific stiffness characteristics (see Figure 7.1). When the structure is set in motion, the inertia forces of the masses ($-m \cdot \gamma$) on the one hand and the elastic forces of the structural web on the other are activated, depending on its stiffness characteristics and acting on the corresponding masses. The solution to the problem of determination of the acceleration, (γ), of the masses results from the requirement of equilibrium for each mass under the action of inertia as well as the elastic forces. However, the difficulty in determining the motion of the masses comes from the fact that, while the elastic forces are generally proportional to the displacements of the masses, the inertia forces are proportional to the second derivative of these displacements with respect to time.

Structures constitute a priori continuous systems in the sense that their mass is continuously distributed and not concentrated at discrete locations. Nevertheless, it is always attempted to consider lumped masses on selected locations equivalently to the real mass distribution based on the structural behaviour as the analytical treatment of the continuous mass distribution is mathematically much more cumbersome. ■

Figure 7.1 Masses in motion and the structural web

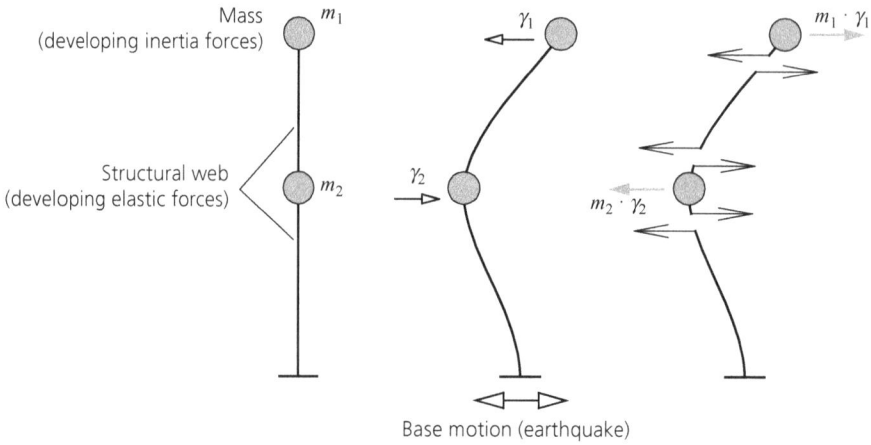

Base motion (earthquake)

Figure 7.2 Effective mass displacement possibilities (degrees of freedom)

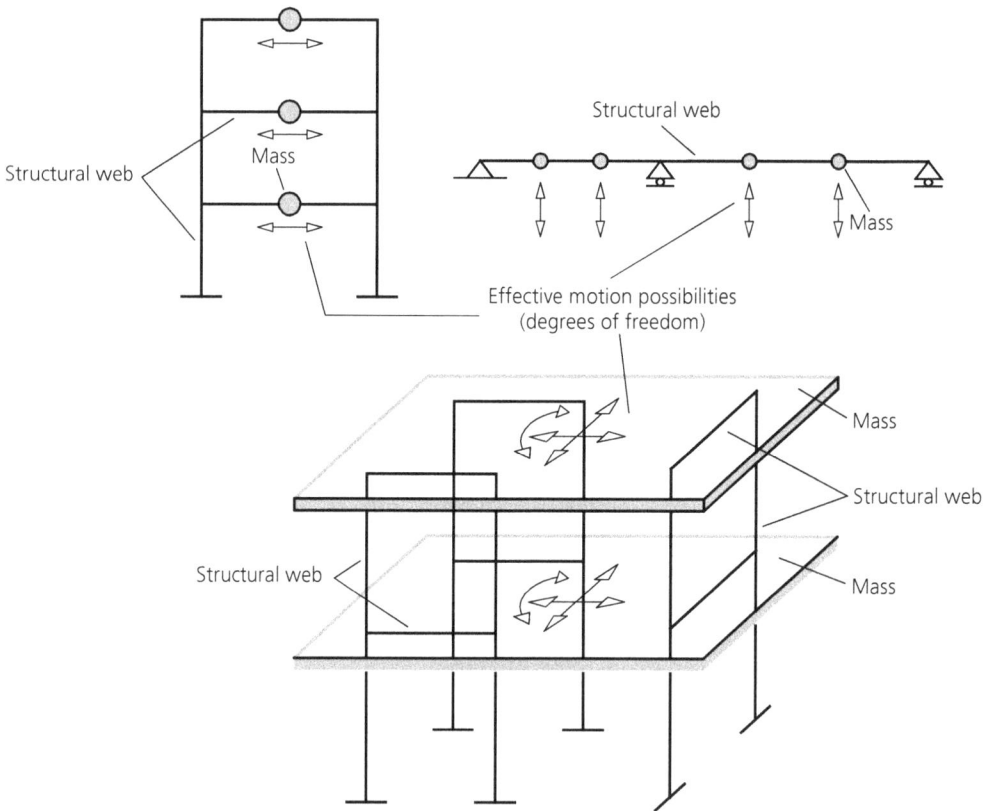

The possible motions of the discrete (lumped) masses generally determine the inertia forces they are subjected to and from this point of view the latter are distinguished between *translational* and *rotational inertia forces*. In the present examination – which is, however, oriented to preliminary design purposes – the inertia forces due to the rotation of the masses are ignored, except for the horizontal diaphragms of multi-storey buildings undergoing an in-plane rotation, $\Delta\omega$, as explained in Chapter 6, Section 6.3.1.

Thus, in the plane frame shown in Figure 7.2, where its masses are assumed to be – approximately – concentrated at each storey, in order to evaluate their lateral excitation, only their horizontal displacement is considered, whereas their rotations about an axis perpendicular to the frame's plane, as well as their vertical displacements, are ignored as a source of inertia forces.

Similarly, in the continuous beam shown in Figure 7.2, where its masses are considered concentrated to some specific points, to evaluate their vertical excitation, only their vertical displacement is considered, whereas their rotation about their own axis is ignored as a source of inertia forces.

Finally, in the two-storey spatial system shown in Figure 7.2, where the mass is assumed equally distributed in each horizontal diaphragm, in order to examine a horizontal seismic excitation, only the horizontal displacement components (Δx, Δy) producing inertia force in the directions x and y, as well as the rotation, $\Delta\omega$, producing an inertia moment at each storey are considered.

In this way, each system is characterised on the basis of its effective displacements, which are usually referred to as *degrees of freedom*. Thus, in the above examples, the frame, the continuous beam and the two-storey system have three, four and six degrees of freedom, respectively.

Despite the fact that load-bearing structures usually exhibit more than one degree of freedom, the study of a single-degree-of-freedom system is absolutely essential in order to understand all the essential characteristics and the dynamic behaviour of multi-degree-of-freedom systems. It is necessary then to prefix the elementary examination and comprehension of the dynamic behaviour of a single mass with only one 'active' displacement, given also that this model applies directly to the practical treatment of more complex systems regarding also those with continuous mass distribution, as will be shown in Sections 7.5 and 7.6.

7.2. Single-degree-of-freedom systems
7.2.1 Dynamic equilibrium

As a single-degree-of-freedom system, a mass, m, connected to the free end of a cantilevered bar of length, L, is examined. Considering that the bending stiffness of the bar is much less than its axial stiffness, the only possible movement of the mass is transversely to the bar's axis. In this sense, the stiffness, k (kN/m), of the bar is considered with respect to the displacement of its free end (see Figure 7.3), being equal to $k = 3 \cdot EI/L^3$ (see Stavridis and Georgiadis, 2025, Section 3.2.3).

It is supposed that the mass is acted on by a horizontal force function $F(t)$ and that at a specific moment, t_1, it undergoes a displacement $u(t_1)$ in the same direction.

Moreover, it is supposed that the mass receives an additional resistance to its motion due to a sort of *internal friction* of the bar's material, which may be considered proportional to the mass's velocity $\dot{u}(t)$, being thus equal to $c \cdot \dot{u}(t)$. The coefficient, c, is called the *damping coefficient*, as by its nature it tends to dampen the motion of the mass.

Figure 7.3 Dynamic equilibrium of a single mass on a cantilevered structural web

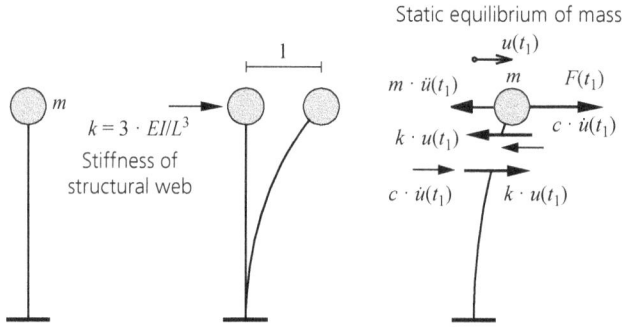

Thus, at the time, t_1, the mass, m, according to d'Alembert's principle, may be considered in equilibrium under the following forces

- the external force $F(t_1)$ in the same direction as $u(t_1)$
- the elastic force $k \cdot u(t_1)$ in the opposite direction to $u(t_1)$
- d'Alembert's inertia force $m \cdot \ddot{u}(t_1)$ in the opposite direction to $u(t_1)$
- the damping force $c \cdot \dot{u}(t_1)$ in the opposite direction to $u(t_1)$

The static equilibrium of the mass requires (see Figure 7.3):

$$m \cdot \ddot{u}(t_1) + k \cdot u(t_1) + c \cdot \dot{u} = F(t_1)$$

It has to be pointed out that the elastic force $k \cdot u(t_1)$ coincides with the force needed to shift the mass, m, by $u(t_1)$, according to the stiffness of its structural web.

7.2.2   Free vibration

7.2.2.1   Undamped vibration

As the dynamic characteristics of the system are independent of the external loading, their investigation is made easier if the force $F(t_1)$ is removed and thus the free vibration of the mass, m, is considered, after an initial arbitrary displacement has been imposed at the rest position. Ignoring the factor of *damping* for the time being, it may be written for any instant of time, t_1, as:

$$m \cdot \ddot{u}(t_1) + k \cdot u(t_1) = 0$$

To determine the displacement function, u, the kinematic conditions of mass, m, at the starting time are needed – that is, its displacement $u_0 = u(t = 0)$, as well as its velocity $\dot{u}_0 = \dot{u}(t = 0)$.

The solution of the above equation shows that the mass oscillates continuously in a periodic manner about the 'rest' position with a deviation $u(t)$ varying sinusoidally with respect to time (see Figure 7.4).

The time needed for a full oscillation of the mass, after which the motion is repeated in exactly the same way, is called the *natural period, T* (s). The inverse magnitude of the period is called the *frequency, f*, and it is $f = 1/T$ (1/s). The frequency, f, represents the number of full oscillations that

Figure 7.4 Free vibration of a single mass

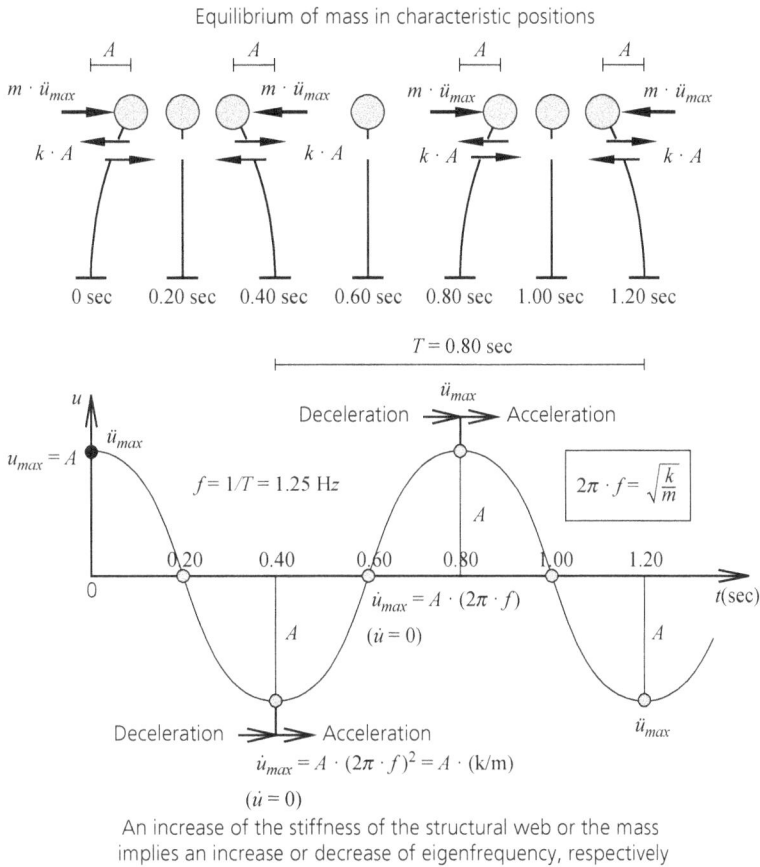

Equilibrium of mass in characteristic positions

$T = 0.80$ sec

$f = 1/T = 1.25$ Hz

$2\pi \cdot f = \sqrt{\dfrac{k}{m}}$

$\ddot{u}_{max} = A \cdot (2\pi \cdot f)$

$(\dot{u} = 0)$

$\ddot{u}_{max} = A \cdot (2\pi \cdot f)^2 = A \cdot (k/m)$

$(\dot{u} = 0)$

An increase of the stiffness of the structural web or the mass implies an increase or decrease of eigenfrequency, respectively

take place within a second, the unit of this quantity being expressed in Hertz. It is also common in vibration studies to use the term *circular frequency* instead of frequency, f, denoting the magnitude, ω, on the basis of the relation:

$$f = \frac{\omega}{(2\pi)}$$

The *natural frequency* of a system depends on its mass and stiffness and determines its dynamic behaviour under the application of external forces. It is found that:

$$\omega = \sqrt{\frac{k}{m}} \quad (\text{rad}/\text{s})$$

and, consequently:

$$f = \frac{1}{2\pi}\sqrt{\frac{k}{m}} \quad (1/\text{s} = \text{Hz})$$

217

Figure 7.5 Dynamic equilibrium of a mass carried by a spring

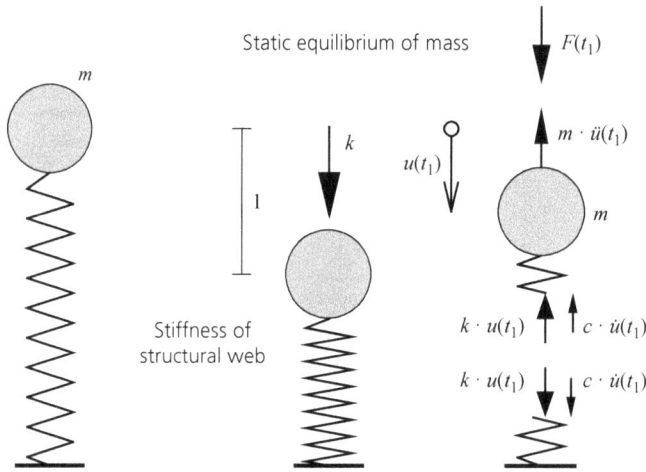

It is pointed out here that the mass unit commonly used for structures is the ton (t) being equal to 1000 kg. It is $1\,t = 1\,kN \cdot s^2/m$ and, given a weight, W, in kN, its mass, m, is $m = W/9.81$ t.

As seen from the last equation, the greater the mass, the fewer vibrations are accomplished in 1 sec, whereas the stiffer the connection of the mass to the structural web, the higher the system's frequency. However, the presence of the square root attenuates the effect of the above parameters.

Of course, the frequency, f, of the system is independent of both the initial deviation, u_0, and the initial velocity, $\dot{u}_0$, of mass, m.

The mass, m, executes a steadily varying, oscillating motion (see Figure 7.4). Its displacement $u(t)$ is expressed as:

$$u(t) = \frac{\dot{u}_0}{\omega} \cdot \sin(\omega t) + u_0 \cdot \cos(\omega t)$$

Its maximum deviation, A, on both sides, also called the *amplitude*, is:

$$A = \sqrt{u_0^2 + \left(\frac{\dot{u}_0}{\omega}\right)^2}$$

At its maximum deviation ($u = \pm A$), the mass has zero velocity and maximum acceleration (or deceleration) $\ddot{u}_{max} = A \cdot \omega^2$; whereas at the *rest position* ($u = 0$), it has the maximum velocity $\dot{u}_{max} = A \cdot \omega$.

Moreover, it should be pointed out that the elastic forces $k \cdot u(t_1)$ have the opposite sense to the corresponding d'Alembert forces $m \cdot \ddot{u}(t_1)$, as is also implied by the basic equation itself. More specifically, as the inertia force $m \cdot \ddot{u}(t_1)$ is opposite to the acceleration vector $\ddot{u}(t_1)$, just before the maximum deviation, A, the inertia force is opposite to the maximum deceleration developed, $\ddot{u}_{max}$,

218

whereas just after this position the inertia force is opposite to the equal acceleration, $\ddot{u}_{max}$. This means that in both cases the inertia force is directed to the right – that is, in the opposite direction to the acting elastic force ($k{\cdot}A$) (see Figure 7.4). Finally, from the equilibrium relation $m{\cdot}\ddot{u}_{max} = k{\cdot}u_{max} = k{\cdot}A$, it may be derived that that $\ddot{u}_{max} = A{\cdot}\omega^2$. ∎

It should be made clear that the selection of the specific cantilever-type single mass system is just indicative of a system with mass, m, and stiffness of structural web, k, subjected to an external force, $F(t)$, corresponding to its active stiffness, k. In this sense, the system consisting, for example, of a mass connected to a spring, as shown in Figure 7.5, may be equally representative, provided of course that the self-weight of the mass, m, is ignored.

However, it should be recognised that while the cantilevered mass is offered as a more illustrative model for the examination of systems consisting of a number of masses arranged in height such as in buildings, the mass–spring model is more representative for the examination of vertical dynamic actions on existing structures like beams or slabs, such as through a machine permanently fixed in place, as shown in Figure 7.6.

Figure 7.6 Free vibration of a mass–spring system

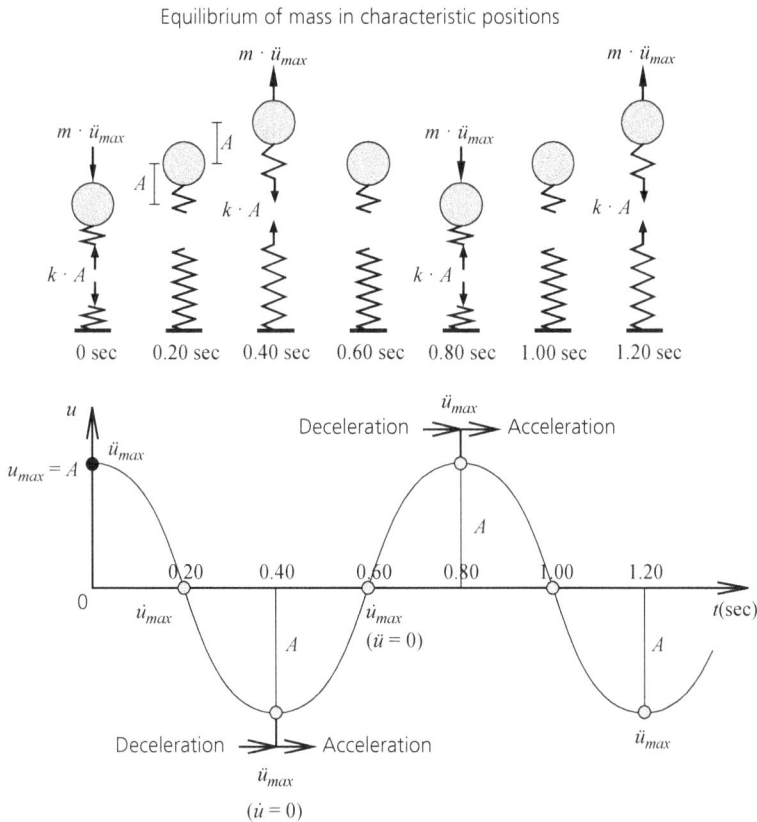

7.2.2.2 Damped vibration

Consideration of the damping coefficient, c, in the following equation:

$$m \cdot \ddot{u}(t) + k \cdot u(t) + c \cdot \dot{u}(t) = 0$$

does not alter the periodicity of the oscillation of the mass. However, it causes a continuous decrease of the amplitude at each vibration cycle. It is pointed out that the system vibrates with a somewhat increased period, T_d, which remains constant (see Figure 7.7). It is found that the corresponding decreased circular frequency, ω_d, is equal to:

$$\omega_d = \sqrt{\omega^2 - \left(\frac{c}{2m}\right)^2}$$

Damping represents the dissipation of energy in each vibration cycle. This energy is absorbed by the internal friction emerging in the means carrying the masses – that is, the structural web – which offers the stiffness to the system. Thus, the damping force $c \cdot \dot{u}(t)$ acts on the mass through the structural web as a restraining force in the same direction as the acting elastic force. Of course, the direct result of damping is, as previously mentioned, the continuous decrease of the vibration's amplitude.

Figure 7.7 Characteristics of damped vibration

The ratio λ of two consecutive amplitudes is constant:
$(A_1/A_3) = (A_2/A_4) = \lambda$
Damping coefficient $\zeta = \ln(\lambda)/2\pi$ $\quad \lambda = e^{2\pi\zeta}$
The presence of damping $\zeta < 10\%$ affects the oscillation period very little
(Practically: $f = \sqrt{k/m}/2\pi$)

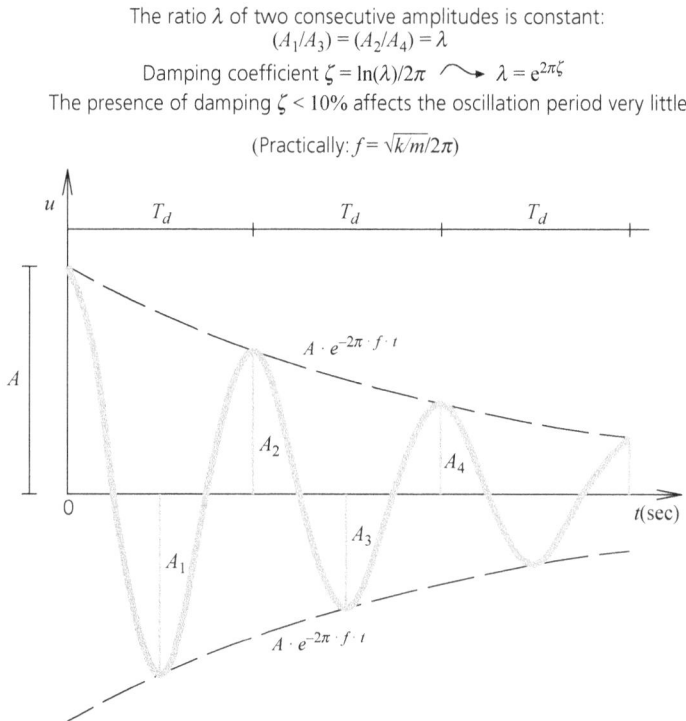

The description of damping as a force proportional to the mass's velocity is not satisfactory from a physical point of view. Nevertheless, it facilitates – as will be shown later – the practically effective quantitative description of damping, thanks to a more convenient mathematical approach involved.

Thus, it is found that if the damping coefficient, c, achieves the critical value $c_{cr} = 2 \cdot m \cdot \omega$, then no vibration can be developed and the mass comes from its maximum deviation position directly to rest. Although it is impossible for such a situation to happen in a structure, its consideration is useful because it represents a reference level for the applied damping coefficient in each case.

At first, the so-called *damping ratio*, ζ, is defined as the ratio of damping coefficient to its critical value $\zeta = c/c_{cr}$. On the basis of this relation, it is:

$$\zeta = c / (2 \cdot m \cdot \omega)$$

and, consequently,

$$\omega_d = \omega \cdot \sqrt{(1 - \zeta^2)}$$

The ratio, ζ, despite the fact that it refers to the unrealistic concept of the coefficient, c, has an essential physical meaning in the reduction of amplitudes. Thus, from the last equation, the magnitude of reduction, δ, between two consecutive amplitudes of the decaying vibration, with a *time distance, T_d*, is:

$$\delta = \ln \frac{u(t)}{u(t + T_d)} = \left(\frac{c}{2m}\right) \cdot \frac{2\pi}{\omega} = \frac{2\pi\zeta}{\sqrt{1 - \zeta^2}}$$

The last equation has a particularly practical importance because the value of ζ may be directly deduced from the measurement of the consecutive amplitudes of the damped vibration of a system exhibiting a constant ratio. As this value does not usually exceed 10% for structural systems encountered practically, it may be considered with a good approximation as:

$$\omega_d = \omega \left(\text{hence } T_d = T\right) \text{ on one side and } \delta = 2 \cdot \pi \cdot \zeta \text{ on the other}$$

In real situations, however, for practical reasons as well as for accuracy, the amplitudes of two vibration cycles are measured having a time distance rather equal to $n \cdot T$, in which case it is (see Figure 7.8):

$$\delta_n = \ln \frac{u(t)}{u(t + n \cdot T)} = n \cdot \delta = 2\pi\zeta \cdot n$$

Hence,

$$\zeta = \frac{\delta_n}{2\pi \cdot n}$$

Finally, the expression of the displacement $u(t)$ of a mass, m, which, after been displaced by A, is left to vibrate freely with a zero initial velocity under a damping ratio ζ ($\omega_d \approx \omega$) is:

$$u(t) = \frac{A}{e^{\zeta\omega t}} \cdot \left[\zeta \cdot \sin(\omega t) + \cos(\omega t)\right]$$

Figure 7.8 Practical determination of damping ratio

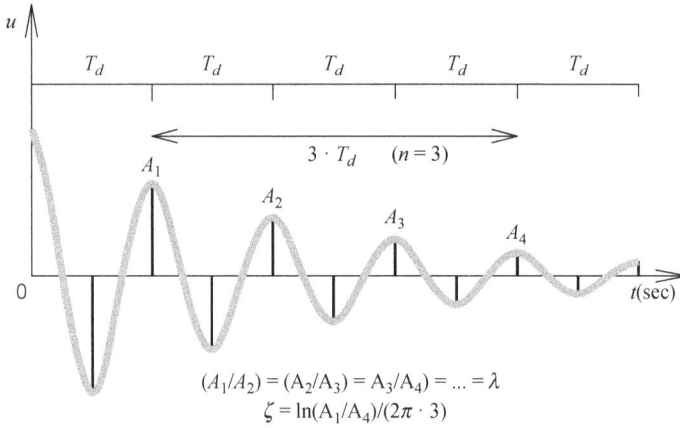

$$(A_1/A_2) = (A_2/A_3) = A_3/A_4) = ... = \lambda$$
$$\zeta = \ln(A_1/A_4)/(2\pi \cdot 3)$$

7.2.3   Forced vibration

The previous analysis of free vibration aimed to depict the basic dynamic characteristics of a single-degree-of-freedom system. Understanding these characteristics is certainly necessary for the comprehension of the dynamic response of the system if the mass is acted on by a force varying in time – that is, exactly as initially considered according to the equation of motion (Müller and Keintzel, 1984):

$$m \cdot \ddot{u}(t) + k \cdot u(t) + c \cdot \dot{u}(t) = F(t)$$

whereby, after the introduction of the damping ratio, ζ, it is:

$$c = 2\zeta\, m \cdot \omega$$

It is useful to express the force $F(t)$ in the form:

$$F(t) = F_1 \cdot f(t)$$

where $f(t)$ describes the variation of the force during the whole period of the examined time, and F_1 is the maximum value of this force.

For the design of a structure, the response of the structural web connecting the mass to the ground is of primary interest. This response depends only on the displacement, u. It is clear that the maximum response corresponds to the maximum displacement, u_{max}, which is clearly greater than that corresponding to the *static* loading of the system with the force, F_1, this being equal to:

$$u_{stat} = F_1 / k$$

Thus, if in a forced vibration the ratio DMF = u/u_{stat} is defined as the *dynamic multiplication factor*, then its maximum value (DMF)$_{max}$ = (u_{max}/u_{stat}) is of great importance for design.

Of course, the response of the system under the loading $F_1 \cdot f(t)$ is fully described by the exact solution of the above differential equation. This solution reflects the fact that the final response is the result of a superposition of the free vibration and the effect of the force imposition on the system, being expressed by the so-called *Duhamel integral*, as shown below. Thus, the general solution of this equation is (Biggs, 1964):

$$u(t) = e^{-\zeta \omega t}\left(u_0 \cdot \cos\left(\omega_d t\right) + \frac{\dot{u}_0 + \zeta \cdot \omega \cdot u_0}{\omega_d}\sin(\omega_d t)\right) + \frac{F_1}{m \cdot \omega_d}\int_0^t f(\tau)\cdot e^{-\zeta\omega(t-\tau)}\cdot\sin\left(\omega_d(t-\tau)\right)d\tau$$

Obviously, the direct use of this solution is not practical. Moreover, the use of special software also dealing with systems of multiple degrees of freedom allows the acquisition of the desired results through a direct numerical approach rather than through the above analytical approach. Nevertheless, the above expression is very useful because it permits the direct analysis of some typical cases and allows particularly important conclusions to be obtained for the response of single-degree and, subsequently, of multi-degree-of-freedom systems.

A few examples of such typical cases are listed below.

- The 'abrupt' application of force, F_1, followed either by its equally 'abrupt' removal after some period, t_d, or by an immediate linearly varied decrease during the same period, t_d.
- The progressive application of the force up to the value F_1 and, consecutively, its decrease down to zero over a total time period, t_d, first in a linear and second in a parabolic manner.
- The progressive linearly increased application of force, F_1, over a time period, t_d, followed by maintenance of this acquired value for an unlimited time.

In all the above cases, the damping effect is ignored for simplicity.

In the curves shown in Figure 7.9, the behaviour of the maximum value of the dynamic magnification factor $(\text{DMF})_{max}$ is depicted for each of the above cases as a function of the ratio of the characteristic time duration, t_d, to the natural period, T, of the vibrating system.

From curve (a) it is seen that in the case of a suddenly applied force, when time reaches $(T/2)$, the factor $(\text{DMF})_{max}$ takes its maximum value of 2, which is not surpassed by any other case. This value is also valid for any t_d greater than $(T/2)$. Otherwise, for a duration, t_d, less than about $(T/5)$, the factor $(\text{DMF})_{max}$ is less than unity – that is, the maximum deviation of the mass is even less than $u_{stat} = F_1/k$. This is attributed to the fact that the analytic expression of DMF for this case for $t < t_d$ comes out as $\text{DMF} = 1 - \cos(\omega t)$ and thus $\text{DMF} < 1$ for $\omega t < \pi/2$.

On the other hand, the immediately occurring linear decrease of the suddenly applied force has more moderate results. Only after an unloading duration, t_d, of about ten times the natural period, T, does the factor $(\text{DMF})_{max}$ begin to approach the value of 2.0, while for a duration less than $(T/3)$ the maximum deviation is less than the static one, u_{stat}.

From curve (b) it is seen that the maximum value of $(\text{DMF})_{max}$ – that is, the maximum mass deviation – occurs when the total loading–unloading duration approaches 90% of the natural period, T. This value equals 1.8 or 1.5 for a parabolic or linear force variation, respectively. For lower or greater durations, t_d, the maximum value of $(\text{DMF})_{max}$ is always lower.

Figure 7.9 Behaviour of a single mass system under a dynamically applied load

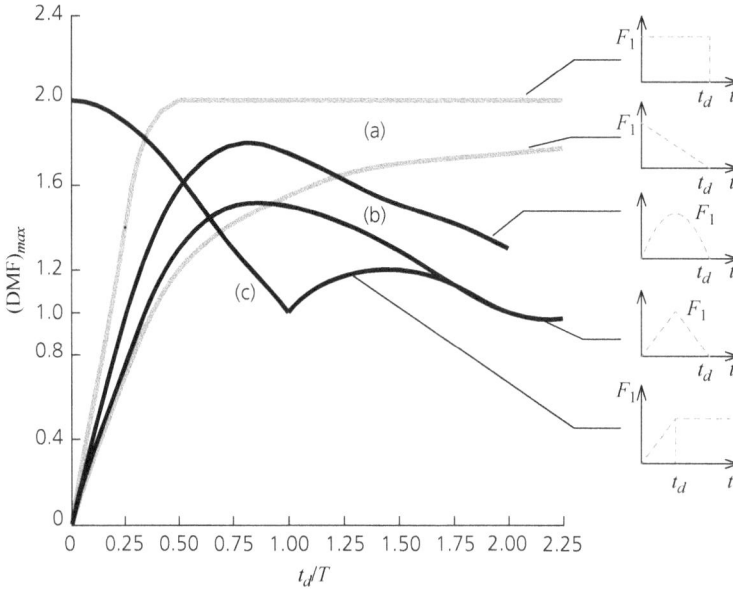

Curve (c) represents the usual force application scheme that is encountered in many practical cases. On the other hand, the more the duration, t_d, falls below T, the higher the value of $(DMF)_{max}$, never exceeding, however, the absolute maximum value of 2.0, no matter how 'abrupt' the loading process is. Nevertheless, for any duration, t_d, greater than T, the maximum value of factor DMF does not exceed the value 1.2. The above results confirm the definition of static loading, as explained in Stavridis and Georgiadis (2025, ref. Chapter 1).

7.2.4   Periodic sinusoidal applied force

Of particular importance for the design of structures is the case in which the acting force $F(t)$ acts on the system in a periodic manner, according to the expression $F(t) = F_0 \cdot \sin(\Omega t)$, where Ω is the frequency of the applied load. The examination of the response under a periodic load permits one to draw useful conclusions about any kind of periodic excitation such as machinery effects, human rhythmic activities, seismic action and so on, as well as about measures required for the attenuation of their effects (dynamic isolation).

By applying the general solution of the previous section in the present case and ignoring the effect of damping, the following expression is obtained:

$$u(t) = U \cdot \sin(\omega t + \varphi) + \frac{F_0 / k}{1 - (\Omega / \omega)^2} \sin(\Omega t)$$

The first term on the right-hand side is essentially identical to the respective expression of the general solution, where:

$$U = A \quad \text{and} \quad \tan \varphi = \frac{u_0}{\dot{u}_0 / \omega}$$

As already mentioned, this term represents that part of the response corresponding to the free vibration of the system, whereas the second term is exclusively due to the externally applied force.

Now, as the consideration of even the least possible damping makes the free vibration (i.e. the first term) vanish quickly, the conclusion may directly be drawn that the maximum value of the dynamic magnification factor $(DMF)_{max}$ is:

$$(DMF)_{max} = \left(\frac{u}{u_{stat}}\right)_{max} = \frac{1}{\left|1-(\Omega/\omega)^2\right|}$$

It is clear that the ratio $r = \Omega/\omega$ plays an important role. For values less than 0.2 and greater than 1.5, the dynamic influence is not important. More specifically, while for values of r tending to zero (static loading) the $(DMF)_{max}$ tends to 1, for high frequency excitations ($r > 2.50$) this factor tends towards zero. These correlations are clearly shown in Figure 7.10. However, as may be deduced from the equation above, when $\Omega = \omega$ then the value of the factor $(DMF)_{max}$ goes to infinity. Actually, what happens in this case is that in each cycle the factor (DMF) steadily increases in time, according to the relation:

$$(DMF)_{\Omega=\omega} = (\sin(\omega t) - \omega t \cdot \cos(\omega t))$$

On the basis of this equation, it is obtained that:

$$(DMF)_{max,\Omega=\omega} = \frac{u_{max}}{F_0/k} = \frac{1}{2}n \cdot \pi$$

where n is the ever-increasing number of cycles.

The situation that occurs in the case in which $\Omega = \omega$ is called *resonance* and is something that should, of course, be avoided. However, as the infinite value of (DMF) corresponds to a system without any damping effect, it should be pointed out that this does not happen in reality, as even a small amount of damping action reduces the corresponding response significantly.

More precisely, for an existing damping factor equal to ζ, it is obtained (Biggs, 1964) that:

$$(DMF)_{max} = \frac{1}{\sqrt{\left(1-(\Omega/\omega)^2\right)^2 + 4\cdot\zeta^2\cdot(\Omega/\omega)^2}}$$

whereby the restrictive contribution of ζ to the response is clear (see Figure 7.10).

Thus, in the case of resonance ($\Omega = \omega$), it is $(DMF)_{max,\,\Omega=\omega} = 1/(2\zeta)$ and, consequently:

$$u_{max} = \frac{1}{2\zeta}\cdot\frac{F_0}{k}$$

However, until the system develops the above maximum deviation of mass, a large number of cycles is needed. ∎

Figure 7.10 Behaviour of the system towards a periodic load action

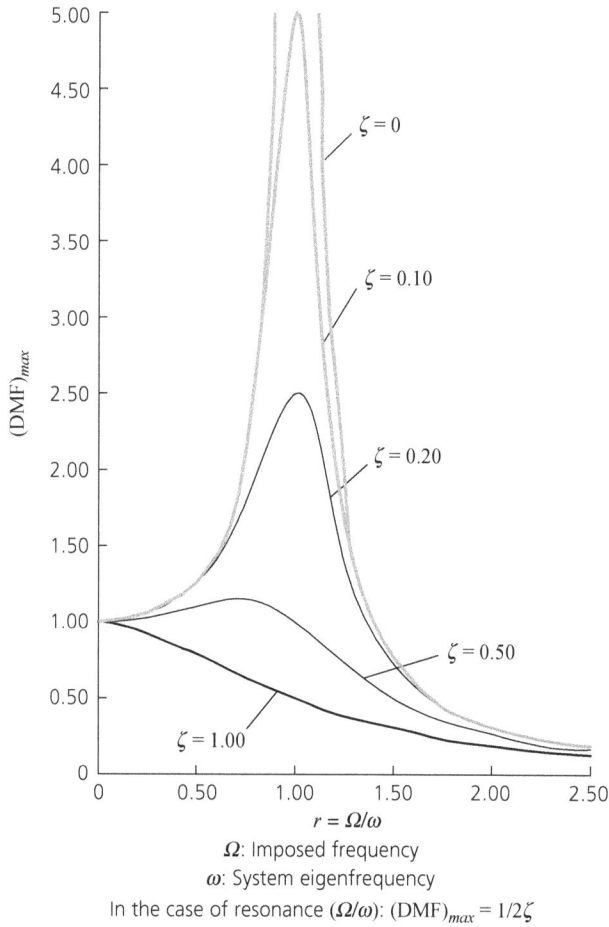

Ω: Imposed frequency
ω: System eigenfrequency
In the case of resonance (Ω/ω): $(DMF)_{max} = 1/2\zeta$

Also of particular interest is the assessment of the forces transmitted to the base of the structure – that is, to the support of the structural web.

As can be seen from the examination of the system's dynamic equilibrium (see Section 7.2.1 and Figure 7.11), the transmitted force, F_g, at the structural web's base consists of its elastic force ($k \cdot u$) plus the damping force ($c \cdot \dot{u}$) acting in the same direction.

It is found that the maximum transmitted force, $maxF_g$, to the base, may be obtained according to the relation:

$$maxF_g = F_0 \cdot \left[\frac{1 + 4 \cdot \zeta^2 \cdot r^2}{(1 - r^2)^2 + 4 \cdot \zeta^2 \cdot r^2} \right]^{(1/2)}$$

Figure 7.11 Transmission of dynamic actions at the base of a structural web

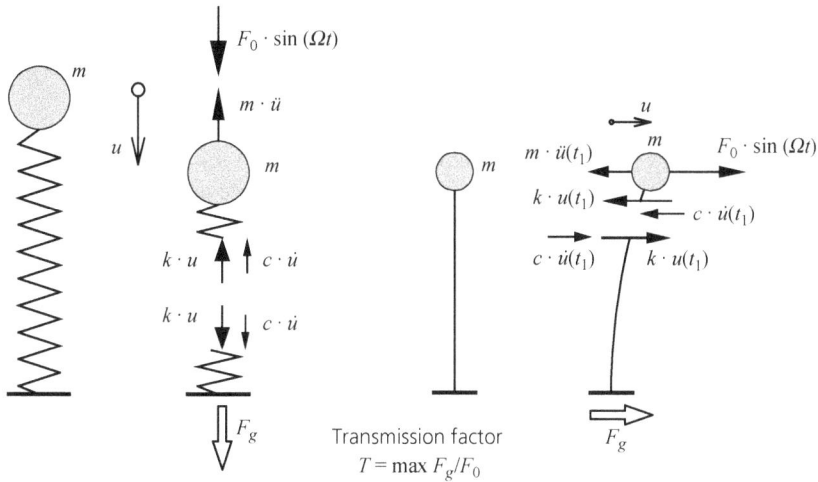

Transmission factor
$$T = \max F_g/F_0$$

The ratio $T = \max F_g/F_0$ is called the *transmission factor* and its dependence on the ratio $r = \Omega/\omega$ as well as on the damping factor is depicted in Figure 7.12.

According to Figure 7.12, given the frequency, Ω, of an imposed load – for example, due to an installed machine – by providing to the system (through appropriate means) a natural circular frequency, ω, which is less than about $\Omega/2.50$, the transmitted force to the base is drastically limited.

Figure 7.12 Transmissibility of dynamic actions to a system's base

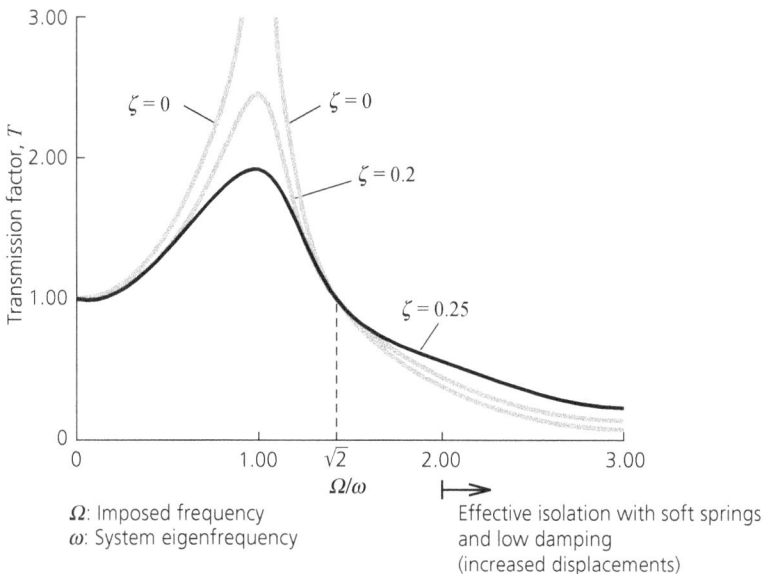

Ω: Imposed frequency
ω: System eigenfrequency

Effective isolation with soft springs and low damping (increased displacements)

7.2.5   Seismic excitation

Seismic excitation consists in an imposed oscillating motion on a structure's support. The cause of this motion is the abrupt (brittle) rupture in some region of an already existing fault in the Earth's crust as a result of accumulated stresses. This rupture is instantly spread out along the fault, causing propagation of a vibration wave within the Earth's crust lasting for a period of about 15 to 30 sec and then coming to a rest.

At first, it is assumed that the imposed horizontal ground motion $u_s(t)$ on the structure's support may be expressed as $u_s(t) = u_{s,0} \cdot f(t)$, where $u_{s,0}$ is the maximum value of the ground displacement in the support's region. It is understood that in the present case of a single-mass system, the direction of u_s corresponds to that of the stiffness, k, of the system (see Figure 7.13).

It is clear that the elastic force acting on mass, m, will be due to the mass's relative displacement to the ground expressed as $(u - u_s)$. Given that mass, m, is not acted on by any external load but is only subjected to the inertia force which is obviously referred to its absolute displacement, u, as well as to the elastic force, it may be written, if the damping force is ignored, as:

$$m \cdot \ddot{u}(t) + k \cdot \left[u(t) - u_s(t) \right] = 0$$

The absolute displacement $u(t)$ of the mass, m, cannot describe the deformation of the structure depending, as pointed out, exclusively on the relative displacement:

$$x(t) = u(t) - u_s(t)$$

Then, substituting the last equation the following is obtained:

$$m \cdot \left(\ddot{x} + \ddot{u}_s \right) + k \cdot x = 0$$

It is clear now that the response $(k \cdot x)$ of the structure depends solely on the ground acceleration $\ddot{u}_s(t)$ (see Figure 7.13).

Thus, the characteristic of an earthquake that is relevant for the design of a structure is only the so-called *accelerogram*, which is recorded by special equipment each time. For a specific earthquake and within a major region, different accelerograms may be recorded for different places, depending on the distance from the respective source of the event (epicentre) as well as on the existing soil properties all along the path followed by the seismic waves to the place recorded.

The accelerogram $\ddot{u}_s(t)$ may be expressed as:

$$\ddot{u}_s(t) = a \cdot f_a(t)$$

where a is the maximum occurring ground acceleration and the function $f_a(t)$ describes its evolution in time.

Now, the last equation may be written as:

$$m \cdot \ddot{x} + k \cdot x = -(m \cdot a) \cdot f_a(t)$$

Figure 7.13 Imposition of seismic ground motion and dynamic equilibrium

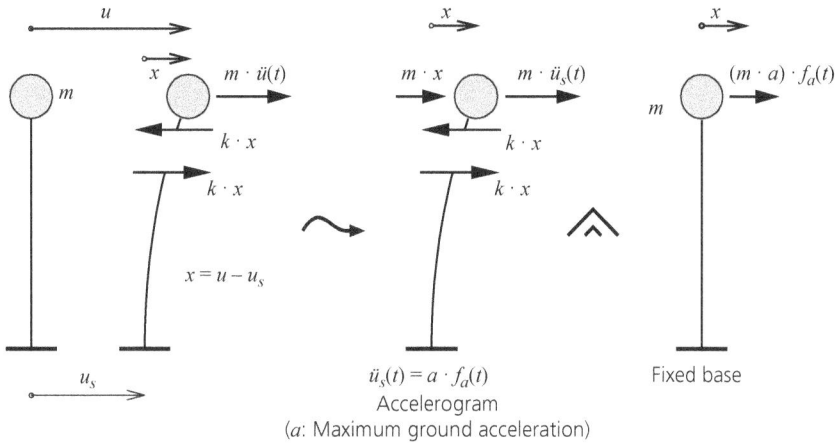

$\ddot{u}_s(t) = a \cdot f_a(t)$
Accelerogram
(a: Maximum ground acceleration)

Imposing a ground motion means essentially imposing an accelerogram,
and this, in turn, means imposing an inertia force on a rigidly supported system

The form of this equation is, according to Section 7.2.3, exactly the same as that of the forced vibration of a single-degree-of-freedom system on a fixed basis (see Figure 7.13) and, since the existing damping obviously concerns the relative velocity, $\dot{x}$, of mass, m, with respect to the ground, the equation may be accordingly modified as follows:

$$m \cdot \ddot{x} + k \cdot x + c \cdot \dot{x} = -(m \cdot a) \cdot f_a(t)$$

It should be noted here that the right-hand term refers merely to a 'fictitious force' that does not really exist (as it does in Section 7.2.3) but represents along with the force $(m \cdot \ddot{x})$ the total inertia force the mass, m, is subjected to.

By assuming initial conditions corresponding to *rest* (i.e. $x = 0$ and $\dot{x} = 0$), the solution of the last equation, analogous to Section 7.2.3, may be written as (Müller and Keintzel, 1984):

$$x(t) = \frac{a}{\omega_d} \left[\int_0^t f_a(\tau) \cdot e^{-\zeta \omega(t-\tau)} \cdot \sin \omega_d (t-\tau) d\tau \right]$$

As previously explained (Section 7.2.2.2), for values of ζ up to 10%, the fundamental circular frequency, ω_d, does not differ substantially from the natural frequency of the system, ω, without damping. However, in any case, for the design of a structure, it is the maximum value, S_d, of the relative displacement that is of paramount interest, as it obviously leads to the maximum internal forces in the structure. It is:

$$S_d = x_{max} = \frac{a}{\omega} \left[\int_0^t f_a(\tau) \cdot e^{-\zeta \omega(t-\tau)} \cdot \sin \omega (t-\tau) d\tau \right]_{max}$$

The time that the maximum deviation, x_{max}, occurs, the acceleration (or deceleration), $\ddot{x}$, also takes its maximum value, $\ddot{x}_{max}$, whereas the velocity, $\dot{x}$ – and, hence, the damping force – vanishes. Thus, at this moment, the elastic force, $k{\cdot}x_{max}$, acting on mass, m, balances its inertia force, depending of course on its absolute acceleration, and is equal to $m \cdot (\ddot{x}_{max} + \ddot{u}_s)$.

Assuming that at this same moment the maximum ground acceleration is practically attained, it may be written (see Figure 7.14):

$$m \cdot (\ddot{x} + \ddot{u}_s)_{max} = k \cdot x_{max} \qquad \text{or} \qquad (\ddot{x} + \ddot{u}_s)_{max} = \omega^2 \cdot x_{max}$$

Thus, the maximum value S_a of the absolute acceleration of the mass is:

$$S_a = (\ddot{x} + \ddot{u})_{max} = a \cdot \omega \cdot \left[\int_0^t f_a(\tau) \cdot e^{-\zeta \omega (t - \tau)} \cdot \sin \omega(t - \tau)d\tau \right]_{max}$$

Of course, this expression is strictly valid if the damping coefficient, ζ, does not exceed 10%. ∎

Thus, it becomes clear that the response of a single-degree-of-freedom system to a specific seismic excitation – that is, with known characteristics regarding time development $f_a(t)$ of the ground acceleration, as well as its maximum value a – depends only on the natural frequency (or natural period) of the system and the respective damping factor.

Thus, based on the last equation and for a given accelerogram, the maximum absolute acceleration, S_a, may be plotted as a function of the natural period, T, for different values of the damping coefficient, ζ. Such a diagram represents the *acceleration spectrum* and it permits, for any oscillator with

Figure 7.14 Position of maximum response in a single-degree-of-freedom system

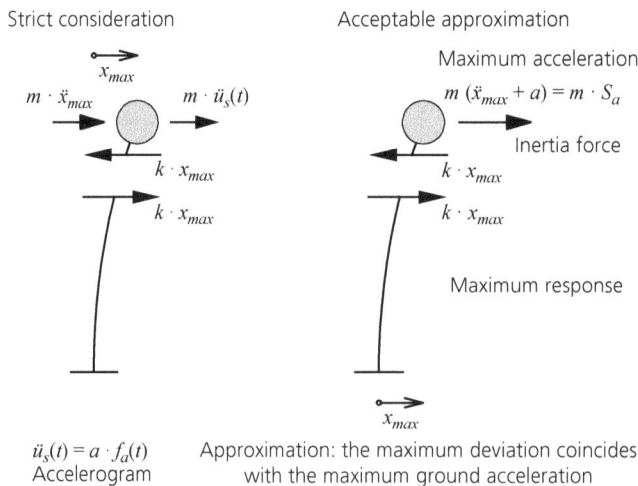

Strict consideration

x_{max}
$m \cdot \ddot{x}_{max}$ $m \cdot \ddot{u}_s(t)$
$k \cdot x_{max}$
$k \cdot x_{max}$

$\ddot{u}_s(t) = a \cdot f_a(t)$
Accelerogram

Acceptable approximation

Maximum acceleration
$m (\ddot{x}_{max} + a) = m \cdot S_a$
Inertia force
$k \cdot x_{max}$
$k \cdot x_{max}$

Maximum response

x_{max}
Approximation: the maximum deviation coincides with the maximum ground acceleration

a known natural period, T, the direct determination of its maximum absolute acceleration. It may be easily recognised that $S_a = S_d \cdot \omega^2$, where S_d is the structure's relative displacement.

It is useful here to point out that S_a – as well as the *spectral displacement* S_d – according to the above expression, is proportional to the maximum ground acceleration, a, characterising the specific accelerogram, so that if divided by a, it represents a spectrum referring to an accelerogram exhibiting the same evolution in time but with respect to a unit maximum acceleration.

In Figure 7.15, a recorded accelerogram of the El Centro earthquake (California, USA, 1940) is illustrated, together with the acceleration spectrum (S_a/a) referred to a unit ground acceleration for two different damping coefficients.

It is clear now that for the practical evaluation of a specific earthquake being described by its accelerogram, the acceleration spectrum, S_a – or the unit one S_a/a together with the ground acceleration, a – represents the examined excitation quite sufficiently. Thus, for a specific system with a known natural period, T, on the basis of the respective spectral value, S_a, the maximum inertia force can be directly obtained as ($m \cdot S_a$). This force represents the maximum shear force, V_a, developed at the structure's base, which is called the *base shear*. It is

$$V_a = m \cdot S_a$$

It is clear that the above inertia force ($m \cdot S_a$) is identical to the maximum elastic force ($k \cdot S_d$) acting on the mass, m (see Figure 7.14). Thus, it is

$$m \cdot S_a = k \cdot S_d = k \cdot S_a / \omega^2$$

In this manner, the use of the acceleration spectrum permits the determination of the maximum response in the structural web through a static loading, which has, of course, a particular practical importance.

From the spectrum of a specific earthquake, it can be deduced that there is always a range of periods, which is more sensitive towards the development of the maximum inertia force or, in other words, of the maximum base shear. As can be seen in Figure 7.15, structures with very small or, particularly, with a high natural period, T – that is, with very low stiffness – develop a clearly limited response.

Regarding now the maximum dynamic magnification factor $(DMF)_{max}$, as defined in Section 7.2.3, the ratio of the maximum relative displacement S_d to the static one $(m \cdot a)/k$ is:

$$(DMF)_{max} = \frac{S_d}{(m \cdot a / k)} = \frac{\omega^2 \cdot S_d}{a} = \frac{S_a}{a} = \omega \cdot \left[\int_0^t f_a(\tau) \cdot e^{-\zeta \omega(t - \tau)} \cdot \sin \omega(t - \tau) d\tau \right]_{max}$$

or, alternatively,

Figure 7.15 Accelerogram and acceleration spectrum (El Centro, California, USA)

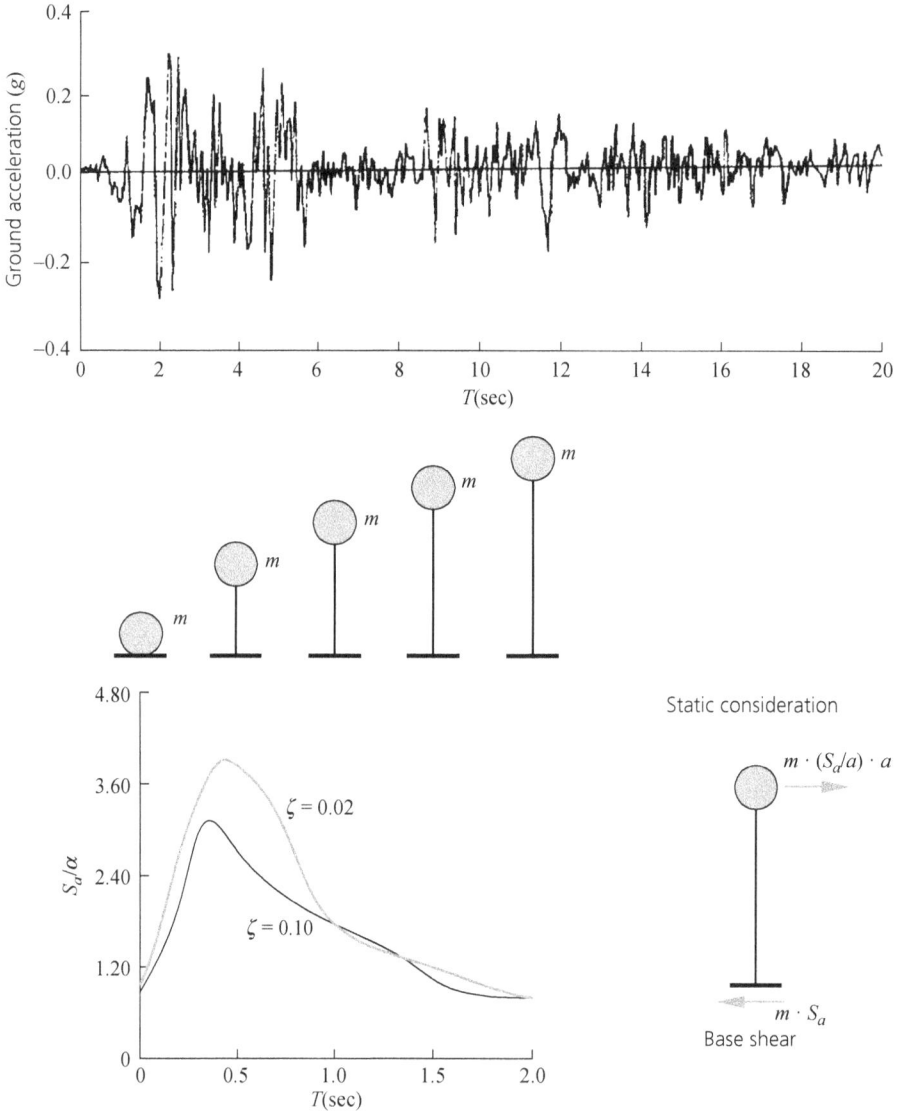

Static consideration

$$S_d = (a/\omega^2)\cdot(\text{DMF})_{max}$$

$$S_a = S_d\cdot\omega^2 = a\cdot(\text{DMF})_{max}$$

It is pointed out that the dynamic factor $(\text{DMF})_{max}$, as can be seen from the above expression, is identical to the acceleration spectral value (S_a/a). ∎

Assuming now in an approximation that the variation of acceleration, a, in time is sinusoidal – that is, $f_a(t) = \sin(\Omega t)$ – then, given that the two last equations are generally valid, the previously found expression of the dynamic factor in Section 7.2.4 may be applied directly:

$$(\text{DMF})_{\Omega,max} = \frac{1}{\sqrt{\left(1-(\Omega/\omega)^2\right)^2 + 4\cdot\zeta^2\cdot(\Omega/\omega)^2}}$$

Thus, it may be written as:

$$S_{d,\Omega} = \left(a/\omega^2\right)\cdot(\text{DMF})_{\Omega,max}$$

and,

$$S_{a,\Omega} = a\cdot(\text{DMF})_{\Omega,max}$$

In this manner, the diagram shown in Figure 7.10 may also be interpreted as an acceleration spectrum – with respect to a unit ground acceleration – for any value of the ratio of frequency Ω – taken as constant – to the natural frequency, ω, of the single-degree-of-freedom system. The magnitude of the assumed frequency, Ω, of maximum ground acceleration definitely depends on the geotechnical characteristics of the foundation soil. Quite indicatively it may be mentioned that it ranges for stiff soils and rocks from 20 to 60 rad/s, for medium stiff soils from 12 to 50 rad/s, and for soft or very soft soils from 10 to 3 rad/s. ∎

Regarding the influence of the foundation soil, it may be seen from Figure 7.16 that its deformability influences, apart from the maximum developed ground acceleration, the response of the structure itself too, due to the variation of the effective inertia force of the mass. This force increases or decreases, according to the region of the respective acceleration spectrum that the natural period of the structure belongs to, by assuming a fixed base (see Figure 7.16).

In design practice, however, the so-called *design spectrum*, which contains the effects of a large number of relevant accelerograms for the region examined, is used as an acceleration spectrum, also making a distinction between the various soil types, given the influence of the foundation soil deformability, as explained above.

Of course, the elaboration of the design spectral curves constitutes a major issue in the design codes of each country. An indicative sketch of these curves is shown in Figure 7.17. The influence of soil type on the maximum effective acceleration, S_a, is obvious. It is worth noting that the initial branch of the spectrum can be conservatively ignored, and the curve can start from the plateau.

7.2.6   Influence of plastic behaviour on seismic response

According to the previous section, the basic equation of motion of the single-degree-of-freedom system under a seismic excitation, with respect to a fixed base, can be written as:

$$m\cdot\ddot{x} + (m\cdot a)\cdot f_a(t) + c\cdot x = -k\cdot x$$

This equation expresses the static equilibrium between the externally applied total inertia force together with the damping force on one side and the opposite acting elastic force induced to the

Figure 7.16 Influence of the deformability of soil on seismic response

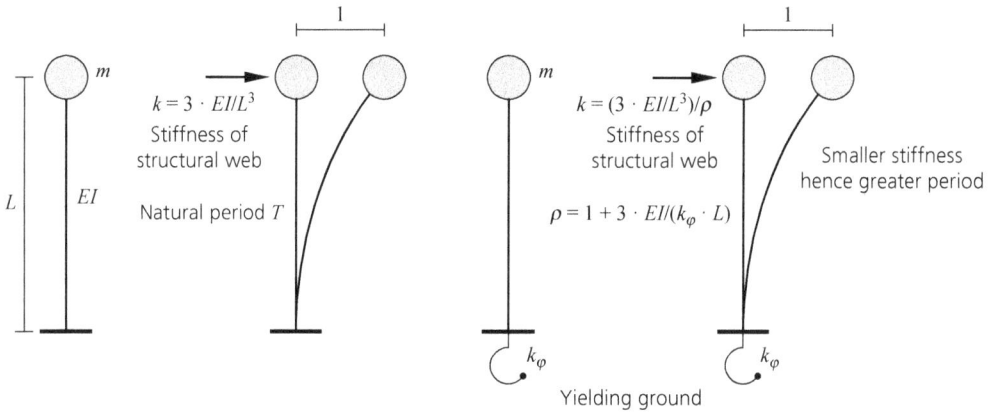

$k = 3 \cdot EI/L^3$
Stiffness of
structural web

Natural period T

$k = (3 \cdot EI/L^3)/\rho$
Stiffness of
structural web

Smaller stiffness
hence greater period

$\rho = 1 + 3 \cdot EI/(k_\varphi \cdot L)$

k_φ

k_φ

Yielding ground

Imposing the same accelerogram on both systems causes in the case of resilient ground
smaller or greater accelerations (inertia forces) depending on the period T itself

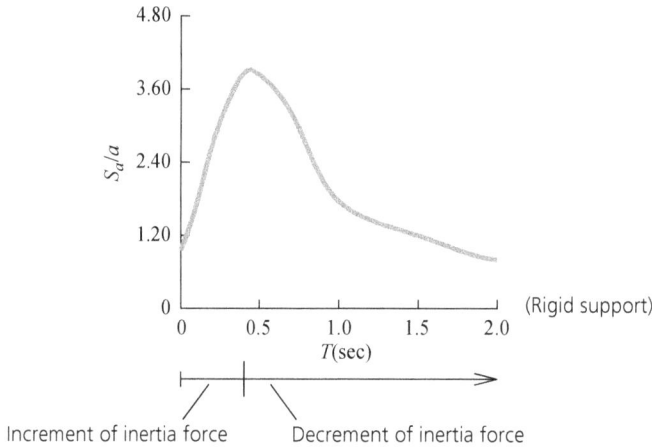

S_d/a

T(sec)

(Rigid support)

Increment of inertia force   Decrement of inertia force

mass by the structural web on the other, due to its imposed deformation (see Figure 7.18). This elastic force $(k \cdot x)$ increases with the displacement, x, of the mass, and it is clear that, regarding the depicted cantilever, the same increase holds also for the bending moment, M, developed at its base – that is, $M = (k \cdot x) \cdot L$. It is clear that the structure works as rigidly clamped at its base and subjected to a varying horizontal force, $m \cdot [\ddot{x} + a \cdot f(t)]$ applied on the mass at the top.

It can be seen that if the last value M reaches the existing bending plastic capacity, M_{pl}, of the cross-section, then a plastic hinge is developed at its base that will not allow further increase of the bending moment, thus setting an upper limit to the development of the transferred force despite the continuing increase of the displacement, x (see Figure 7.18), this limit being obviously equal to $F_{pl} = M_{pl}/L$. The displacement, x, can, of course, continue to increase up to a certain value if the specific cross-section of the structural element can provide the required rotational deformability (see Section 7.2.7.1 and Stavridis and Georgiadis (2025, Section 4.1.2)).

Figure 7.17 Design spectrum

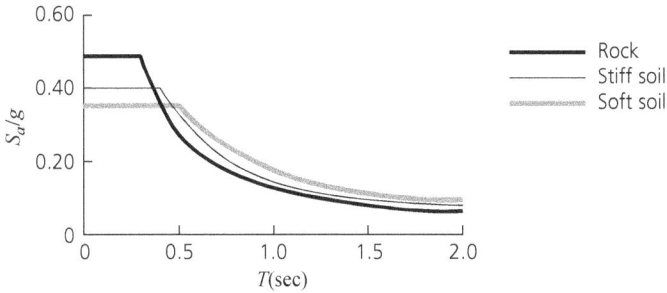

Figure 7.18 Limitation of the inertia force due to plastification

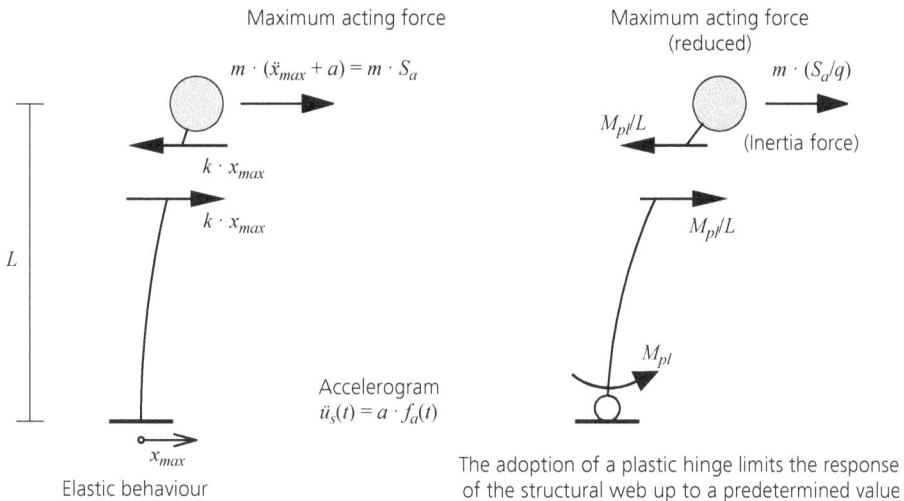

Maximum acting force

$$m \cdot (\ddot{x}_{max} + a) = m \cdot S_a$$

$k \cdot x_{max}$

$k \cdot x_{max}$

L

x_{max}

Accelerogram
$\ddot{u}_s(t) = a \cdot f_a(t)$

Elastic behaviour

Maximum acting force
(reduced)

M_{pl}/L

$m \cdot (S_a/q)$

(Inertia force)

M_{pl}/L

M_{pl}

The adoption of a plastic hinge limits the response
of the structural web up to a predetermined value

It is clear that the above equation can be satisfied with lower inertia forces than would be required if no plastic hinges were allowed to form – that is, if the structure behaved in a purely elastic manner. This implies that when plastic hinges are permitted, the required bending capacity of the structure is reduced compared with the case where the bending moments develop entirely according to the elastic response of the structural web under the imposed (dynamic) load $m \cdot a \cdot f_a(t)$.

Thus, the design of a structure for an imposed accelerogram $a \cdot f_a(t)$ can be dealt with either by (a) adopting a cross-section with a flexural strength, $M_{pl}^{(A)}$, that will remain in the elastic state throughout the earthquake or (b) adopting a cross-section with a lower flexural strength, $M_{pl}^{(B)}$, being able to satisfy the equilibrium during the whole dynamic response, provided the required rotation of the plastic joint can be achieved. Since $M_{pl}^{(B)} < M_{pl}^{(A)}$, it is clear that in order to apply this economical solution to the seismic problem, a special check for rotational deformability (or ductility) is required.

7.2.7 Ductility and seismic design

7.2.7.1 Characteristics of plastic bending deformation

The concept of ductility is crucial for understanding the influence of plastic behaviour on the seismic design of structures. To introduce it, it is useful to consider first the basic characteristics of plastic flexural deformation. As has been clarified in previous chapters, the curvature, φ, at any point of a member under bending can be expressed – according to Stavridis and Georgiadis (2025, Section 4.1.1.7) – in terms of the strains, ε_1 (shortening) and ε_2 (elongation) of the end fibres of the cross-section at that point as:

$$\varphi = \frac{1}{r} = \frac{\varepsilon_1 + \varepsilon_2}{h} = \frac{\Delta\varepsilon}{h} \qquad \text{(see Figure 7.19)}$$

In the case of steel sections and since the limiting value of ε can be considered at least equal to $\varepsilon_u = 15 \cdot \varepsilon_y$, where ε_y is the yielding strain, the limiting value, φ_u, is estimated as:

$$\varphi_u = \frac{2 \cdot 15 \cdot \varepsilon_y}{h} = 30 \cdot \frac{f_y}{E_s} \cdot \frac{1}{h}$$

where f_y is the yielding strength of steel and E_s its modulus of elasticity.

Figure 7.19 Curvature development due to applied bending moment

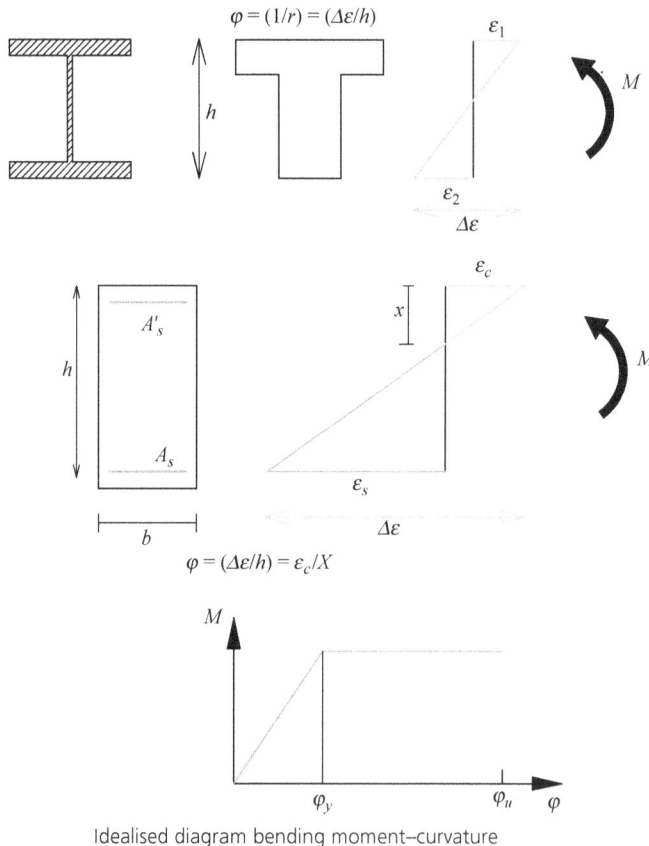

Idealised diagram bending moment–curvature

Since the linear strain distribution of the cross-section generally implies a neutral line, where $\varepsilon = 0$, the above expression of the curvature can also be written as:

$$\varphi = \frac{1}{r} = \frac{\varepsilon_1}{x}$$

where x is the height of the considered compressive zone.

This expression is more appropriate for reinforced concrete sections in the form:

$$\varphi = \varepsilon_c/x$$

A rectangular reinforced concrete cross-section is now considered having reinforcement in both the tension and compression zones $\rho = A_s/(b \cdot h)$ and $\rho' = A_s'/(b \cdot h)$, respectively, as shown in Figure 7.19. By assuming that under the ultimate bending moment both the tensile and compressive reinforcement yield, the equilibrium of the tensile forces of the reinforcement with the compressive force of the concrete allows the estimatation of the depth of the compressive zone of the cross-section as:

$$x_u = \frac{h \cdot (\rho - \rho')}{0.81} \cdot \frac{f_y}{f_c}$$

and since the cracking strain of concrete $\varepsilon_c = 0.0035$ (see Stavridis and Georgiadis, 2025, Section 4.2.2.1), the final expression of the limiting curvature is:

$$\varphi_u = \frac{0.00284}{h \cdot (\rho - \rho')} \cdot \frac{f_c}{f_y}$$

As the presence of a compressive force generally decreases $\Delta\varepsilon$ and increases x (see Stavridis and Georgiadis, 2025, Section 6.6.2), it may be concluded from both the above expressions of φ that an acting compressive force has an adverse influence on the plastic deformability of the cross-section since the value of φ_u decreases. Clearly, the arrangement of dense stirrups (the so-called *tightening*) helps to increase the available value of φ_u (see Stavridis and Georgiadis, 2025, Section 6.6.2). ∎

A statically applied force, F, at the top of the cantilever considered in the previous paragraph will cause the linear bending moment diagram shown in Figure 7.20. The progressive application of force, F, up to the value $F_{pl} = M_{pl}/L$ will cause the moment, M_{pl}, at the base of the cantilever together with the maximum elastic displacement of the cantilever ($u_y = F_{pl}/k$). Then, under the constant load, F_{pl}, a further displacement, u_{pl}, at the top will occur, according to Figure 7.20. Assuming now that the cross-section of the cantilever exhibits the moment–curvature diagram shown in Figure 7.19 with a maximum curvature capacity equal to φ_u, it may be concluded that the maximum allowable displacement, u_u, of the cantilever top corresponds to this ultimate curvature.

By analogy with the picture of the progressive elastoplastic response presented in Stavridis and Georgiadis (2025, Section 4.1.2) concerning a simply supported beam, it can be assumed that in the present case a certain plastification takes place at the base of the cantilever over a length, L_{pl}, under a characteristic curvature equal to φ_u. The plastification length, L_{pl}, can generally be considered to be in the order of the height of the cross-section in the case of steel cross-sections, whereas for reinforced concrete the empirical relationship:

$$L_{pl} = 0.08 \cdot L + 0.022 \cdot d \cdot f_y$$

Figure 7.20 Deformation in the plastified cantilever base

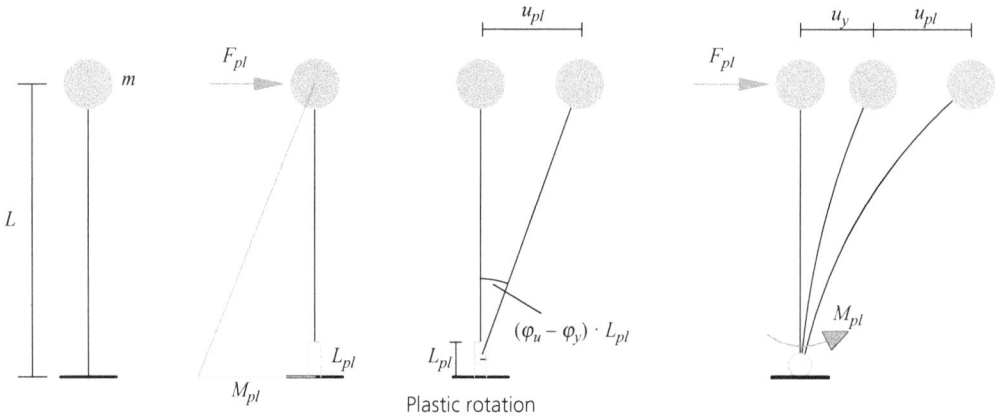

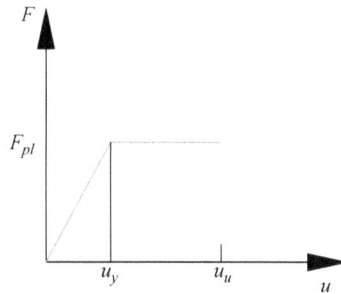

Diagram $F - u$

Displacement ductility

$$\mu_\delta = u_u/u_y$$

may be applied, where d is the diameter of the longitudinal reinforcement and f_y is its yield stress in MPa.

The developing displacement, u_u, at the top of the cantilever is thus composed of an elastic part, u_y, which may be expressed as $u_y = \varphi_y \cdot L^2/3$ (as found by following the principle of virtual work procedure described in Stavridis and Georgiadis, (2025, Section 2.3.3)) and a plastic part, u_{pl}, due to the rotation as a rigid body of the cantilever length $(L - 0.5 \cdot L_{pl})$ by the angle of plastic rotation $(\varphi_u - \varphi_y) \cdot L_{pl}$ (see Figure 7.20) – that is, $u_{pl} = (\varphi_u - \varphi_y) \cdot L_{pl} - (L - 0.5 \cdot L_{pl})$. Thus, it is $u_u = u_y + u_{pl}$ and, finally:

$$u_u = \varphi_y \cdot L^2/3 + (\varphi_u - \varphi_y) \cdot L_{pl} \cdot (L - 0.5 \cdot L_{pl}).$$

The total chord rotation, θ_u, at the ultimate state is thus equal to $\theta_u = u_u/L$, which may be written as:

$$\theta_u = \varphi_y \cdot \frac{L}{3} + (\varphi_u - \varphi_y) \cdot L_{pl} \cdot (1 - 0.5\frac{L_{pl}}{L})$$

actually consisting of the elastic part:

$$\theta_y = \varphi_y \cdot \frac{L}{3}$$

and the remaining plastic part:

$$\theta_{pl} = (\varphi_u - \varphi_y) \cdot L_{pl} \cdot \left(1 - 0.5 \frac{L_{pl}}{L}\right)$$

The last expression may be considered applicable to any frame element where the length, L, measures the distance between the plastic joint and the point where the bending moment vanishes. This value of the angle of chord rotation represents the angle up to which the rotation of a plastic joint is allowed to develop.

7.2.7.2 The concept and role of ductility in seismic design

The concept of ductility refers to the non-linear elasto-plastic pattern of the development of a specific deformation magnitude induced in a member by a progressively increasing static load. The ductility, μ, is the ratio of the ultimate value, Δ_u, of the selected deformation to the value, Δ_y, of this deformation at the end of the elastic phase during the progressive increase of the loading. In the present case of the cantilever beam, given that the displacement, u, of its top is the most characteristic deformation, the ductility, μ_δ, is considered as $\mu_\delta = u_u/u_y$.

It may now be considered that the base of the cantilever carrying a mass, m, as shown in Figure 7.21 is subjected to a specific seismic accelerogram (e.g. El Centro, see Section 7.2.5). It is also assumed that during the application of the accelerogram, the cross-section of the cantilever at the base attains its flexural strength, M_{pl}, whenever needed, thus developing plastic deformations with a maximum displacement of the mass equal to Δ_u.

The displacement of mass, m, at any instant of time is due to the static action of the horizontal inertia force acting on it. If this force reaches the value $F_{pl} = M_{pl}/L$, the corresponding elastic displacement will be $\Delta_y = (M_{pl}/L) \cdot L^3/(3 \cdot EI)$, according to the elastic stiffness of the cantilever $k = 3 \cdot EI/L^3$. Given the maximum displacement, Δ_u, the ductility, μ_δ, according to the above is $\mu_\delta = \Delta_u/\Delta_y$.

Supposing now that the single-mass system is subjected to the same accelerogram but depicts throughout elastic behaviour, if F_{el} is the maximum transmitted shear force from the mass and Δ_{el} is the corresponding displacement of the top, it will be $F_{el} = (3 \cdot EI/L^3) \cdot \Delta_{el}$.

The imposition of the accelerogram on the system following elastic behaviour requires the undertaking of the force, F_{el}, whereas in order to withstand the same earthquake through the development of plastic deformations, a clearly smaller force, F_{pl}, has to be taken up, of course, with the additional requirement for the cross-section to provide these plastic deformations. The ratio of these two forces $q = F_{el}/F_{pl}$, known as the *behavioural coefficient* or *behavioural factor*, is important for the seismic design of structures and in general is related to the existing ductility $\mu_\delta = \Delta_u/\Delta_y$. It has been found that for values of the fundamental period, T, greater than the characteristic limit, T_c, of the applicable design spectrum (see Figure 7.21), it may be approximated that $\Delta_u = \Delta_{el}$ and this, of course, implies $\mu_\delta = q$, as shown in Figure 7.21. However, if the fundamental period, T, is smaller, then Δ_u results generally larger than Δ_{el} and μ_δ can be expressed approximately by the empirical relationship:

$$\mu_\delta = (q-1) \cdot \left(\frac{T_c}{T}\right) + 1$$

Figure 7.21 Elastoplastic behaviour of a single-degree-of-freedom system in an imposed seismogram

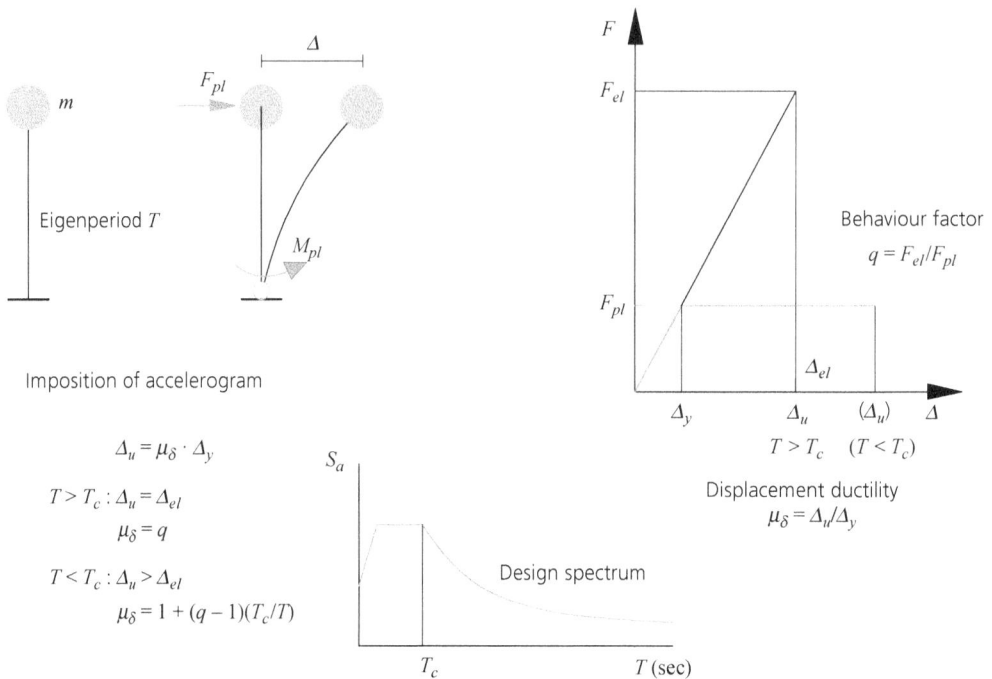

Eigenperiod T

F_{pl}

m

Δ

M_{pl}

Imposition of accelerogram

Behaviour factor
$q = F_{el}/F_{pl}$

F_{el}

F_{pl}

Δ_{el}

$\Delta_y \qquad \Delta_u \quad (\Delta_u) \quad \Delta$

$T > T_c \quad (T < T_c)$

Displacement ductility
$\mu_\delta = \Delta_u/\Delta_y$

$\Delta_u = \mu_\delta \cdot \Delta_y$

$T > T_c : \Delta_u = \Delta_{el}$
$\mu_\delta = q$

$T < T_c : \Delta_u > \Delta_{el}$
$\mu_\delta = 1 + (q - 1)(T_c/T)$

S_a

Design spectrum

$T_c \qquad\qquad T\ (\text{sec})$

7.2.7.3 Design practice

The problem of seismic design always lies in checking whether a structure, such as the above cantilever with a given lumped mass, 'survives' in a specific seismic environment described generally by an adopted acceleration spectrum. As noted earlier (see Section 7.2.5), this spectrum is an envelope of the acceleration spectra of the possible specific accelerograms to which the structural system is likely to be subjected at a given place.

In principle, two options are available for the seismic design of a structure.

▨ The structure is dimensioned with the aim of keeping it in the elastic region during the earthquake. Based on the stiffness, k, of the structural web and the adopted mass, m, the corresponding period, T, is determined on the basis of which a value of the spectral acceleration, S_a, is assigned to the mass. The design control consists of whether the structure (cantilever) is able to take up the response under the static action of the inertia force ($m \cdot S_a$).

▨ The second option assumes that the structure (cantilever) will behave elastically up to a certain acting inertia force $F_{pl} = M_{pl}/L$ according to the selected flexural strength, M_{pl}, with a respective displacement, Δ_y; after this limit, the structure will further yield until the maximum displacement, Δ_u, has been reached according to the force–displacement diagram, which may be set out on the basis of the selected cross-section (see Figure 7.21). Assuming now that the structure would exhibit unbounded elastic behaviour, then it is clear that the active inertia force according to the respectively calculated period, T, would be equal to $F_{el} = (m \cdot S_a)$. The existing ratio $q = F_{el}/F_{pl}$, as explained above, leads to a certain value of the ductility, μ_δ, depending on the period, T. The additional requirement for the structure is that it must be capable of deforming by the displacement $(\mu_\delta \cdot \Delta_y)$, according to the above force–displacement diagram, whereby the value of Δ_y as previously mentioned is equal to $(M_{pl} \cdot L^2 / 3 \cdot EI)$.

This design procedure consists in accepting in advance a certain behaviour coefficient, q, and dimensioning the structure (cantilever) on the basis of the inertial force $(S_a \cdot m/q)$, where S_a is derived according to the adopted spectrum. The verification involves estimating whether the dimensioned structure is able to develop a deflection, Δ_u, greater than or equal to the value $\mu_\delta \cdot \Delta_y$ through the elasto-plastic behaviour.

It is clear that the selection of the coefficient, q, is essentially left to the choice of the designer, depending on the estimated (or intended) ductility of the structure. A value of q equal to 1.0 means acceptance of elastic behaviour and no reduction of the inertial seismic forces. The structure should simply have the required strength without any particular ductility requirement. A higher value of q, on the other hand, means a reduced statically applied seismic force and therefore a reduced strength requirement, but on the condition that the development of the required plastic deformation is possible. A steel structure may allow the consideration of q up to a value of 4.0, while for reinforced concrete, a suitable arrangement of the reinforcement may allow a value of q up to 3.0.

7.3.   Multi-degree-of-freedom systems

Multi-degree-of-freedom systems contain more than one discrete mass or one mass with multiple degrees of freedom. These masses are 'carried' by a massless structural web exhibiting a clearly determinable behaviour regarding its deformability and stiffness. Within this frame of behaviour, the masses show displacements or rotations involving either negligible or non-negligible inertia forces. These last displacements are considered as effective for the corresponding masses and their number represents the so-called *effective degrees of freedom* of the system. Examples of such multi-degree-of-freedom systems are mentioned in the introductory section of this chapter (see Figure 7.2).

7.3.1   The use of the stiffness matrix

7.3.1.1   The concept of the stiffness matrix

Of decisive importance for the description of the behaviour of a multi-degree-of-freedom system is the so-called *stiffness matrix*, which is a set of magnitudes concerning the stiffness of its structural web, which are determined as shown in the example of a three-storey plane model shown in Figure 7.22. Here, it is assumed that masses m_1, m_2 and m_3 develop inertia forces in a horizontal excitation (e.g. earthquake) with respect only to their horizontal displacements. Thus, the effective displacements of the system are the lateral storey drifts Δ_1, Δ_2 and Δ_3.

The stiffness matrix consists in the horizontal juxtaposition of three group forces that are determined as follows (see Figure 7.22).

The first group consists of the three – horizontal – forces (k_{11}, k_{21}, k_{31}) that are required to be applied correspondingly to the effective displacements in order to produce a unit displacement in level 1 and zero shift in the two others (i.e. $\Delta_1 = 1$, $\Delta_2 = 0$, $\Delta_3 = 0$).

Analogously, the second group consists of the forces (k_{12}, k_{22}, k_{32}) needed to act in the direction of the effective displacements in order to produce a unit displacement in level 2 and zero shift in the two others (i.e. $\Delta_1 = 0$, $\Delta_2 = 1$, $\Delta_3 = 0$). Finally, the third group contains those forces (k_{13}, k_{23}, k_{33}), which, acting along the effective displacements, produce $\Delta_1 = 0$, $\Delta_2 = 0$, $\Delta_3 = 1$.

Figure 7.22 The elements of a stiffness matrix

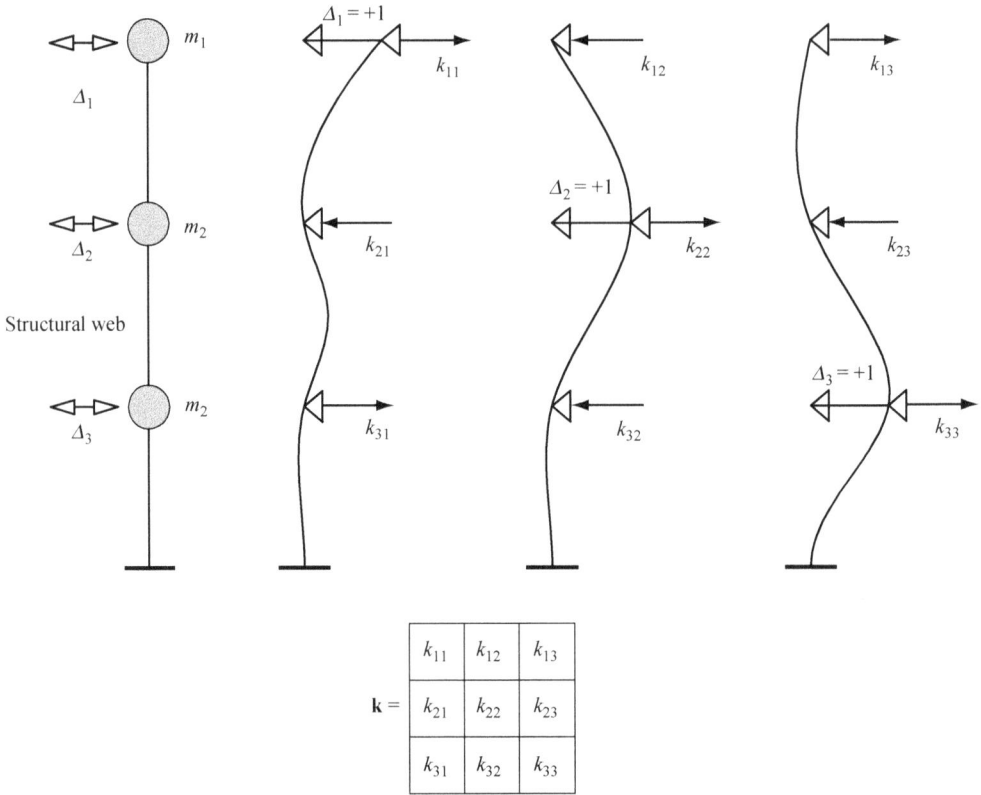

$$k = \begin{array}{|c|c|c|} \hline k_{11} & k_{12} & k_{13} \\ \hline k_{21} & k_{22} & k_{23} \\ \hline k_{31} & k_{32} & k_{33} \\ \hline \end{array}$$

The stiffness matrix represents the stiffness of the structural web
with respect to the effective mass displacements

Each of the above groups consists of the reactions developed if, at first, fictitious simple supports are inserted into the system in order to restrain all its effective displacements, and then the respective unit support settlement (1 m) is imposed, as shown in Figure 7.22.

Defining on purpose the positive senses of these forces as coinciding with those of displacements (Δ_1, Δ_2 and Δ_3) (i.e. to the right), it becomes clear that the forces of each group will exhibit alternating signs.

The structural web that carries the three masses may be represented by the stiffness matrix k by juxtaposing horizontally the above three force groups, as previously mentioned, according to the following formulation:

$$\mathbf{k} = \begin{bmatrix} k_{11} & k_{12} & k_{13} \\ k_{21} & k_{22} & k_{23} \\ k_{31} & k_{32} & k_{33} \end{bmatrix}$$

It is useful to be familiar with the treatment and the essence of matrix operations. However, it must be pointed out here that the use of matrices contributes very little, if at all, to the understanding of the static or dynamic behaviour of any system – at least at this introductory level. Nevertheless, it constitutes a necessary tool for the compact writing of equilibrium equations and the relevant relations, thus facilitating their methodical numerical treatment through computer programs.

7.3.1.2 Matrix operations

In the present context, of main interest are either *square matrices* – having an equal number of rows and columns – or *column matrices* – consisting of a vertical juxtaposition of numerical elements.

A square matrix **k** in the order $(N \times N)$, has N rows and N columns (see Figure 7.23). The elements of the square matrix along the diagonal starting from the upper left element constitute the *diagonal elements*. A column matrix $\boldsymbol{a}$ of the order $(N \times 1)$ has N rows and 1 column.

Interchange of a matrix's rows and columns – or vice versa – leads to the *transposed matrix* being denoted by the letter T. Thus, while matrix $\mathbf{k}^T$ still has N rows and N columns (i.e. $N \times N$), matrix $\boldsymbol{a}^T$ has 1 row and N columns (i.e. $1 \times N$).

The basic physical concept of a matrix is directly related to the column matrix. This is the recording of quantities corresponding to specific magnitudes – for example, the storey drifts of the previous structure or the acting forces at each storey level. The appropriate horizontal juxtaposition of such column matrices leads to square matrices, as the stiffness matrix has been previously defined.

The sum of two column matrices is a matrix of the same order and comprises the sums of their corresponding terms. Although the addition also applies to two square matrices of the same order, the physical meaning of this operation is clearer and better understood in this context with respect to column matrices.

The multiplication of two matrices is of fundamental importance and the way it is accomplished is shown in Figure 7.23. The multiplication of two matrices is feasible only when the number of columns in the 'left' matrix is equal to the number of rows in the 'right' matrix. This may also be

Figure 7.23 Layout for the multiplication of matrices

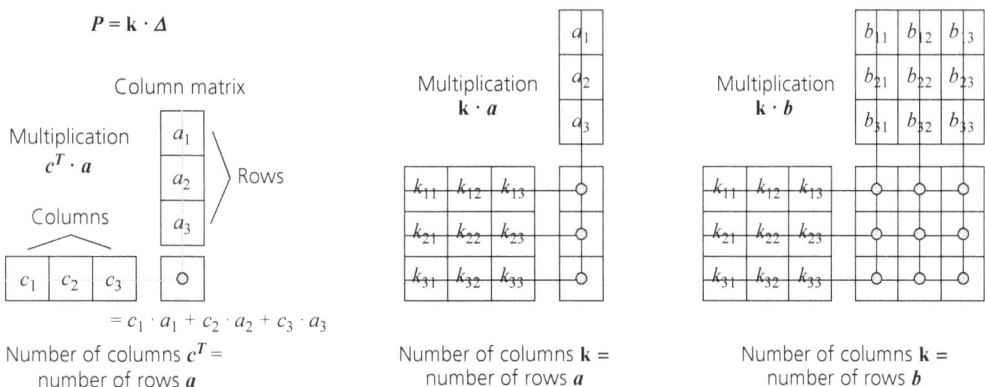

understood from the rule that the final order of the product matrix is found by eliminating the 'interior' dimensions of the matrices, as illustrated below.

Thus, product $\mathbf{k} \cdot \mathbf{a}$ is written as $(N \times N) \cdot (N \times 1)$ and is a matrix of order $(N \times 1)$, whereas product $\mathbf{a} \cdot \mathbf{k}$, which is written as $(N \times 1) \cdot (N \times N)$, does not make sense. Of course, product $\mathbf{a}^{\mathrm{T}} \cdot \mathbf{k}$ does have a meaning since it is written as $(1 \times N) \cdot (N \times N)$, which leads to a matrix of order $(1 \times N)$, whereas it can be seen that product $\mathbf{a}^{\mathrm{T}} \cdot \mathbf{a}$, written as $(1 \times N) \cdot (N \times 1)$, leads to a matrix of order (1×1) – that is, a single number.

It is thus clear, taking also into account the schematic layout of Figure 7.23, that the multiplication of two matrices consists in the consecutive formation of matrix products of the last type $(1 \times N) \cdot (N \times 1)$, which, of course, represent a single number.

The square matrix, the diagonal elements of which consist of unities while all other elements are null, is called the *identity matrix* and is denoted by $\mathbf{I}$. The consideration of the identity matrix allows reference to the basic concept of the so-called *inverse matrix* of a square matrix $\mathbf{A}$ of the order $(N \times N)$, which is denoted by $\mathbf{A}^{-1}$. The inverse matrix is one that satisfies the relation:

$$\mathbf{A} \cdot \mathbf{A}^{-1} = \mathbf{I} \text{ or (identically)} \, \mathbf{A}^{-1} \cdot \mathbf{A} = \mathbf{I}$$

Obviously, both matrix $\mathbf{A}^{-1}$ and matrix $\mathbf{I}$ are of the order $(N \times N)$.

The concept of the inverse matrix is directly related to the solution of a linear system of N equations with N unknowns. Such a system may be represented through the following matrix notation

$$\mathbf{A} \cdot X = C$$

Matrix $\mathbf{A}$ consists of the coefficients of the unknowns, while column matrices X and C contain the unknowns and the known terms of the right-hand side of the equations, respectively. Multiplying both sides by the inverse matrix $\mathbf{A}^{-1}$ results in:

$$\mathbf{A}^{-1} \cdot \mathbf{A} \cdot X = \mathbf{A}^{-1} \cdot C$$

that is,

$$\mathbf{I} \cdot X = \mathbf{A}^{-1} \cdot C \text{ and finally } X = \mathbf{A}^{-1} \cdot C$$

Of course, the inversion of matrix $\mathbf{A}$ requires a numerical procedure equally cumbersome as that of the solution to the linear system, so that the advantage of using the matrix notation concerns only the possibility of a compact formulation and not the actual computing effort required, which in any case should be undertaken using a computer.

7.3.2 Correlation of loading and displacements

The usefulness of the matrix notation in describing in a compact manner the behaviour of a system is shown through the stiffness matrix $\mathbf{k}$ (3×3) of the structural model in Figure 7.24. After the physical definition of $\mathbf{k}$ in Section 7.3.1.1, it is easily understood that the three forces (P_1, P_2, P_3) required to be applied to the three storeys in order to produce the displacements $(\Delta_1, \Delta_2, \Delta_3)$ can be obtained from the relation:

$$P = \mathbf{k} \cdot \Delta \tag{a}$$

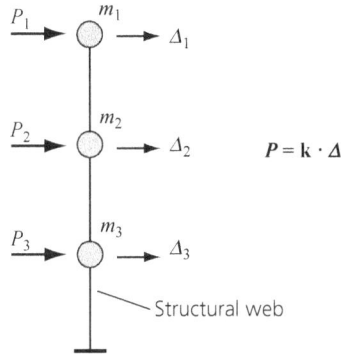

Figure 7.24 Correlation of loading and deformation in a multi-degree-of-freedom system

where matrix P (3 × 1) represents the forces (P_1, P_2, P_3) and matrix Δ (3 × 1) represents the displacements (Δ_1, Δ_2, Δ_3).

According to the introduced 'matrix language', it can be seen that if loads P are known and the displacements Δ caused by them are sought, these will be obtained by multiplying both sides of the above equation by the inverse k^{-1} of the stiffness matrix k:

$$k^{-1} \cdot P = k^{-1} \cdot k \cdot \Delta$$

so that, finally:

$$\Delta = k^{-1} \cdot P$$

Clearly, this result represents nothing more than the solution of the linear system (a), without this compact matrix formulation helping the solution itself.

7.3.3 Multi-storey systems

At this point, it is useful – for later use too – to describe using the compact matrix method the overall response of a three-storey spatial system under horizontal loads, according to the previous examination in Chapter 6 being analogously referred to a fixed orthogonal system Oxy (see Figure 7.25).

Chapter 6 showed how the structural web of the multi-storey spatial system consists of elements of plane stiffness. Each plane element is represented by its own stiffness matrix S_i, as previously defined in Section 7.3.1.1 and shown in Figure 7.26.

The overall stiffness matrix K of the order 9 × 9 (each K_i within K is of the order 3 × 3) referring to the whole system relates the loads P_x and P_y (order 3 × 1) with the developed displacements Δ_x, Δ_y and Δ_ω at each diaphragm. The assemblage of matrix K is made according to the following layout:

$$K = \begin{bmatrix} K_1 & K_2 & K_3 \\ K_2 & K_4 & K_5 \\ K_3 & K_5 & K_6 \end{bmatrix}$$

Figure 7.25 Layout of a multi-storey system

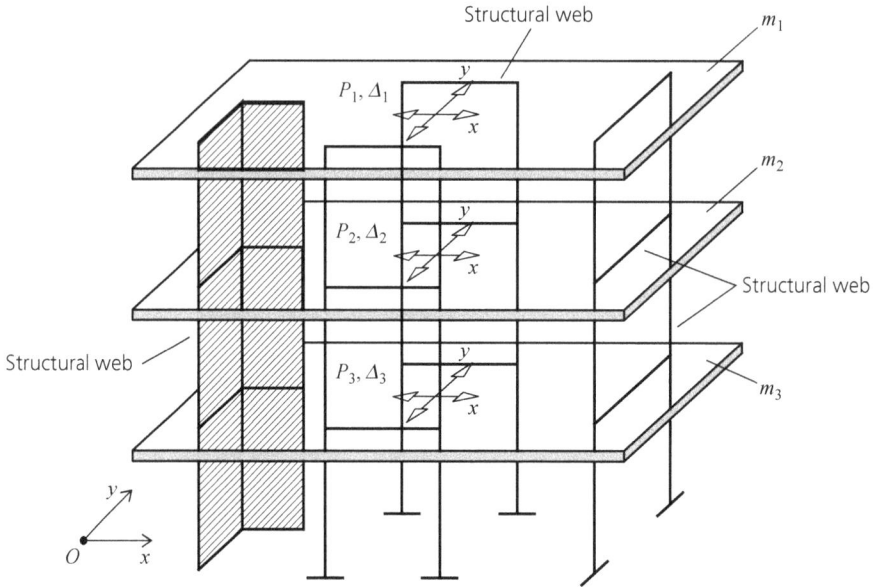

Figure 7.25 Layout of a multi-storey system

where

$$K_1 = \Sigma S_i \cdot \cos^2 \beta_i$$
$$K_2 = \Sigma S_i \cdot \sin\beta_i \cdot \cos\beta_i$$
$$K_3 = \Sigma S_i \cdot e_i \cdot \cos\beta_i$$
$$K_4 = \Sigma S_i \cdot \sin^2 \beta_i$$
$$K_5 = \Sigma S_i \cdot e_i \cdot \sin\beta_i$$
$$K_6 = \Sigma S_i \cdot e_i^2$$

all matrices K_i being of the order 3×3 (Stavridis, 1986).

The correspondence with the relevant expressions given in Chapter 6, Section 6.3.3 – which refer to a single storey – is absolute.

Thus, the relation between loading and deformation of the system is written as follows:

$$\begin{Bmatrix} P_x \\ P_y \\ P_\omega \end{Bmatrix} = \begin{bmatrix} K_1 & K_2 & K_3 \\ K_2 & K_4 & K_5 \\ K_3 & K_5 & K_6 \end{bmatrix} \cdot \begin{Bmatrix} \varDelta_x \\ \varDelta_y \\ \varDelta_\omega \end{Bmatrix}$$

It is pointed out that the matrices P_x and $\varDelta_x$ consist of the components in the x-direction of the magnitudes (P_1, P_2, P_3) and $(\varDelta_1, \varDelta_2, \varDelta_3)$, respectively (see Figure 7.22). This is also analogous for matrices P_y and $\varDelta_y$. Matrix P_ω consists of the components $(P_y \cdot x_e - P_x \cdot y_e)$ in accordance to section 6.3.3, while matrix $\varDelta_\omega$ consists of the diaphragm rotations (see Chapter 6, Section 6.3.1).

Figure 7.26 Determination of matrix of lateral stiffness for a plane element

Figure 7.26 Determination of matrix of lateral stiffness for a plane element

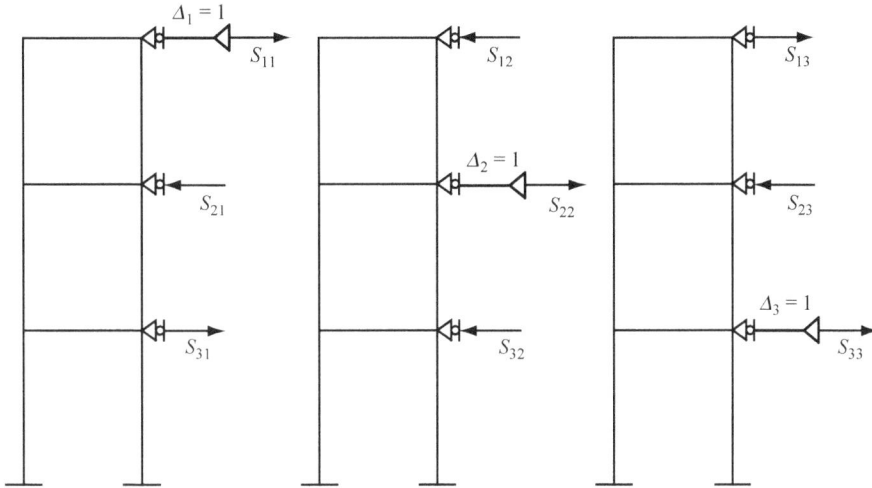

The lateral stiffness matrix is assembled from the reactions
developed at the lateral (imaginable) supports of the plane element

It is, of course, clear that given a multi-storey system with N diaphragms, the above equation allows the determination of the storey displacements through the inversion of matrix $\mathbf{K}$ of the order $(3N \times 3N)$ as examined above, by using suitable computer software. However, in the context of examination of dynamic behaviour from a preliminary design point of view, of prime interest is the orthogonal layout of plane elements about the x and y axes. Then, the constituent matrices of the global stiffness matrix $\mathbf{K}$ are expressed as follows (see Chapter 6, Section 6.3.4):

$$\mathbf{K}_1 = \Sigma \mathbf{S}_x, \quad \mathbf{K}_2 = 0, \quad \mathbf{K}_3 = -\Sigma(\mathbf{S}_x \cdot y)$$
$$\mathbf{K}_4 = \Sigma \mathbf{S}_y, \quad \mathbf{K}_5 = \Sigma(\mathbf{S}_y \cdot x_i), \quad \mathbf{K}_6 = \Sigma(\mathbf{S}_x \cdot y_i^2) + \Sigma(\mathbf{S}_y \cdot x_i^2)$$

Thus, the above condensed equation may be split up as follows:

$$P_x = \mathbf{K}_1 \cdot \Delta_x + \mathbf{K}_3 \cdot \Delta_\omega$$
$$P_y = \mathbf{K}_4 \cdot \Delta_y + \mathbf{K}_5 \cdot \Delta_\omega$$
$$P_\omega = \mathbf{K}_3 \cdot \Delta_x + \mathbf{K}_5 \cdot \Delta_y + \mathbf{K}_6 \cdot \Delta_\omega$$

If the arrangement of the plane elements results in the nullification – or even the approximate elimination – of the matrices $\mathbf{K}_3$ and $\mathbf{K}_5$, then from the preceding equation it follows that $\Delta_\omega = 0$. This implies that the diaphragms do not rotate, which is indeed the intended behaviour, as discussed in Chapter 6. Taking all of the above into account, the following relation can be written as:

$$P_x = \mathbf{K}_1 \cdot \Delta_x = (\Sigma \mathbf{S}_x) \cdot \Delta_x$$

Figure 7.27 Equivalent plane system in both directions

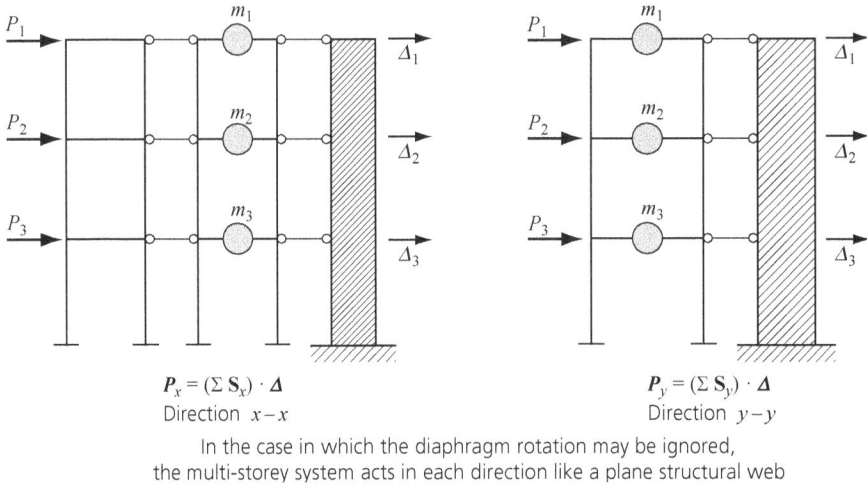

$$P_x = (\Sigma\, S_x) \cdot \Delta$$
Direction $x-x$

$$P_y = (\Sigma\, S_y) \cdot \Delta$$
Direction $y-y$

In the case in which the diaphragm rotation may be ignored,
the multi-storey system acts in each direction like a plane structural web

and,

$$P_y = K_4 \cdot \Delta_y = (\Sigma S_y) \cdot \Delta_y$$

The above relations mean that, in the examined case, the displacements in x and y are independent of each other. They are obtained by applying loads P_x and P_y to the two separate systems, which are formed through coupling of all the corresponding plane elements in each direction, carrying at each level the corresponding mass (see Figure 7.27). This coupling is accomplished at each level through hinged undeformable bars, as shown in Figure 7.27. The above conclusion has a particularly practical importance for both the assessment of the dynamic characteristics of a multi-storey mass system and the determination of the equivalent static loading because of a seismic excitation, as will be examined later in this chapter.

7.3.4   Free vibration

7.3.4.1   Plane systems

The study of multi-degree-of-freedom systems will be continued on an illustrative basis using a model with three degrees of freedom, without thereby limiting the general validity of any conclusions for systems with more degrees of freedom.

At this point, it is noted that the rectilinear structural web shown in Figure 7.28 should not necessarily be considered merely as a straight member carrying at some points masses m_1, m_2 and m_3, but as a general structural system described by its stiffness matrix **k**. Consequently, the model may also be considered as a plane frame like the one shown in Figure 7.28 having the same stiffness matrix **k** (see Figure 7.26). This is clarified because the lateral movement of a vertical cantilever is quite different from the one of a frame, as pointed out in Stavridis and Georgiadis (2025, Section 6.4.3, Figure 6.22).

Figure 7.28 Simulation of a multi-degree-of-freedom frame with a rectilinear model

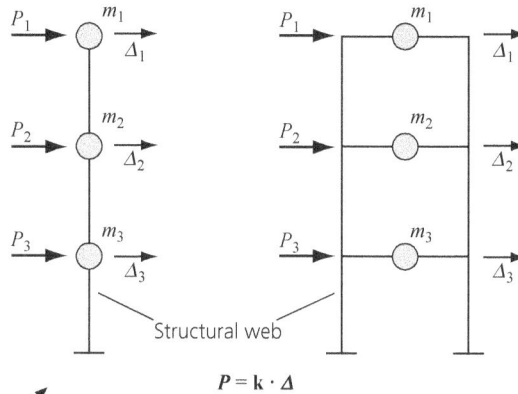

$$P = k \cdot \Delta$$

The 'rectilinear' model is used merely to
represent a structural web having a stiffness matrix [k]
The two structural systems exhibit a totally different behaviour under lateral loads

First, it is assumed that mass, m_i, excited by an external force $F_i(t)$ in the direction of its own degree of freedom – that is, horizontally – undergoes a displacement $x_i(t)$. The positive sense for all the forces and displacements is considered common (see Figure 7.29).

Both the set of forces $F_i(t)$ and the set of displacements $x_i(t)$ may be represented by the column matrices $F(t)$ and x, respectively. For reasons of matrix manipulation practice only, the system of masses, m_i, is represented by the square diagonal matrix $\mathbf{m}$ of order $(N \cdot N)$. In this matrix, masses m_i constitute the diagonal elements with all the other elements being zero. It may be written as:

$$\mathbf{m} = \mathbf{I} \cdot m$$

Figure 7.29 Dynamic equilibrium in the multi-degree-of-freedom system under external loading

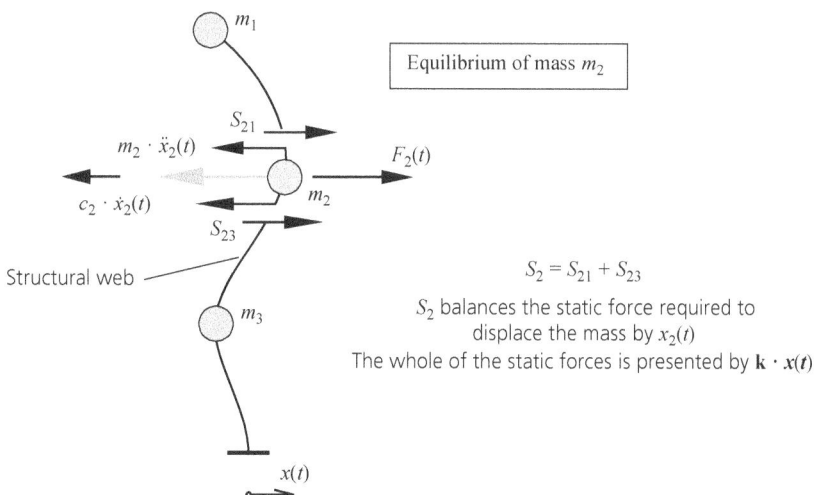

Just like in a single-degree-of-freedom system, a specific mass, m_i, may be considered as being always in equilibrium under the following forces (see Figure 7.29):

(a) the external force, $F_i(t)$
(b) the total of the elastic forces, S_i, transmitted from the structural web to the mass, as a consequence of its deformational configuration $x(t)$
(c) the inertia force, $m_i \cdot \ddot{x}_i(t)$
(d) the damping force, $c_i \cdot \dot{x}_i(t)$

As forces (b), (c) and (d) have the opposite sense to force $F_i(t)$, for the equilibrium of each mass, m_i, it must be (see Figure 7.29):

$$m_i \cdot \ddot{x}_i(t) + c_i \cdot \dot{x}_i(t) + S_i = F_i(t)$$

It is clear that the deformational configuration $x(t)$ for the specific time instant (t) may be produced by a unique set of external forces depending on the stiffness of the structural web. According to the previous explanation, this set of forces can be given by the expression $k \cdot x(t)$ (column matrix $N \cdot 1$) and it is clear that each of these forces equals the corresponding sum, S_i, of the elastic forces acting directly on mass, m_i.

The simultaneous consideration of equilibrium of all masses may now be expressed by the following matrix formulation:

$$\mathbf{m} \cdot \ddot{x} + \mathbf{c} \cdot \dot{x} + \mathbf{k} \cdot x = \mathbf{F}(t)$$

The damping coefficients, c_i, are represented by the square diagonal matrix $\mathbf{c}$. ∎

Figure 7.30 Eigenforms and dynamic characteristics of a multi-degree-of-freedom system

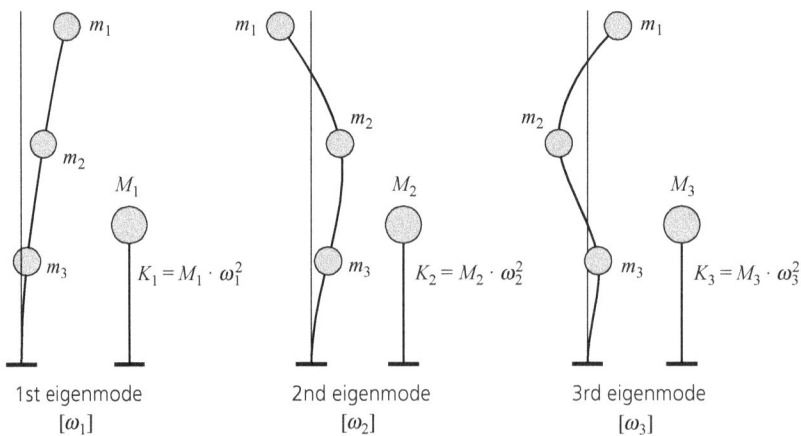

1st eigenmode
$[\omega_1]$

2nd eigenmode
$[\omega_2]$

3rd eigenmode
$[\omega_3]$

If a structural web is deformed according to some eigenmode and then left free, then all the masses are vibrating in phase under the corresponding eigenfrequency
The whole system acts then as an ideal mass with an ideal stiffness

As for a single-degree-of-freedom system, it is expedient here also to seek the basic dynamic characteristics of a multi-degree-of-freedom system through the examination of its undamped free vibration.

The free vibration is evoked by imposing an arbitrary displacement at a certain mass, according to its corresponding degree of freedom (e.g. a horizontal displacement $\Delta_3 = \delta$) and then leaving the system free to vibrate.

In this case, the global equilibrium condition is written as:

$$\mathbf{m} \cdot \ddot{x} + \mathbf{k} \cdot x = 0$$

By trying out the *harmonic* solution $x = \boldsymbol{\varphi} \cdot \sin(\omega t)$, where the magnitude ω is, of course, unknown and, moreover, the column matrix $\boldsymbol{\varphi}$ also consists of unknown quantities, given that $\ddot{x} = -\omega^2 \cdot \boldsymbol{\varphi} \cdot \sin(\omega t)$, the above equation is written as (Tedesko *et al.*, 1999):

$$[\mathbf{m}^{-1} \cdot \mathbf{k} - \omega^2 \cdot \mathbf{I}] \cdot \boldsymbol{\varphi} = 0 \qquad \text{or} \qquad [\mathbf{k} - \omega^2 \mathbf{m}] \boldsymbol{\varphi} = 0$$

This equation, known as the *eigenvalue equation* or *frequency equation* of the system, represents a linear system of N equations with N unknowns φ_i. According to the laws of linear algebra, in order for a non-zero solution to exist in a linear system having all its right-side members equal to zero, the determinant of the coefficients of the unknowns must be equal to zero. This determinant, as can be concluded from the last equation, consists in a polynomial of Nth degree with respect to ω^2, for which a number of N values (roots) can be determined. It is, of course, clear that for each *eigenvalue* ω^2, the corresponding solution of the above linear system does not simply consist of a well-defined set of values φ_i, but rather of an infinite group of values $\lambda \cdot \varphi_i$, where λ is an arbitrary number.

It is clear that because of the computational effort required for the determination of the above magnitudes, these can be obtained using appropriate computer software.

From a physical point of view, the above signifies that the N-degree-of-freedom system has N different *eigenfrequencies, ω*, and to each one of them corresponds a certain displacement configuration, $\boldsymbol{\varphi}$, referring to each degree of freedom, which may be considered as multiplied by an arbitrary factor, positive or negative. This displacement configuration is called the *eigenmode* or *natural mode* (see Figure 7.30). The physical meaning of any of the eigenmodes, $\boldsymbol{\varphi}$, is that if it is somehow imposed on the system (i.e. through an appropriate application of horizontal displacements) and the system is left free, then every mass of the system will vibrate like a single-mass oscillator under the same natural frequency, ω, which corresponds to the selected eigenmode.

An additional analytical property of the eigenmodes, $\boldsymbol{\varphi}$, is pointed out here. For any pair of different eigenmodes $\boldsymbol{\varphi}_r$ and $\boldsymbol{\varphi}_s$, it is:

$$\boldsymbol{\varphi}_r^T \cdot \mathbf{m} \cdot \boldsymbol{\varphi}_s = 0 \quad \text{and} \quad \boldsymbol{\varphi}_r^T \cdot \mathbf{k} \cdot \boldsymbol{\varphi}_s = 0$$

which is known as the *orthogonality condition* of the eigenmodes, $\boldsymbol{\varphi}$.

The fact that the whole system may vibrate at any of the N eigenfrequencies, ω_r, exhibiting for each mass amplitudes represented by the eigenmode, $\boldsymbol{\varphi}_r$, meaning that all masses are found at the same time either in the rest position or in their maximum deviation, suggests the idea of considering an ideal (fictitious) mass, M_r, connected to an ideal (fictitious) stiffness, K_r (see Figure 7.30). It is indeed found that for the corresponding eigenmode, φ_r, the last magnitudes may be determined as:

$$M_r = \varphi_r^T \cdot \mathbf{m} \cdot \varphi_r = \sum_1^N m_r \cdot \varphi_r^2 \quad \text{and} \quad K_r = \varphi_r^T \cdot \mathbf{k} \cdot \varphi_r$$

According, of course, to the characteristics of a typical single-degree-of-freedom system, it must be (Biggs, 1964):

$$\omega_r^2 = K_r / M_r$$

In this way, for a certain multi-degree-of-freedom system, N single-mass oscillators may be considered, each one of them with specific characteristics of mass, stiffness and frequency. The lowest frequency, ω, of them all is called the *fundamental frequency* and often plays the governing role of greater physical importance than the others, as will be discussed later.

More specifically, if an arbitrary horizontal displacement is imposed on any mass of any of the two systems shown in Figure 7.31, and then the structure is left free, every mass of the respective system will vibrate at a frequency that practically coincides with the fundamental one. This fact, which may apply to any multi-degree-of-freedom system, leads to an important result that allows the determination of its fundamental frequency.

If on every mass of a multi-degree-of-freedom system an arbitrary static force, F_r, is applied, then the masses will be displaced by Δ_r, respectively (see Figure 7.32) and, if the system is then left free, a harmonic motion for every mass with a natural frequency, ω, will be immediately established, whereby all masses take both their rest position and their maximum deviation, Δ_r, at the same time instant.

Figure 7.31 Causing the fundamental frequency to appear

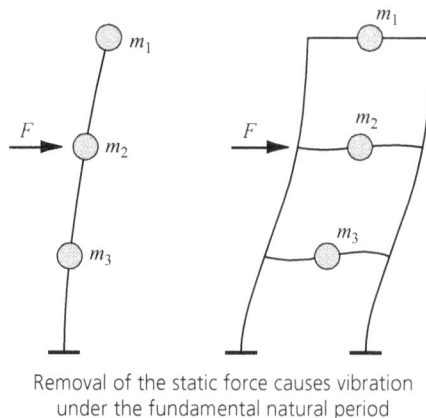

Removal of the static force causes vibration under the fundamental natural period

Figure 7.32 Practical evaluation of the fundamental eigenfrequency

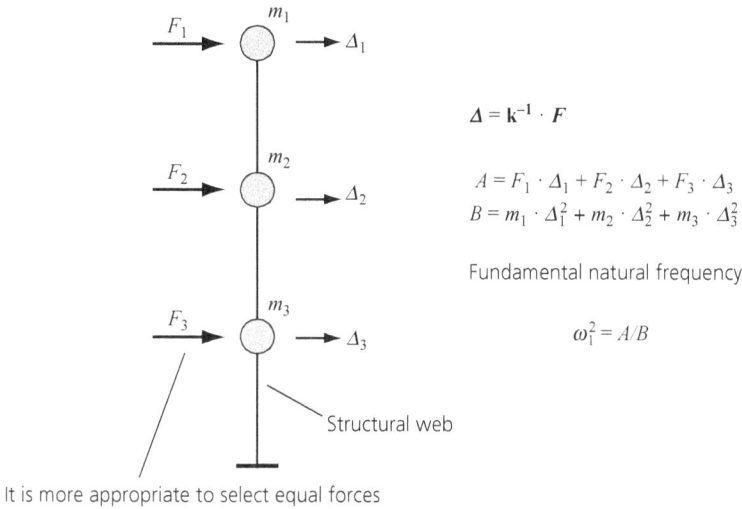

$$\Delta = k^{-1} \cdot F$$

$$A = F_1 \cdot \Delta_1 + F_2 \cdot \Delta_2 + F_3 \cdot \Delta_3$$
$$B = m_1 \cdot \Delta_1^2 + m_2 \cdot \Delta_2^2 + m_3 \cdot \Delta_3^2$$

Fundamental natural frequency

$$\omega_1^2 = A/B$$

Structural web

It is more appropriate to select equal forces

According to the characteristics of the single-degree-of-freedom systems, the maximum velocity of every mass at rest position will be ($\omega \cdot \Delta_r$) and, consequently, the kinetic energy of the whole of masses will be $\sum (1/2) \cdot m_r \cdot (\omega \cdot \Delta_r)^2$.

Clearly, this kinetic energy should be equal to the total work done by forces, F_r, which is $\sum (1/2) \cdot F_r \cdot \Delta_r$. Thus, the value of the fundamental frequency, ω, may be readily obtained as (Biggs, 1964):

$$\omega^2 = \frac{\sum F_r \cdot \Delta_r}{\sum m_r \cdot (\Delta_r)^2}$$

The practical computation is particularly facilitated if forces, F_r, are selected as equal (see Figure 7.32). In this case, the displacements, Δ_r, may be also used to represent approximately the displacement components of the first (fundamental) eigenmode vector φ_1. ∎

The damping action is logically expressed through the damping coefficient, ζ, with regard to each eigenmode. To the first eigenmode corresponding to the fundamental frequency, ω, the value of ζ usually applied is assigned as directly related to the nature of the structural web material, according to Section 7.2.2.2.

For the experimental determination of the damping coefficient, as well as of the eigenfrequency of each eigenmode, the same procedure may be followed as in the case of a single-degree-of-freedom system – that is, by creating through appropriate forces the corresponding displacement configuration φ_r and following the corresponding free vibration of the system (see Section 7.2.2.2).

The measured eigenfrequencies for each eigenmode should not differ practically from those obtained theoretically, as explained above, provided that the damping coefficients do not exceed 10%.

7.3.4.2   Approximate evaluation of the fundamental period in multi-storey systems

For a multi-storey system consisting of a certain number (N) of horizontal diaphragms connected together and to the ground through plane frame elements as well as shear walls in orthogonal layout, it is possible to make an assessment of the fundamental period of the system on the basis of the result given in Section 7.3.4.1.

Considering that all diaphragms are subjected to the same lateral force, F, acting on the centre of stiffness of each diaphragm along a selected direction (x or y), according to Chapter 6, Section 6.3.5, the displacement Δ_i for level (i) can be approximately assessed by the relation:

$$\Delta_i = \Sigma \frac{V_i}{K^i}$$

where (see Section 6.3.5):

$$K^i = \Sigma k \cdot \frac{h}{h_i} + (3 \cdot E / h_i^3) \cdot \Sigma I$$

referring to the selected direction x or y, respectively.

Thus, by applying the final equation in Section 7.3.4.1, it may be written as:

$$\omega^2 = F \cdot \frac{\Sigma \Delta_i}{\Sigma m \cdot \Delta_i^2} \quad \text{(see Figure 7.33)}$$

allowing the assessment of the fundamental period $T = 2 \cdot \pi / \omega$ of the system. It is clear that both orthogonal stiffness directions must be examined according to the following procedure (see Figure 7.33).

Figure 7.33 Practical determination of the fundamental eigenfrequency of a multi-storey system

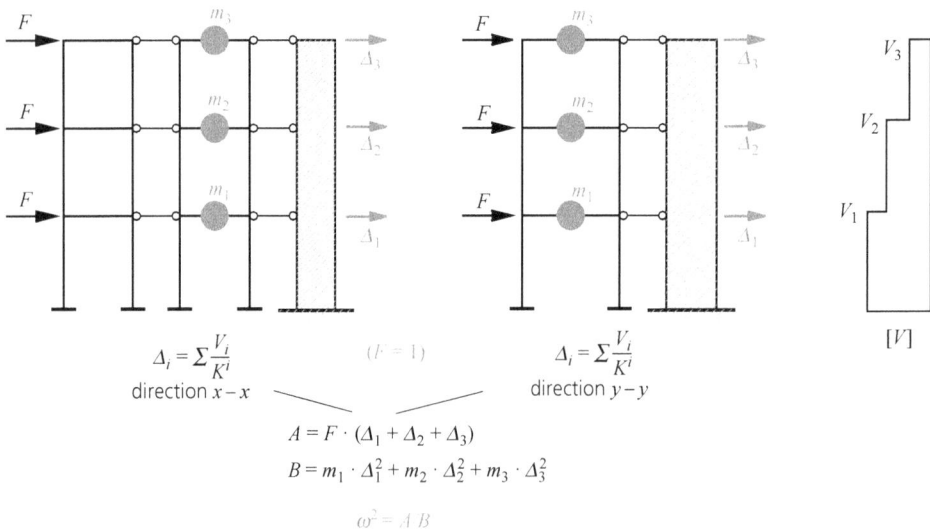

$$\Delta_i = \Sigma \frac{V_i}{K^i}$$
direction $x - x$

$$(F = 1)$$

$$\Delta_i = \Sigma \frac{V_i}{K^i}$$
direction $y - y$

$$A = F \cdot (\Delta_1 + \Delta_2 + \Delta_3)$$
$$B = m_1 \cdot \Delta_1^2 + m_2 \cdot \Delta_2^2 + m_3 \cdot \Delta_3^2$$

$$\omega^2 = A \, B$$

- A plane formation is created separately for the *x*- and *y*-directions, consisting of the respective plane elements in each direction,
- Unit loads are applied to the different levels of the structural web, for which the corresponding displacements are calculated. The fundamental eigenfrequency for the considered direction (*x* or *y*) is obtained by direct application of the above expression of ω^2.

Worked example

The plan view shown in Figure 7.34 applies for three consecutive storeys, each having a height of 3.50 m and a slab thickness of 25 cm. All the frame girders have a width and depth of 30 cm and 80 cm, respectively, while all the columns have a section of 50/30 cm. The wall elements have a thickness of 20 cm.

Considering as effective mass, *m*, only the one corresponding to the slab's self-weight, it is:

Slab self-weight $= 15.40 \cdot 8.0 \cdot 0.25 \cdot 25.0 = 770$ kN and the mass is $m = 770/9.81 = 78.5$ t

The distinction of the two plane systems is made as shown in Figure 7.33.

X-DIRECTION

The lateral stiffness of each frame-storey is equal to 41 840 kN/m. The total stiffness, *K*, for each level is:

$$K^1 = (41840 + 41840) \cdot (3.50/3.50) + 3 \cdot 10^7 \cdot 2.0^3 \cdot 0.20 / (3.5^3 \cdot 12) = 176974 \text{ kN/m}$$

$$K^2 = (41840 + 41840) \cdot (3.50/7.00) + 3 \cdot 10^7 \cdot 2.0^3 \cdot 0.20 / (7.0^3 \cdot 12) = 53501 \text{ kN/m}$$

$$K^3 = (41840 + 41840) \cdot (3.50/10.5) + 3 \cdot 10^7 \cdot 2.0^3 \cdot 0.20 / (10.5^3 \cdot 12) = 31348 \text{ kN/m}$$

By applying horizontal unit loads at the corresponding levels of each plane system (*F* = 1 kN), the following displacements can be obtained:

$$\Delta_1 = 3.0/176974 = 0.017 \text{ mm}$$
$$\Delta_2 = 3.0/176974 + 2.0/53501 = 0.0543 \text{ mm}$$
$$\Delta_3 = 3.0/176974 + 2.0/53501 + 1.0/31348 = 0.086 \text{ mm}$$

Figure 7.34 Plan view layout of vertical elements of stiffness

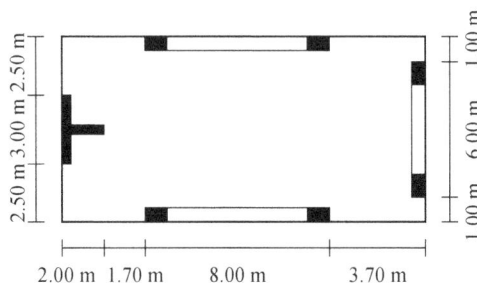

whereby the numbering of levels increases upwards to conform with the notation in Figure 7.33 and, according to the expression given in Section 7.3.4.2, it is:

$$\omega^2 = \frac{1.0}{78.5} \cdot \frac{0.000017 + 0.0000543 + 0.000086}{0.000017^2 + 0.0000543^2 + 0.0000862^2} = 188.4 \, sec^{-2}$$

with $\omega = 13.7$/sec, the period is:

$$T = 2 \cdot \pi / \omega = 0.46 \, sec$$

Y-DIRECTION
The lateral stiffness of each frame-storey is equal to 43 922 kN/m

The total stiffness, K, for each level is:

$$K^1 = (43\,922) \cdot (3.50 / 3.50) + 3 \cdot 10^7 \cdot 3.0^3 \cdot 0.20 / (3.5^3 \cdot 12) = 358791 \, kN/m$$
$$K^2 = (43\,922) \cdot (3.50 / 7.00) + 3 \cdot 10^7 \cdot 3.0^3 \cdot 0.20 / (7.0^3 \cdot 12) = 61320 \, kN/m$$
$$K^3 = (43\,922) \cdot (3.50 / 10.5) + 3 \cdot 10^7 \cdot 3.0^3 \cdot 0.20 / (10.5^3 \cdot 12) = 26303 \, kN/m$$

By applying horizontal unit loads at the corresponding levels of each plane system ($F = 1$ kN), the following displacements are obtained:

$\Delta_1 = 3.0 / 358791 = 0.0084 \, mm$
$\Delta_2 = 3.0 / 358791 + 2.0 / 61320 = 0.041 mm$
$\Delta_3 = 3.0 / 358791 + 2.0 / 61320 + 1.0 / 26303 = 0.079 \, mm$

whereby the upwards numbering of levels is again emphasised, conforming to the notation in Figure 7.33, and, according to the expression given in Section 7.3.4.2, it is determined as:

$$\omega^2 = \frac{1.0}{78.5} \cdot \frac{0.0000084 + 0.000041 + 0.000079}{0.0000084^2 + 0.000041^2 + 0.000079^2} = 204.6 \, sec^{-2}$$

with $\omega = 14.3$/sec, the period is:

$$T = 2 \cdot \pi / \omega = 0.44 \, sec$$

7.3.4.3   Multi-storey spatial system
In the case of a multi-storey building with horizontal diaphragms arranged vertically over each other, and that are carried by the structural web consisting of vertical plane elements, as examined in Chapter 6, it becomes clear that the dynamic equilibrium of the diaphragm masses is related to displacements Δx, Δy of each diaphragm, as well as to its rotation $\Delta \omega$. While displacements Δx and Δy activate the inertia forces ($m \cdot \Delta \ddot{x}$) and ($m \cdot \Delta \ddot{y}$) of the mass of each diaphragm, the rotation $\Delta \omega$ activates the rotational moment of inertia ($\Theta \cdot \Delta \ddot{\omega}$) on the basis of the rotational moment of inertia Θ of the diaphragm's mass with respect to its centroid.

In the case of an equally distributed mass in the diaphragm, it is

$$\Theta = m \cdot (I_x + I_y)$$

where I_x and I_y are the moment of inertia of the diaphragm area with respect to the x and y axes, respectively.

By taking into account the *expanded* mass matrix ($3N \times 3N$):

$$\bar{m} = \begin{bmatrix} m & 0 & 0 \\ 0 & m & 0 \\ 0 & 0 & \Theta \end{bmatrix}$$

the eigenfrequency equation is obtained analogously to that obtained previously (Stavridis, 1986)

$$\left[\bar{m}^{-1} \cdot K - \omega^2 \cdot I \right] \cdot \varphi = 0$$

The eigenfrequencies, ω, can be determined using computer software.

In the case in which the diaphragm rotations are negligible, even approximately, the fundamental eigenfrequency of the whole system may be assessed with practically satisfactory accuracy on the basis of the expression of ω^2 given in Section 7.3.4.2.

7.3.5   Forced vibration

The response of a multi-degree-of-freedom system subjected to external forces is governed, as analysed in Section 7.3.2.1, by the system of differential equations that is expressed in a condensed matrix form as (Müller and Keintzel, 1984):

$$m \cdot \ddot{x} + c \cdot \dot{x} + k \cdot x = F(t)$$

As in free vibration, in the present case it is also found that the multi-degree-of-freedom system behaves as a system of N discrete single-mass systems that are independent of each other, having a fictitious mass, M, as well as a fictitious stiffness, K, according to the expressions given in Section 7.3.4.1 (see Figure 7.35). Each such system is governed by the equation:

$$M_r \cdot \ddot{q}_r + 2\zeta_r M_r \omega_r \cdot \dot{q}_r + K_r \cdot q_r = \varphi_r^T \cdot F(t)$$

which, for the case usually encountered where $F(t) = P \cdot f(t)$ – that is, when all loads exhibit the same variation in time – it may be further written as:

$$M_r \cdot \ddot{q}_r + 2\zeta_r M_r \omega_r \cdot \dot{q}_r + K_r \cdot q_r = \varphi_r^T \cdot P \cdot f(t) = \left(\sum_1^N \varphi_i^{(r)} \cdot P_i \right) \cdot f(t)$$

Although such an equation may be analytically treated according to Section 7.2.3, the numeric solution obtained by suitable computer software is more practical, provided that the time steps used in the solution are smaller than $1/10$ of the corresponding eigenperiod $T_r = 2\pi/\omega_r$.

It is noted that the determination of the displacements $q_r(t)$ of the N fictitious single-degree-of-freedom system allows the calculation of the truly developed displacements $x(t)$ of the masses

Figure 7.35 Dynamic response of a multi-degree-of-freedom system under external loads

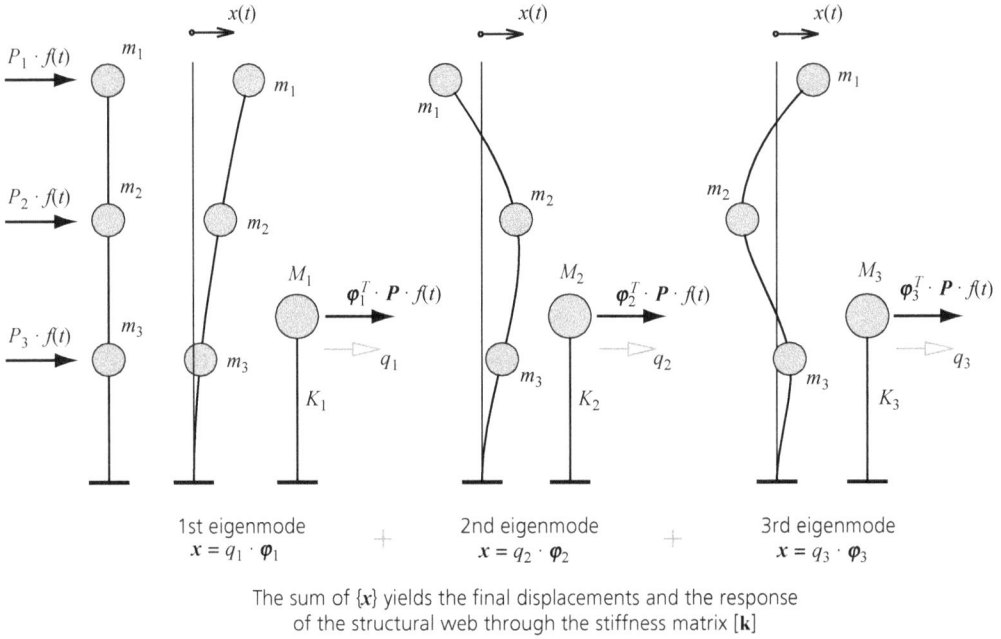

1st eigenmode
$x = q_1 \cdot \boldsymbol{\varphi}_1$

+

2nd eigenmode
$x = q_2 \cdot \boldsymbol{\varphi}_2$

+

3rd eigenmode
$x = q_3 \cdot \boldsymbol{\varphi}_3$

The sum of {x} yields the final displacements and the response
of the structural web through the stiffness matrix [k]

in the multi-degree-of-freedom system, according to relation $x(t) = \boldsymbol{\Phi} \cdot q(t)$, where $\boldsymbol{\Phi}$ is a square matrix $(N \cdot N)$, obtained from the juxtaposition of the N column matrices $\boldsymbol{\varphi}_r$ (see Figure 7.35). Moreover, the fact that each eigenmode may be considered multiplied by an arbitrary factor, λ, does not affect the resulting displacements $x(t)$ at all, as this factor, according to the above equations, is finally eliminated.

7.4. Seismic excitation
7.4.1 Dynamic analysis
The governing factor for the response of a system due to seismic excitation is, just as in the single-degree-of-freedom case (see Section 7.2.5), the accelerogram $\ddot{u}_s(t)$, expressed as:

$$\ddot{u}_s(t) = a \cdot f_a(t)$$

where a represents the maximum ground acceleration (see Figure 7.36).

It is obvious that the inertia mass forces are caused by their absolute acceleration, whereas the elastic as well as the damping forces acting on the masses depend only on their relative displacements to the ground. Given that no external forces act on the system, the condensed equation of dynamic equilibrium according to the previous section is written as:

$$\mathbf{m} \cdot \left[\ddot{x} + \boldsymbol{I} \cdot a \cdot f_a(t) \right] + \mathbf{c} \cdot \dot{x} + \mathbf{k} \cdot x = \mathbf{0}$$

This equation may also be written as:

$$\mathbf{m} \cdot \ddot{x} + \mathbf{c} \cdot \dot{x} + \mathbf{k} \cdot x = -\mathbf{m} \cdot \mathbf{I} \cdot a \cdot f_a(t)$$

where $\mathbf{I}$ represents a column matrix containing N unit elements.

Thus, the problem of seismic excitation is posed as a problem of forced vibration according to the previous section. In other words, the multi-degree-of-freedom system supported on fixed ground under a seismic excitation is equivalent to the same system subjected to the external loads expressed by the right-hand side of the last equation (see Figure 7.36).

As discussed previously, the multi-degree-of-freedom system can be 'split' into N single-mass oscillators, each of them corresponding to an eigenfrequency, ω, with the accompanying eigenmode, φ. On this basis, the corresponding fictitious mass, M_r, and the fictitious stiffness, K_r, may be determined. The equation of motion for the r-th eigenfrequency is written as (Müller and Keintzel, 1984):

$$M_r \cdot \ddot{q}_r + 2\zeta_r M_r \omega_r \cdot \dot{q}_r + K_r \cdot q_r = -\varphi_r^T \cdot \mathbf{m} \cdot \mathbf{I} \cdot a \cdot f_a(t)$$

and dividing it by M_r:

$$\ddot{q}_r + 2\zeta\omega_r \cdot \dot{q}_r + \omega_r{}^2 \cdot q_r = -\varphi_r^T \cdot \mathbf{m} \cdot \mathbf{I} \frac{1}{M_r} \cdot a \cdot f_a(t)$$

Figure 7.36 Seismic loading of a multi-degree-of-freedom system

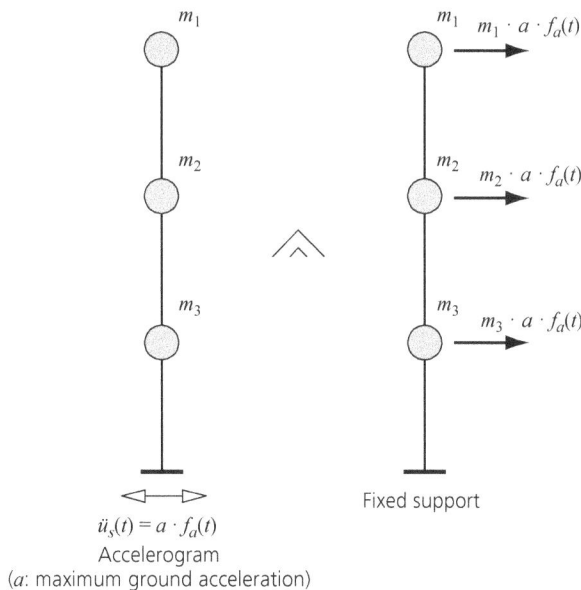

m_1

m_2

m_3

$\ddot{u}_s(t) = a \cdot f_a(t)$
Accelerogram
(a: maximum ground acceleration)

m_1 $m_1 \cdot a \cdot f_a(t)$

m_2 $m_2 \cdot a \cdot f_a(t)$

m_3 $m_3 \cdot a \cdot f_a(t)$

Fixed support

This equation is identical to the equation of motion of the single-mass system in Section 7.2.5, which is:

$$\ddot{x} + (2 \cdot \zeta \cdot \omega) \cdot \dot{x} + \omega^2 \cdot x = -a \cdot f_a(t)$$

It is observed that the maximum ground acceleration, a, is now multiplied by the factor:

$$\Gamma_r = \frac{\varphi_r^T \cdot \mathbf{m} \cdot \mathbf{I}}{M_r} = \frac{\sum m_i \cdot \varphi_i^{(r)}}{\sum m_i \cdot [\varphi_i^{(r)}]^2}$$

which as a pure number is characteristic of the eigenmode r and is called the *participation factor*. If multiplied by the quantity $\sum m_i \cdot \varphi_i^{(r)}$, it leads to the mass $M_{eff}^{(r)}$, which can be considered to 'represent' the r-th eigenmode in the dynamic response of the system (see also Section 7.5.2). Its relation to the total mass $M = \sum m_i$ of the system is expressed by the ratio $\varepsilon_r = M_{eff}^{(r)} / M$, which shows the degree of participation of the r-th eigenmode in the overall system response. It is:

$$M_{eff}^{(r)} = \Gamma_r \cdot \sum m_i \cdot \varphi_i^{(r)} = \left(\sum m_i \cdot \varphi_i^{(r)}\right)^2 / M_r = \left(\sum m_i \cdot \varphi_i^{(r)}\right)^2 / \sum m_i \cdot \left(\varphi_i^{(r)}\right)^2$$

$M_{eff}^{(r)}$ is called the *effective mass of the r-th eigenmode*. The degree, ε_r, of its contribution to the total mass decreases with increasing number of the eigenmode. It can be shown that the sum of the effective masses $M_{eff}^{(r)}$ for all eigenmodes equals the total mass $\sum m_i$ of the system – that is, $\sum \varepsilon_r = 1$. However, a practically acceptable approximation of the problem is obtained if the analysis is restricted to that number, s, of eigenmodes, which ensures that:

$$\sum_1^s \varepsilon_i \geq 0.90$$

Attention is drawn here to not confuse the concept of the *effective mass*, $M_{eff}^{(r)}$, with that of the *eigenmode mass*, M_r, as defined in Section 7.3.4.1. It should be emphasised that in contrast to the effective masses, $M_{eff}^{(r)}$, the sum of the eigenmode masses, M_r, is not equal to the total mass of the system.

According now to the results of the single-mass oscillator, it is possible to determine the corresponding $(q_r)_{max}$ through the acceleration spectrum for the earthquake on the basis of the value $S_a^{(r)}$ for the r-th eigenfrequency, considering as maximum ground acceleration the value $(\Gamma_r \cdot a)$ instead of a. Hence, according to Section 7.2.5, it is:

$$(q_r)_{max} = \frac{1}{\omega^2} \cdot (\ddot{q}_r + \ddot{u}_s)_{max} = \frac{1}{\omega^2} \cdot \Gamma_r \cdot a \cdot \omega \cdot \left[\int_0^t f_a(\tau) \cdot e^{-\zeta\omega(t-\tau)} \cdot \sin\omega(t-\tau)d\tau \right]_{max}$$

that is,

$$(q_r)_{max} = \frac{S_a^{(r)}}{\omega_r^2} \cdot \Gamma_r$$

Figure 7.37 Seismic analysis of a multi-degree-of-freedom system

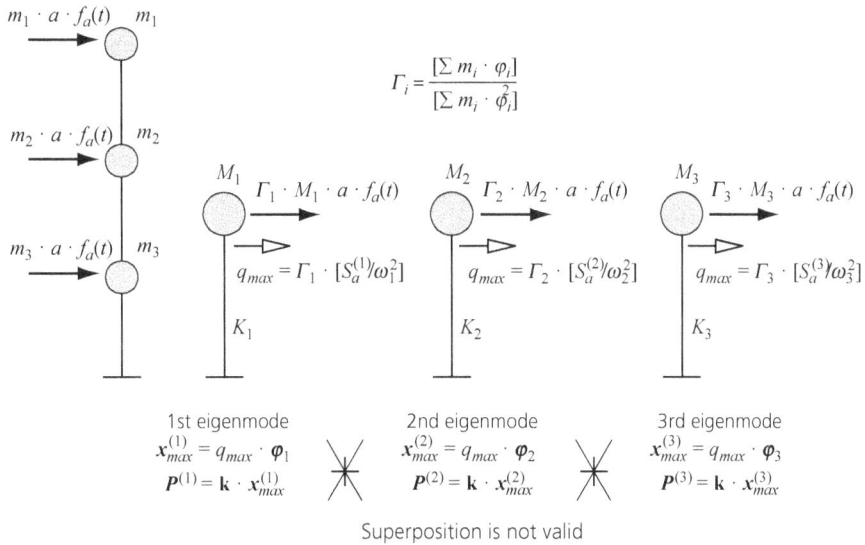

$$\Gamma_i = \frac{[\sum m_i \cdot \varphi_i]}{[\sum m_i \cdot \varphi_i^2]}$$

$m_1 \cdot a \cdot f_a(t) \quad m_1$

$m_2 \cdot a \cdot f_a(t) \quad m_2$

$m_3 \cdot a \cdot f_a(t) \quad m_3$

$M_1 \quad \Gamma_1 \cdot M_1 \cdot a \cdot f_a(t)$

$q_{max} = \Gamma_1 \cdot [S_a^{(1)}/\omega_1^2]$

K_1

$M_2 \quad \Gamma_2 \cdot M_2 \cdot a \cdot f_a(t)$

$q_{max} = \Gamma_2 \cdot [S_a^{(2)}/\omega_2^2]$

K_2

$M_3 \quad \Gamma_3 \cdot M_3 \cdot a \cdot f_a(t)$

$q_{max} = \Gamma_3 \cdot [S_a^{(3)}/\omega_3^2]$

K_3

1st eigenmode
$x_{max}^{(1)} = q_{max} \cdot \varphi_1$
$P^{(1)} = k \cdot x_{max}^{(1)}$

2nd eigenmode
$x_{max}^{(2)} = q_{max} \cdot \varphi_2$
$P^{(2)} = k \cdot x_{max}^{(2)}$

3rd eigenmode
$x_{max}^{(3)} = q_{max} \cdot \varphi_3$
$P^{(3)} = k \cdot x_{max}^{(3)}$

Superposition is not valid

Now, for each oscillator (r) the mass displacements $(x_{max})_r$ of the multi-degree-of-freedom system may be determined on the basis of the computed $(q_r)_{max}$. It is (see Figure 7.37):

$$(x_{max})_r = \varphi_r \cdot (q_r)_{max} = \frac{\Gamma_r \cdot S_a^{(r)}}{\omega_r^2} \cdot \varphi_r$$

Thus, a maximum deviation is assigned to each mass, corresponding to each of the N fictitious oscillators. Of course, the actually developed displacement of a specific mass may not be deduced as the sum of the above maximum deviations $(x_{max})_r$ on the basis of the N oscillators, because the corresponding $(q_r)_{max}$ does not refer to the same time instant. Nevertheless, according to a practically acceptable probabilistic consideration, this mass deviation may be obtained as the square root of the sum of the squares of the above components.

Regarding now the determination of the maximum response in the structural web, it is clear that for the development of the above displacements $(x_{max})_r$ of the N masses, the action of those forces P_r is required, which are suggested by the stiffness of the structural web itself, according to the relation:

$$P_r = k \cdot (x_{max})_r = \frac{\Gamma_r \cdot S_a^{(r)}}{\omega_r^2} \cdot k \cdot \varphi_r$$

For the determination of the stiffness matrix k, Section 7.3.1.1 may be followed.

It is thus possible to determine the N systems of external forces P to the maximum (real) deviations x_{max} corresponding to the N (fictitious) oscillators. However, as these forces do not correspond to simultaneously developed displacements, their superposition is not meaningful and one has to refer to the above-mentioned treatment through the square root of the sum of their squares in order to obtain a practically acceptable approximation for loading of the structural system.

7.4.2   Equivalent static loads

In order to take into account the contribution of all the eigenmodes of a system in its seismic excitation, is difficult to manage the bulk of calculations involved in Section 7.4.1, particularly for preliminary design purposes. This inconvenience leads to the search for such a static loading acting on the masses of the system so that the resulting response approximates to that due to the 'exact' dynamic analysis. The practical suggestion to that purpose is given by the fact that, for a regular distribution of mass and stiffness in a structural web, it is generally the first eigenmode that plays the governing role in its dynamic response (Müller and Keintzel, 1984).

Thus, in seeking these equivalent static loads, it is approximately but conservatively assumed that they may be represent by the maximum inertia forces H that the system's masses are subjected to. For the r-th eigenmode, and according to the previous equation in Section 7.4.1 and Figure 7.38 as well as the eigenvalue equation in section 7.3.4.1, it is:

$$H_r = k \cdot (x_{max})_r = k \cdot \varphi_r \cdot (q_r)_{max} = \omega_r^2 \cdot m \, \varphi_r \cdot (q_r)_{max}$$

On the basis of the expression of $(q_r)_{max}$ in Section 7.4.1 (see also Section 7.3.5):

$$H_r = m \cdot \varphi_r \cdot \frac{\Sigma m_i \cdot \varphi_i^{(r)}}{\Sigma m_i \cdot \left(\varphi_i^{(r)}\right)^2} \cdot S_a^{(r)}$$

Looking now for a more convenient expression for these forces, their sum (ΣH) over the whole H_r height of the structural web is considered. This force called the *base shear* constitutes the sum, V_r, of the actions which the masses are subjected to and it is transmitted directly to the ground. It thus comes out that:

$$\Sigma H_r = V_r = \frac{\left(\Sigma m_i \cdot \varphi_i^{(r)}\right)^2}{\Sigma m_i \cdot \left(\varphi_i^{(r)}\right)^2} \cdot S_a^{(r)}$$

Here, it should be noted that while strictly speaking $S_a^{(r)}$ applies to a certain accelerogram, for design purposes the greatest spectrum design value should be considered; therefore, the notation $S_a^{(r)}$ will be considered as referring to the corresponding spectrum.

Recognising in the above equation the expression of the active mass $M_{eff}^{(r)}$ of the r-th eigenmode as defined in Section 7.4.1, it can be written as:

$$V_r = M_{eff}^r \cdot S_a^{(r)}$$

This result confirms the role of the effective mass $M_{eff}^{(r)}$ of the r-th eigenmode, as mentioned in Section 7.4.1, according to the degree of participation of ε_r in the total mass as $\varepsilon_r = M_{eff}^{(r)} / M$.

The base shear force, V_r, is distributed over the levels of the structure according to the above expression of the forces, $H_i^{(r)}$:

$$H_i^{(r)} = V_r \cdot \frac{m_i \cdot \varphi_i^{(r)}}{\sum m_i \cdot \varphi_i^{(r)}}$$

It can now be assumed that the number, s, of eigenmodes required to satisfactorily approximate the state of stress of the structure through the equivalent static loads is that which ensures that the sum of the corresponding participation degrees, ε, is greater than 90% (see Section 7.4.1).

By assuming now that the degree of participation ε_1 of the first eigenmode is generally of the order of 0.9, then the effective mass $M_{eff}^{(1)}$ can be considered approximately equal to the total mass, M, of the system as mentioned above and thus the base shear may be expressed as:

$$V = M \cdot S_a^{(1)}$$

being distributed at each level (i) according to the relation:

$$H_i = V \cdot \frac{m_i \cdot \varphi_i^{(1)}}{\sum m_i \cdot \varphi_i^{(1)}}$$

as has been adopted by most seismic codes.

The last expression for equivalent static loads may be further simplified for the purposes of preliminary design. Considering the first eigenmode as a straight line, the relations of the eigenmode coefficients $\varphi^{(1)}$ can be replaced by the relations of the heights of the different levels of the model according to the expression:

$$H_i = V \cdot \frac{m_i \cdot h_i}{\sum m_i \cdot h_i} \quad \text{(see Figure 7.38)}$$

More specifically, for a multi-storey building – as considered in Section 7.3.3 and from a dynamic viewpoint in Section 7.4.1 – which is subjected to a certain ground acceleration, in order to determine the equivalent horizontal static load for each level, the following procedure should be followed.

▨ The value $S_a^{(1)}$ and then base shear, V, are determined from the acceleration spectrum used, separately for the two directions, on the basis of the fundamental eigenfrequency in x and y, (e.g. calculated according to Section 7.3.4.2), as well as of the maximum adopted ground acceleration.

▨ The corresponding equivalent static loads are obtained through direct application of one of the two last expressions. In the case in which the eigenmode values φ_i are used, these may be considered as the already determined displacements of the equivalent plane system under the horizontal unit loads, applied in order to assess its fundamental frequency, as described in Section 7.3.4.2. It is clear that, for a uniform distribution of masses, these static loads increase toward higher storeys.

▨ On the basis now of the forces acting on each storey's diaphragm, the response of the vertical – as well as the horizontal – elements of the multi-storey system may be readily assessed, according to Chapter 6, Sections 6.3.1 and 6.3.2.

Figure 7.38 Determination of equivalent static loading in a multi-storey system

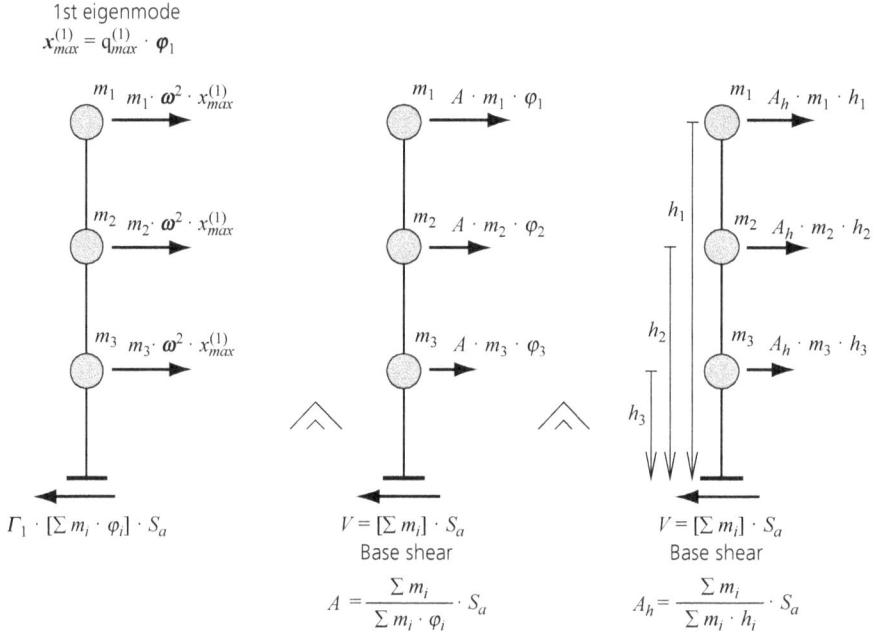

1st eigenmode
$$x_{max}^{(1)} = q_{max}^{(1)} \cdot \varphi_1$$

m_1 $m_1 \cdot \omega^2 \cdot x_{max}^{(1)}$

m_2 $m_2 \cdot \omega^2 \cdot x_{max}^{(1)}$

m_3 $m_3 \cdot \omega^2 \cdot x_{max}^{(1)}$

$$\Gamma_1 \cdot [\Sigma \, m_i \cdot \varphi_i] \cdot S_a$$

m_1 $A \cdot m_1 \cdot \varphi_1$

m_2 $A \cdot m_2 \cdot \varphi_2$

m_3 $A \cdot m_3 \cdot \varphi_3$

$$V = [\Sigma \, m_i] \cdot S_a$$
Base shear

$$A = \frac{\Sigma \, m_i}{\Sigma \, m_i \cdot \varphi_i} \cdot S_a$$

m_1 $A_h \cdot m_1 \cdot h_1$

h_1 m_2 $A_h \cdot m_2 \cdot h_2$

h_2 m_3 $A_h \cdot m_3 \cdot h_3$

h_3

$$V = [\Sigma \, m_i] \cdot S_a$$
Base shear

$$A_h = \frac{\Sigma \, m_i}{\Sigma \, m_i \cdot h_i} \cdot S_a$$

Worked example

Following the example given in Section 7.3.4.2 and based on the acceleration spectrum of the current Greek Seismic Code (ELOT, 2003), a maximum spectral acceleration $S_a = 2.50$ is calculated in soil class A, both for the x–x direction with fundamental period $T = 0.32$ sec and for the y–y direction with fundamental period $T = 0.26$ sec. This means that for a maximum ground acceleration for example of 0.16g, the total seismic force (base shear) is:

$$V = 3 \cdot 78.5 \cdot 0.7 \cdot 9.81 \cdot 2.50 = 924 \, \text{kN}$$

In the case of an earthquake in the x-direction, the equivalent static loads applied to the centre of gravity of each of the three diaphragms are calculated on the basis of the deviations, φ_i taken as the estimated corresponding displacements in the above example. It is:

$$\Sigma m_i \cdot \varphi_i = 78.5 \cdot (0.017 + 0.0543 + 0.086) = 12.35 \, \text{t}$$

Thus, the forces exerted in the x-direction are (see Figure 7.38):

$P_1 = 924 \cdot 78.5 \cdot 0.017 / 12.35 = 100 \, \text{kN}$
$P_2 = 924 \cdot 78.5 \cdot 0.0543 / 12.35 = 319 \, \text{kN}$
$P_3 = 924 \cdot 78.5 \cdot 0.086 / 12.35 = 505 \, \text{kN}$

In the case of an earthquake in the y-direction, the equivalent static loads, which are also applied to the centre of gravity of each of the three slabs, are calculated accordingly. It is:

$$\sum m_i \cdot \varphi_i = 78.5 \cdot (0.0084 + 0.041 + 0.079) = 10.08 \text{ t}$$

The forces exerted in the y-direction are (see Figure 7.38):

$P_1 = 924 \cdot 78.5 \cdot 0.0084 / 10.08 = 60 \text{ kN}$

$P_2 = 924 \cdot 78.5 \cdot 0.041 / 10.08 = 295 \text{ kN}$

$P_3 = 924 \cdot 78.5 \cdot 0.079 / 10.08 = 568 \text{ kN}$

that is, with practically the same distribution as in the x-direction.

The assessment of the state of stress of a multi-storey system under these loads in both the x-and y-directions can now be carried out in the way discussed in Chapter 6, Section 6.3.1.

7.4.3   Consideration of plastic behaviour in the seismic design of frames

7.4.3.1   Introduction

The dynamic analysis of a frame with a certain mass at each level subjected to a specific accelerogram imposed at its base, as described previously, leads to the determination of its stress state by assuming elastic behaviour. As in the case of the single-mass system, it is again clear that the preset exhaustion of the flexural strength of any member (beam or column) implies a limit on the application of the corresponding forces and moments involved in the equilibrium of the level in question. Thus, equilibrium can be achieved with these limited internal forces, allowing, of course, for greater deformations (displacements) than those in the elastic state. The uptake of the accelerogram by the structure can thus be carried out by smaller sectional forces than in the elastic state, as long as the structure provides the required deformability due to the forthcoming plastic behaviour.

The seismic design of a frame is based on the equivalent static loads defined in Section 7.4.2. These loads represent the maximum inertial forces on the masses and can be used as a substitute for the dynamic design analysis, no matter whether plastic behaviour is taken into account or not. However, in the next sections the consideration is focused on handling these equivalent loads by taking into account the plastic deformability of the structure. This is in direct correspondence with what has been discussed in this respect in Sections 7.2.6 and 7.2.7 for the single-mass system.

7.4.3.2   Equivalent static lateral loads: push-over analysis

In order to directly assess the effects of plastic behaviour on a multi-degree-of-freedom system with concentrated masses on the floors – represented by specific loads (usually distributed ones) on the beams – it is necessary to consider the plastic deformability of frame under statically applied horizontal inertia forces deduced on the basis of those defined in Section 7.4.2 and according to Chapter 6, Section 6.3.1. These forces are assumed to be applied gradually at the levels of the frame following an incremental pattern suggested by the distribution of the base shear over the various levels, as already established in Section 7.4.2. During this loading, successive plastic hinges are developed that behave in the way described in Stavridis and Georgiadis (2025, Sections 4.1.2 and 6.6.1). This process stops when at some point no further load increment can be taken up by the frame, either because the rotational capacity of any joint has reached its limit or because the structure becomes a mechanism. This process is illustrated by the example of a two-storey frame shown in Figure 7.39, which follows the successive stages of joint plastification.

Figure 7.39 Gradual frame plastification due to continuous increase in lateral loading

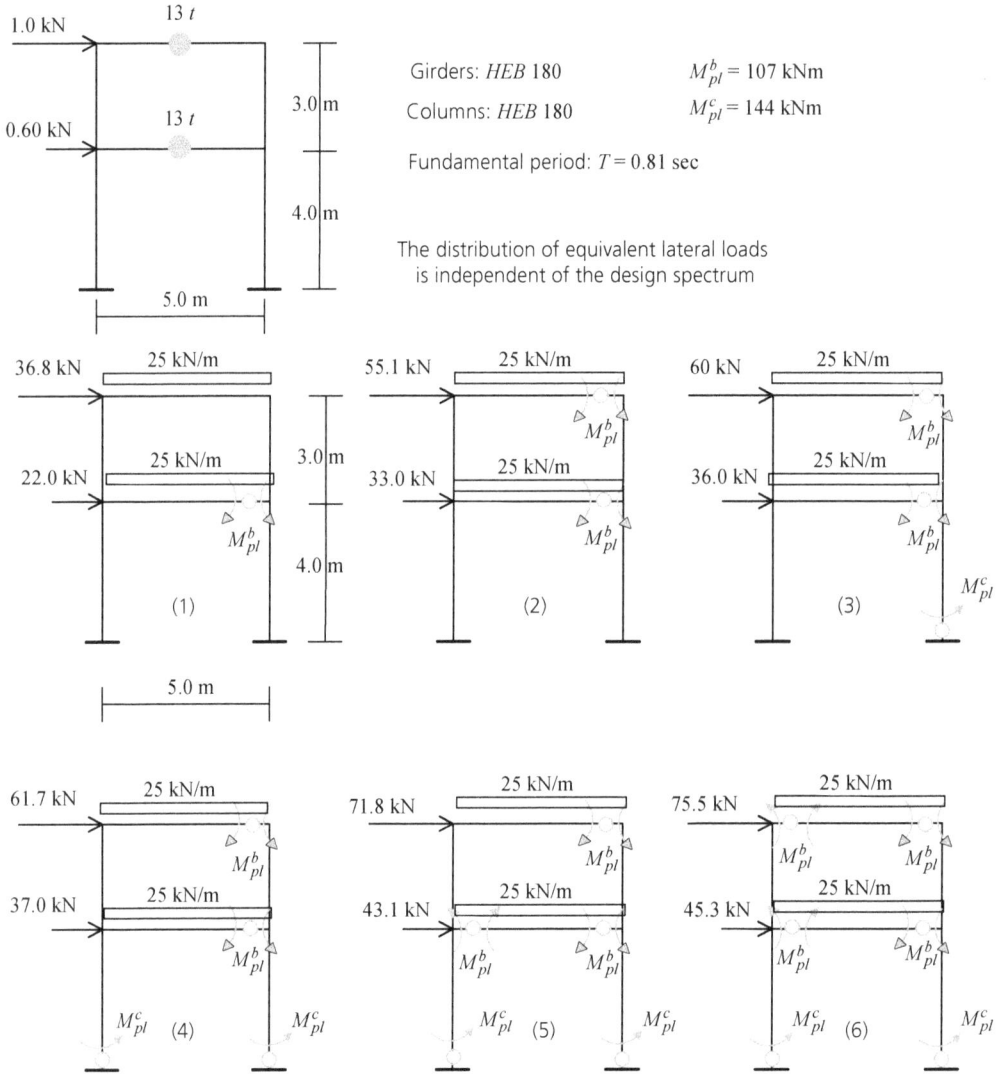

Girders: *HEB* 180 $M_{pl}^b = 107$ kNm

Columns: *HEB* 180 $M_{pl}^c = 144$ kNm

Fundamental period: $T = 0.81$ sec

The distribution of equivalent lateral loads
is independent of the design spectrum

In every plastification phase under the external loads,
an elastic state of stress is considered

It should be pointed out here that the allowable rotation of each plastic joint is determined (L/2) the basis of the expression of θ_{pl} (see Section 7.2.7.1), which, by considering the half-length of the element as relevant because of the induced antisymmetric deformation, can be written as:

$$\theta_{pl} = (\varphi_u - \varphi_y) \cdot L_{pl} \cdot \left(1 - \frac{L_{pl}}{L} \right)$$

Where L represents the length of the element and L_{pl} is as defined in section 7.2.7.1

The mapping of the deformational behaviour of the frame during this gradual loading can be done by plotting the horizontal displacement of the top level, Δ_{top}, against the corresponding total horizontal force, F (base shear), applied in a diagram, as shown in Figure 7.40, called the *push-over curve*. The effect of the successive development of plastic hinges on the frame's deformability (i.e. stiffness) is apparent from the ever-changing slope (F/Δ) of the curve.

When designing the members of a frame, the engineer should aim to maximise the deformability of the structure under the relevant static equivalent seismic forces. In other words, the push-over curve should extend as much as possible in terms of both the total horizontal force the frame can resist and the maximum displacement at the top level.

A safe and proven design strategy for sizing the elements of a frame with the above objective is to allow the creation of plastic joints only at the ends of the beams and at the base of the ground floor columns. This is achieved, bearing in mind the pattern of the bending moment diagram, if at each node the sum of the plastic bending moments of the adjacent beams is less than the sum of the plastic bending moments of the adjacent columns (see Figure 7.41). In this way, as long as the required deformability of the plastic hinges at the ends of the beams is ensured, all columns remain in an elastic state throughout the progressive loading except for their base joints that normally get plastified last.

Figure 7.40 Push-over curve

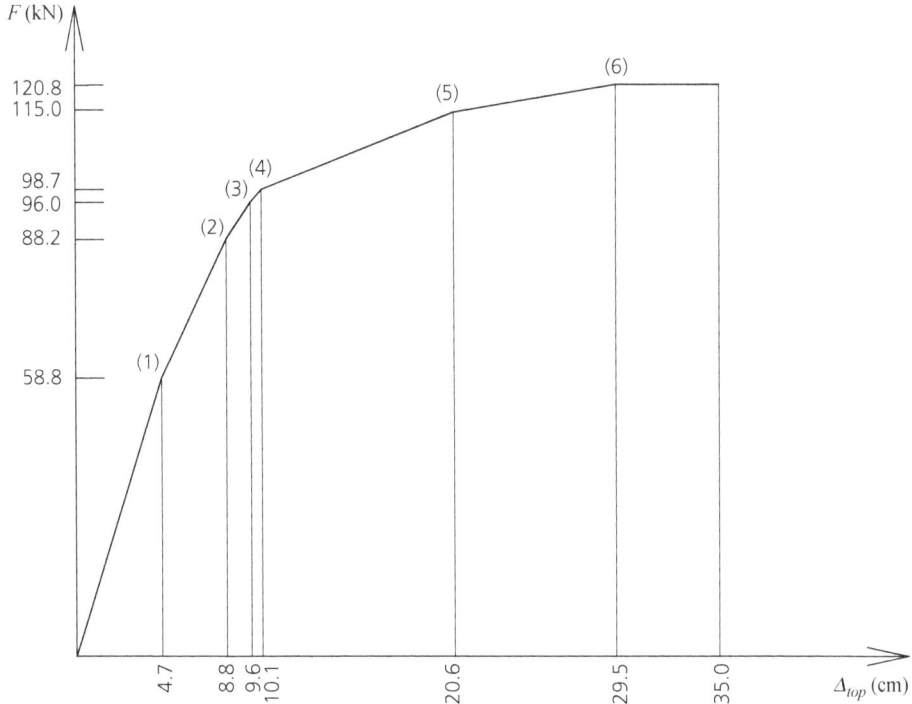

Capacity curve of the frame (push-over curve)

Figure 7.41 Design goal for each frame node

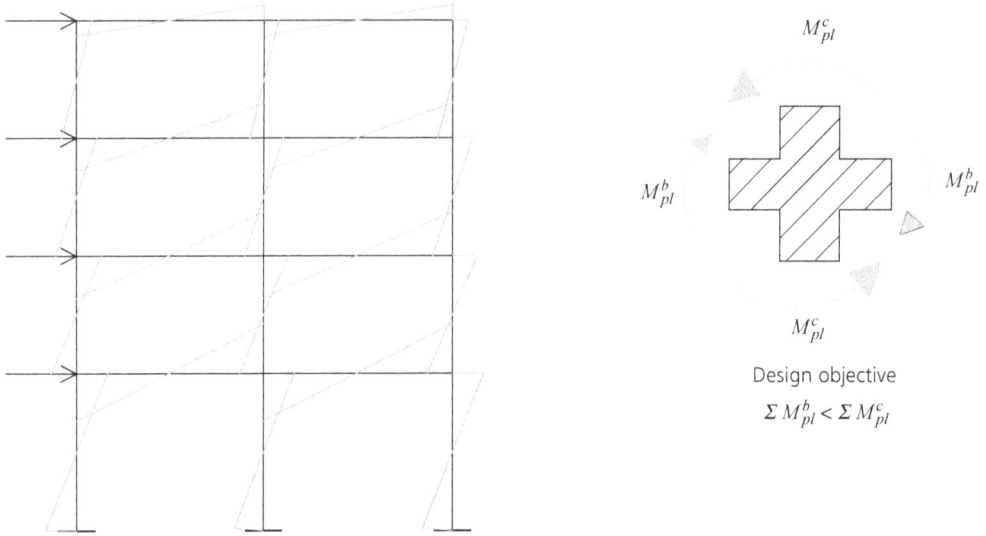

M^c_{pl}

M^b_{pl}

M^b_{pl}

M^c_{pl}

Design objective

$\Sigma M^b_{pl} < \Sigma M^c_{pl}$

7.4.3.3   Formation and use of the push-over curve for design

It should be recognised that, although the push-over curve refers to the multi-degree-of-freedom system of the frame, the fact that it refers to the relationship between a horizontal loading and a selected horizontal displacement allows it to be considered as referring essentially to a single-mass system (see Section 7.2.7.1). This is of particular importance for the comprehension of its use in the design.

In this sense, the relationship between horizontal force and induced displacement considered in the system discussed in Section 7.2.7.1 (see Figures 7.19 and 7.20) constitutes the push-over curve for the single-degree-of-freedom system. As this push-over curve is a bilinear diagram, this allows the push-over curve of the frame to be considered in the same way (see Figure 7.42).

Now, it can be considered that the enclosed area below the push-over diagram of the frame repre-sents the work produced by the horizontal forces during the induced deformation on the structure. Thus, in order to accomplish the set out of the fictitious bilinear diagram for the multi-storey frame, the extent of the elastic branch represented by the force, F_y, as well as the corresponding deforma-tion, Δ_y, should be so selected that the further extension of the horizontal branch up to the value of the limit displacement, Δ_u, ensures the equality of the deformation work, E, produced by the actual push-over process (i.e. the area measured under the curve itself) with that of the respectively equivalent bilinear curve. This equal area condition leads to the relation:

$$F_y \cdot \left(\Delta_u - \frac{\Delta_y}{2} \right) = E$$

A choice of the displacement, Δ_y, leads directly to the determination of F_y. A reasonable sugges-tion that has been made in the literature is that the *ascending* elastic branch of the bilinear diagram should intersect the push-over curve at a point corresponding to a horizontal force equal to 60% of the yield force, F_y. This point should have coordinates F^* and Δ^* satisfying the relations:

Figure 7.42 Drawing an equivalent bilinear push-over diagram

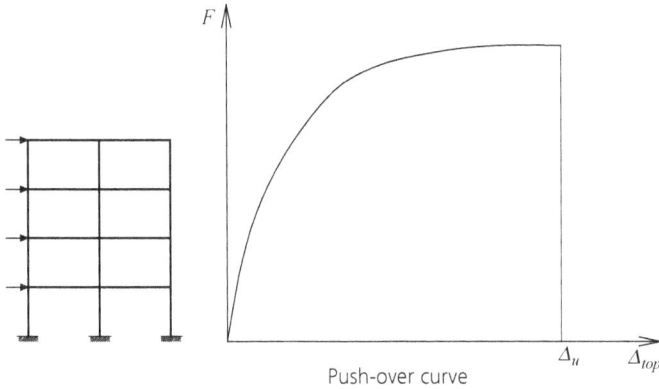

Push-over curve

Bilinearisation of the push–over curve

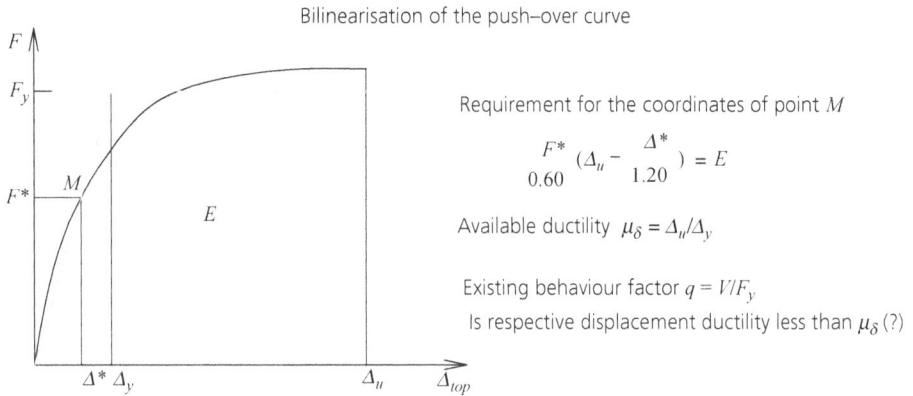

Requirement for the coordinates of point M

$$\frac{F^*}{0.60}(\Delta_u - \frac{\Delta^*}{1.20}) = E$$

Available ductility $\mu_\delta = \Delta_u/\Delta_y$

Existing behaviour factor $q = V/F_y$
Is respective displacement ductility less than μ_δ (?)

$$F_y = \frac{F^*}{0.60}$$

and,

$$\Delta_y = \frac{\Delta^*}{0.60}$$

However, any other reasonable estimation of F_y and Δ_y could also be applied.

As explained above, the push-over curve is established for the specific distribution of equivalent static loads resulting from the given dynamic behaviour of the frame (see Section 7.4.2).

A normal distribution of masses and strengths in a frame generally leads to an approximate value of participation, ε_1, of the first (fundamental) eigenmode of 0.9 and this allows the system to be 'represented' by a single-mass system with a mass equal to the active mass of the first eigenmode $M_{eff}^{(1)}$ as practically equal to the total mass, M, and of an appropriate stiffness to provide the fundamental period of the dimensioned frame. Then, the total base shear stress based on the elastic design spectrum is $V = M \cdot S_a$ and, consequently, the coefficient of behaviour, q, of the frame is $q = V / F_y$.

The thus implied value of ductility, μ_δ, according to the relation presented in Section 7.2.7.2, determines the displacement requirement, Δ_{top}, of the frame roof $\Delta_{top} = \mu_\delta \cdot \Delta_y$, and if the latter value is less than Δ_u the frame is considered seismically adequate.

It is clear that if an *elastic* design is followed for the structure, the response under the equivalent loads, H_i, (see Section 7.4.2) must be taken into account. The dimensioning of the frame should then simply provide the necessary strength to the members, without any further requirement concerning plastic deformability.

7.4.3.5 Summary of the seismic check procedure

The seismic design check of an already dimensioned frame can in principle be based on the assumption of either elastic or plastic behaviour. In both cases, a static load, equivalent to the dynamic load, is basically assumed, with horizontal loads distributed at the different levels of the frame, according to Section 7.4.2 on the base of the adopted design spectrum.

Following the elastic analysis, the frame is designed to provide simply the strength required by the induced sectional forces due to the above loads.

Following the plastic analysis, it should be checked whether the frame can provide the deformability (ductility) that is determined on the basis of the existing coefficient of behaviour, q.

The whole procedure is carried out in the following way.

- Perform the eigenvalue analysis and calculate the fundamental period, $T_{(1)}$, and the corresponding first eigenmode displacements φ_1. Determine the active mass, $M_{eff}^{(1)}$, and its ratio, ε_1, to the total mass. Check whether $\varepsilon_1 \approx 0.90$.
- Determine the applicable design spectrum $[S_a, T]$ based on the maximum ground acceleration.
- Determine the equivalent static loads for each level corresponding to the first eigenmode and calculate the base shear force, V.
- Construct the push-over curve.
- Bilinearise the push-over curve. Determine the values $(F_y, \Delta_y, \Delta_u)$.
- Calculate the behavioural coefficient $q = V / F_y$.
- Determine the required ductility, μ_δ.
- Determine the targeted deformation $\Delta_{top} = \mu_\delta \cdot \Delta_y$.
- Check whether the frame can develop this deformation according to the push-over curve.

Worked example

The indicative plan view, as shown in Figure 7.43, is used for a four-storey reinforced concrete structure. The first floor has a height of 4.50 m from the ground and the other three floors have a uniform height of 3.0 m. All four slabs have a thickness of 20 cm for each floor and apart from their self-weight they are assumed to bear an additional permanent load $g_1 = 1.0$ kN/m^2 and a live load $p = 2.0$ kN/m^2.

All beams have a width of 30 cm and a depth of 60 cm, while all the columns have a 50/50 cm cross-section. The beams have 5ϕ16 bottom and top reinforcement. The columns have 16ϕ20 reinforcement, equally spaced around the perimeter with ϕ10/25 stirrups. The concrete is assumed to have an $f_c = 28$ MPa and the steel an $f_y = 420$ MPa.

Figure 7.43 Plan view of the skeletal layout

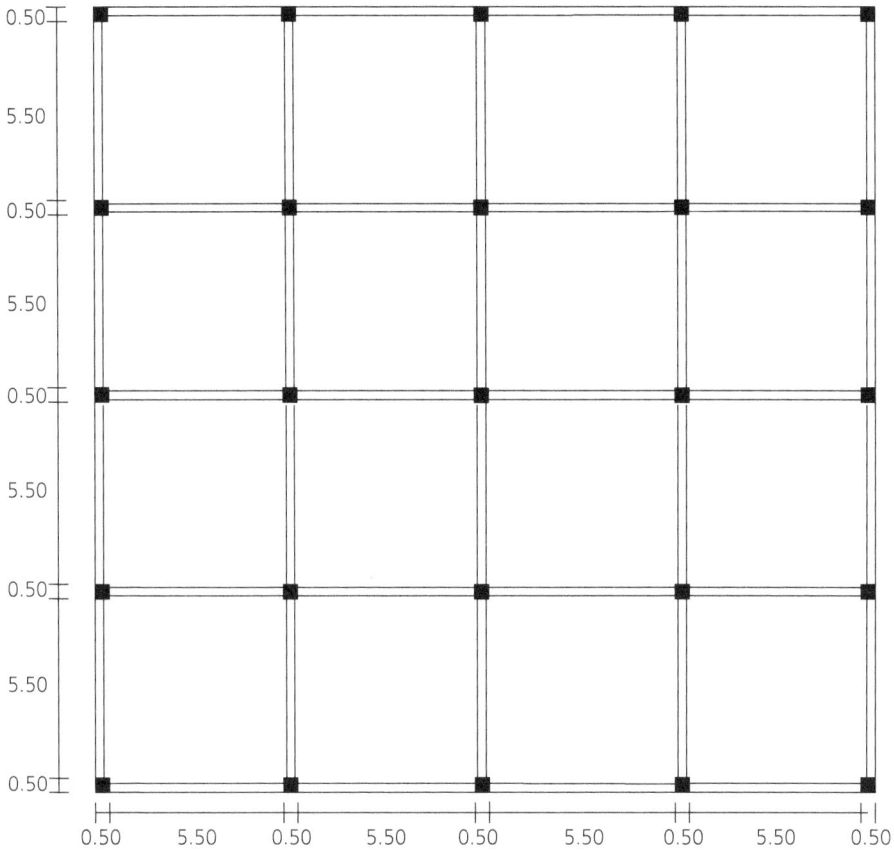

The structure is checked for a seismic excitation with a maximum ground acceleration of 0.24g, in soil class B, according to European seismic code (i.e. BS EN 1998-1:2004: Eurocode 8 (BSI, 2004)).

A typical frame is considered as shown in Figure 7.44. The dynamic analysis gives a fundamental period equal to 0.71 sec, while the active mass $M_{eff}^{(1)}$ is equal to 396 t with participation $\varepsilon_1 = 0.86$, so just the first mode is considered appropriate ($\varepsilon_1 \approx 0.9$).

The push-over curve is constructed and bilinearised, as shown in Figure 7.44, with characteristic values F_y, Δ_y and Δ_u equal to 726 kN, 0.065 m and 0.258 m, respectively.

The spectral acceleration is determined on the basis of the fundamental period $T = 0.71$ sec, which lies between $T_C = 0.5$ sec and $T_D = 2.0$ sec, as well as the soil factor for class B that is equal to 1.2.

It is $S_a = 0.24 \cdot 9.81 \cdot 1.2 \cdot 2.5 \cdot (0.50/0.71) = 4.97$ m/sec^2 and, subsequently, a base shear $V = 396.0 \cdot 4.97 = 1968$ kN is specified. On the basis of the bilinearised push-over curve, a behavioural coefficient $q = 1968.0/726.0 = 2.71$ is determined and thus, according to Section 7.2.7.2, a ductility index $\mu_\delta = q = 2.71$ is derived with a subsequent target displacement at the top equal to $\Delta_{top} = \mu_\delta \cdot \Delta_y = 2.71 \cdot 0.065 = 0.18$ m, being ensured by the push-over deformability curve of the frame, showing an ultimate value $\Delta_u = 0.258$ m (see Figure 7.44).

271

Figure 7.44 Formation layout and push-over curve of a typical frame

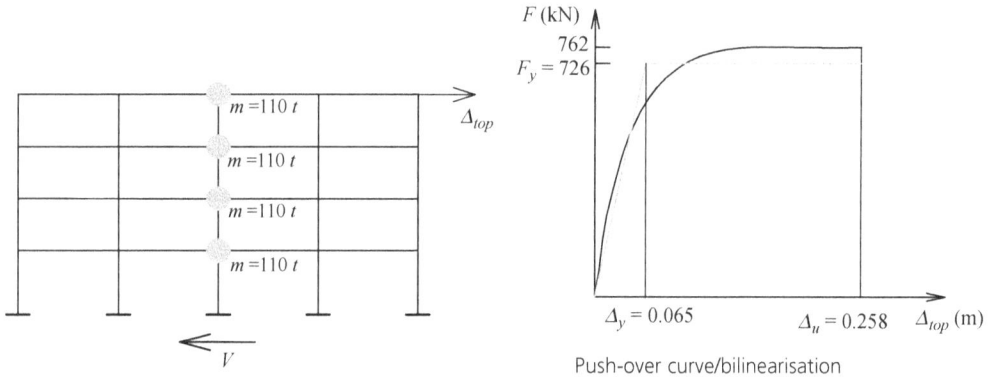

Push-over curve/bilinearisation

7.5.   Approximate treatment of continuous systems

As pointed out at the beginning of this chapter, structures have a continuous mass distribution. However, the inertia forces that have to be considered in order to express the equilibrium of each part of a structure lead to practically unmanageable differential equations. Therefore, this approach is not practically possible in structural design.

On the other hand, the discrete systems examined so far, although they are not identical to real structures, are able to express their dynamic behaviour with acceptable accuracy. In this way, the mass can be considered lumped in as many places of the structure as desired by the engineer.

The total (continuous) mass, m, of a beam, for example, may be divided into five lumped equidistant masses ($m/5$) distributed over the beam, thus creating a system of five degrees of freedom, if only those inertia forces that correspond to vertical displacements are accounted for (see Figure 7.45).

In the same sense, the mass, m, of a slab may be divided in 25 equidistant masses ($m/25$), corresponding to a system with 25 degrees of freedom based only on the vertical deflections (see Figure 7.45).

Nevertheless, analysis of the above discrete systems certainly requires the use of appropriate computer programs. This does not directly meet the needs of preliminary design in what concerns basic design parameters, such as, for instance, the fundamental frequency of the system, or its maximum deformation, or even its maximum acceleration under specific dynamic loads. Thus, it is appropriate – wherever possible and at least for usual cases – to pursue the equivalent dynamic behaviour of the continuous systems through single-mass systems, exhibiting the same frequency as the fundamental frequency of the examined system. This idea is based on the fact that in most practical cases, the fundamental eigenfrequency of a system plays the most governing role in its dynamic behaviour, as already pointed out in Section 7.3.4.2.

Figure 7.45 shows the 'substitution' of a beam and slab by a corresponding 'equivalent' single-mass system, of which the mass, stiffness and equivalent dynamic loading are to be determined (Biggs, 1964).

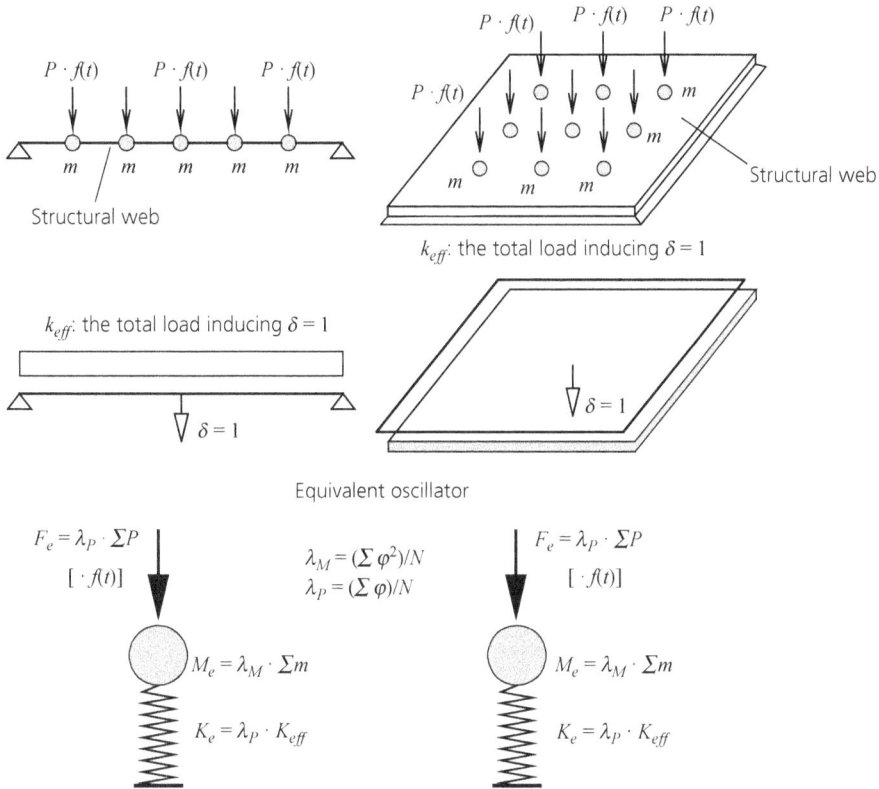

Figure 7.45 Equivalent substitution of a continuous structure through a single-mass system

The determination of the above parameters is based on the fact that, as discussed in Sections 7.3.2 and 7.3.3, the dynamic analysis of an N-degree-of-freedom system goes back to the analysis of N independent single-mass oscillators and, as explained above, it is the one with the lowest frequency that may more accurately represent the whole system.

As discussed in Section 7.3.2, the mass, M_e, of this oscillator can be expressed as:

$$M_e = \boldsymbol{\varphi}_1^{\mathbf{T}} \cdot \mathbf{m} \cdot \boldsymbol{\varphi}_1 = \sum_{s=1}^{N} m_s \cdot \varphi_s^{2} = \lambda_M \cdot \sum_{s=1}^{N} m_s$$

where φ_s are the deviation values of the first eigenmode of the N-degree system and, in the case of a uniformly distributed total mass, M, in the discrete model, the *mass factor*, λ_M, is:

$$\lambda_M = \frac{\sum\limits_{s=1}^{N} \varphi_s^{2}}{N} \qquad \text{(dimensionless number)}$$

Thus, the mass, M_e, of the equivalent oscillator is equal to $M_e = \lambda_M \cdot M$ being expressed as a part of the total mass, M, of the system (as $\lambda_M < 1$) (see Figure 7.45).

273

Assuming now that the discrete system is subjected to the dynamic loading:

$$F(t) = P \cdot f(t)$$

where matrix P consists of the specific loads acting on the N masses of the system, according to Section 7.3.3, the loading term of the single-mass oscillator with the lowest frequency (i.e. the *fundamental oscillator*) of the N-degree system is:

$$F_1 = \varphi_1^T \cdot P \cdot f(t) = \left(\sum_{s=1}^{N} P_s \cdot \varphi_s \right) \cdot f(t)$$

Considering that the load $F_e \cdot f(t)$ will act on the equivalent oscillator, it is:

$$F_e = \sum_{s=1}^{N} P_s \cdot \varphi_s = \lambda_P \cdot \sum_{s=1}^{N} P_s \quad \text{(see Figure 7.45)}$$

and in the case of an equally distributed total load to the N masses of the discrete system, the load factor λ_P is:

$$\lambda_P = \frac{\sum\limits_{s=1}^{N} \varphi_s}{N} \quad \text{(dimensionless number)}$$

In the case of a uniform distribution of load, p (kN/m), on a beam of length, L, the total load is $\sum P_s = (p \cdot L)$, whereas for a uniformly distributed load, q (kN/m²), on an orthogonal slab with dimensions (a, b), the total load is $\sum P_s = (q \cdot a \cdot b)$.

It should be noted here that the aim of seeking an equivalent single-mass oscillator is basically to determine the kinematic behaviour (displacement, velocity, acceleration) of a certain critical point of the initial system for preliminary design purposes. This point, in the case of a single span beam or a slab, is obviously the centre of mass of the system. For this purpose, and in extension of the initially adopted notion of stiffness in Stavridis and Georgiadis (2025, Section 2.3.7), the concept of effective stiffness, k_{eff}, is introduced as the total load that, under a certain distribution, causes a unit displacement at the reference point of the structure – that is, at its mid-point (see Figure 7.46). It is then clear that the deflection, δ, of that point due to the total load, $\sum P_s$, for the considered distribution is:

$$\delta = \frac{\sum P_s}{k_{eff}}$$

For a uniform load distribution, in the case of a beam of length, L, it is $k_{eff} = 76.8 \cdot EI/L^3$, whereas in the case of a simply supported square slab of side, L, it is $k_{eff} = 271 \cdot EI/L^2$.

Thus, the sought-after stiffness, K_e, of the equivalent oscillator results from the requirement that the transverse displacement, δ, at the considered reference point of the continuous system, due to the static loads, P_s, should be identical to the displacement of the equivalent oscillator under the load $F_e = \lambda_P \cdot \sum P_s$. Thus,

Figure 7.46 Equivalent substitute parameters of single-mass systems for beams and slabs

Type of loading and beam Length L	Mass factor λ_M	Load factor λ_P	Effective stiffness k_{eff} $(\cdot\,EI/L^3)$	Uniform load on plate	a/b	Mass factor λ_M	Load factor λ_P	Effective stiffness k_{eff} $(\cdot\,Eh^3/12\cdot a^2)$
(simply supported beam, distributed load)	0.5	0.637	76.80	Plate perimetrically simply supported	1.0	0.31	0.45	271
					0.9	0.33	0.47	248
(simply supported beam, point load)	0.5	1.0	48		0.8	0.35	0.49	228
				Plate thickness: h	0.7	0.37	0.51	216
(propped cantilever, distributed load)	0.479	0.595	185		0.6	0.39	0.53	212
					0.5	0.41	0.55	216
(propped cantilever, point load)	0.479	1.0	107	Plate perimetrically fixed	1.0	0.21	0.33	870
					0.9	0.23	0.34	798
(fixed beam, distributed load)	0.396	0.523	384		0.8	0.25	0.36	757
					0.7	0.27	0.38	744
(fixed beam, point load)	0.396	1.0	192	Plate thickness: h	0.6	0.29	0.41	778
					0.5	0.31	0.43	866

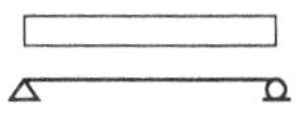

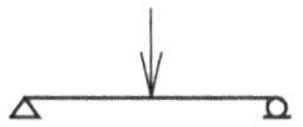

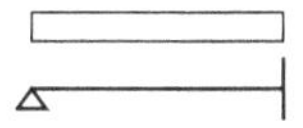

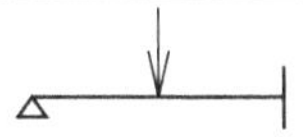

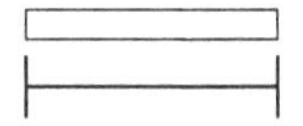

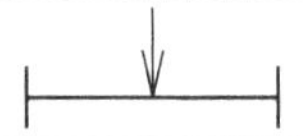

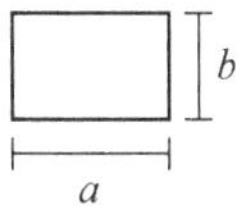

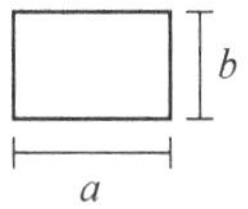

(a)   (b)

$$\delta = \frac{\sum P_s}{k_{eff}} = \frac{\lambda_P \cdot \sum P_s}{K_e}$$

and, hence,

$$K_e = \lambda_P \cdot k_{eff} \qquad \text{(see Figures 7.45 and 7.46)}$$

The fundamental frequency, f_1, of the continuous system, being equal to that of the equivalent oscillator, is obtained as:

$$f_1 = \frac{1}{2\pi}\sqrt{\frac{K_e}{M_e}} = \frac{1}{2\pi}\sqrt{\frac{\lambda_P}{\lambda_M}}\cdot\sqrt{\frac{k_{eff}}{M}}$$

where factors λ_P and λ_M refer to a uniform distribution of load and mass, respectively.

The tables in Figure 7.46 give the values of the mass and load factors λ_P and λ_M, respectively, for various cases of beams with distributed or concentrated load, as well as for perimetrically supported slabs (Biggs, 1964).

7.6. Design to avoid annoying vibrations
7.6.1 Human activities
Apart from the seismic excitation of multi-storey systems (Section 7.3.4), checking of vertical vibrations of horizontal load-carrying elements, such as beams or slabs, induced during their normal use by humans constitutes a major design issue in several cases (Bachmann and Ammann, 1987). A pedestrian bridge, for example, or an office floor, or even a floor designed for sports or

dance activities, should in no case undergo vibrations that are felt by the people using them in a disturbing way. This actually happens when the frequency with which the foot strikes the floor lies near the fundamental frequency of the structure.

Generally, footbridges should be designed with a fundamental frequency out of the range of 1.6 to 2.4 Hz and, if possible, not lower than the value of 4.50 Hz. This is because the normal walking frequency lies within that range, increasing the risk of resonance. For dance floors and floors intended for sports activities, this frequency should not be less than 6.0 Hz and 6.8 Hz, respectively. In the case of office floors, where smaller acceleration limits have to be attained, the fundamental frequency should be above 7.50 Hz (Bachmann et al., 1995).

Defining human tolerance to vibrations (comfort criteria) is a complex and highly subjective issue that differs among individuals. Even the same person may react differently to the same vibration on different days. Moreover, human posture and activity (walking, running, sitting or standing), the structure's location (e.g. for bridges near hospitals users may be more sensitive to vibrations than for bridges along hiking trails) and appearance, surrounding environment, height above the ground, as well as the previous experience of the users and the people around them, are critical for the perception of the vibration. Although the comfort criteria differ between different design standards and guidelines, it is generally suggested to use a limit of vertical acceleration of 0.5 m/s^2 and 0.3 m/s^2 for moving and sitting people, respectively.

Finally, it is worth mentioning that, apart from vertical accelerations, human activities can also cause lateral accelerations. Lateral accelerations, although they are generally smaller, should be considered appropriately in the design as the human body is more sensitive to them. Even a small lateral acceleration can affect humans walking and synchronise them with the structure, increasing disproportionally the lateral response of the structure, as happened in the case of the Millennium Bridge in London.

7.6.2   Vibrations induced by machines

The use of machinery generally causes vibrations in the structures on which they are placed, which may annoy people standing near them. Machines produce continuous periodical sinusoidal loading, characterised by the maximum value of force, F_0, under a frequency, f_0, according to the data supplied by the vendor.

In addition to what was discussed in the previous section regarding comfort appraisal, as a measure of the acceptable level of accelerations felt by a person, the so-called *sensitivity factor*, K, may be considered. This is calculated by the empirical formula (Bachmann and Ammann, 1987):

$$K = d\frac{0.80 \cdot f^2}{\sqrt{1+0.032 \cdot f^2}}$$

where d is the maximum amplitude of vibration (mm) and f is the fundamental eigenfrequency (Hz) of the structural part (e.g. a floor slab) supporting the machine.

Machines are generally supported on a structure either directly or through appropriate springs. The design goal consists usually in the limitation of deflection, $\delta_{max} = u_{max}$, and/or velocity, $v_{max} = \dot{u}_{max}$, of the supporting part at values tolerable by the machine, respectively, as well as in limiting the sensitivity factor, K, under the value of 2.0, so that the floor vibration is tolerable by the people standing nearby. ∎

Figure 7.47 Direct support of a machine on a continuous system

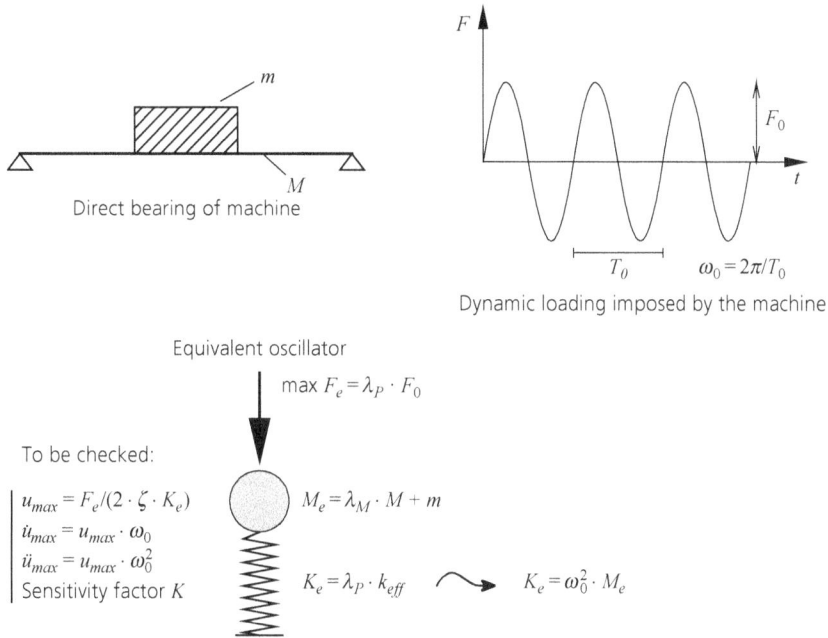

Direct bearing of machine

Dynamic loading imposed by the machine

Equivalent oscillator

max $F_e = \lambda_P \cdot F_0$

To be checked:

$u_{max} = F_e/(2 \cdot \zeta \cdot K_e)$
$\dot{u}_{max} = u_{max} \cdot \omega_0$
$\ddot{u}_{max} = u_{max} \cdot \omega_0^2$
Sensitivity factor K

$M_e = \lambda_M \cdot M + m$

$K_e = \lambda_P \cdot k_{eff}$ $\quad K_e = \omega_0^2 \cdot M_e$

In the case in which a machine is directly supported on the floor, the whole system may be described by a fictitious oscillator with a mass, M_e, and spring stiffness, K_e, under an equivalent dynamic loading, F_e (Bachmann and Ammann, 1987). This oscillator exhibits a damping coefficient identical to that of the floor supporting structure (see Figure 7.47).

The mass, M_e, of the equivalent oscillator is determined as the sum of the equivalent vibrating mass $\lambda_M \cdot M$ of the supporting structural part (usually a slab) having mass, M, and the mass, m, of the machine itself. So, it is:

$$M_e = \lambda_M \cdot M + m$$

The spring stiffness, K_e, is expressed, according to Section 7.5, through the effective stiffness, k_{eff}, of the supporting part, multiplied by the factor λ_P (see Figure 7.47).

If the fundamental eigenfrequency, f_1, of the supporting structure being calculated, according to Section 7.5:

$$f_1 = \frac{1}{2\pi}\sqrt{\frac{\lambda_P}{\lambda_M}} \cdot \sqrt{\frac{k_{eff}}{M}}$$

is lower than the operation frequency, f_0, of the machine, then instead of calculating the stiffness, K_e, of the oscillator as $\lambda_P \cdot k_{eff}$, it is more useful to consider it equal to:

$$K_e = \omega_0^2 \cdot M_e = (2 \cdot \pi \cdot f_0)^2 \cdot M_e$$

277

in order to account for higher frequencies applied to the supporting structural part, which may lead to resonance.

On the basis of the adopted damping coefficient, ζ, for the structure, the determination of the maximum deflection, u_{max}, of the structure and, consequently, of the machine itself, is possible according to Section 7.2.4 as:

$$u_{max} = \frac{1}{2 \cdot \zeta} \cdot \frac{F_e}{K_e} = \frac{1}{2 \cdot \zeta} \cdot \frac{\lambda_p \cdot F_0}{K_e}$$

which may be directly compared with the maximum, δ_{max}, supplied by the machinery vendor. Moreover, the maximum velocity, as well the maximum acceleration developed can be checked, according to Section 7.2.2.1, as:

$$\dot{u}_{max} = u_{max} \cdot \omega = (2 \cdot \pi \cdot f_0) \cdot u_{max}$$

and

$$\ddot{u}_{max} = u_{max} \cdot \omega^2 = (2 \cdot \pi \cdot f_0)^2 \cdot u_{max}$$

However, of decisive importance for the acceptance or not of the dynamic behaviour of the system is the value of the sensitivity factor, K, according to the foregoing formula:

$$K = u_{max} \frac{0.80 \cdot f_0^2}{\sqrt{1 + 0.032 \cdot f_0^2}}$$

This factor should not exceed the value of 2.0. ∎

If the above solution does not lead to acceptable results, then, in order to limit the response of the supporting structure, the machine should be placed on springs with appropriately selected stiffness. The two masses of the machine and the structure are separated, as shown in Figure 7.48, thus constituting a two-degrees-of-freedom system consisting of an 'upper' mass, m_M, supported through a spring having stiffness, k_M, on the 'lower' mass, m_S, being in turn supported through a spring of stiffness, k_S, on a solid base.

The quantities m_S and k_S concerning the structure may be considered as known quantities, but the characteristics of the 'upper' system have to be determined, namely *machine mass*, m_M, and *spring stiffness*, k_M. The machine mass is not necessarily the mass, m, only, but also any appropriately attached mass, m_B, to it – for example, through a concrete pedestal – required to fulfil the dynamic requirements (see Figure 7.48).

The eigenfrequencies f_1 and f_2 of the two-degrees-of-freedom system may be directly determined through the expression (Bachmann and Ammann, 1987):

$$f_{1/2} = \left[\frac{1}{8 \cdot \pi^2} \left(\frac{k_M}{m_M} + \frac{k_M + k_S}{m_S} \pm \sqrt{\left(\frac{k_M}{m_M} + \frac{k_M + k_S}{m_S} \right)^2 - \frac{4 \cdot k_M \cdot k_S}{m_M \cdot m_S}} \right) \right]^{1/2}$$

Figure 7.48 Supporting a machine on a continuous system through springs

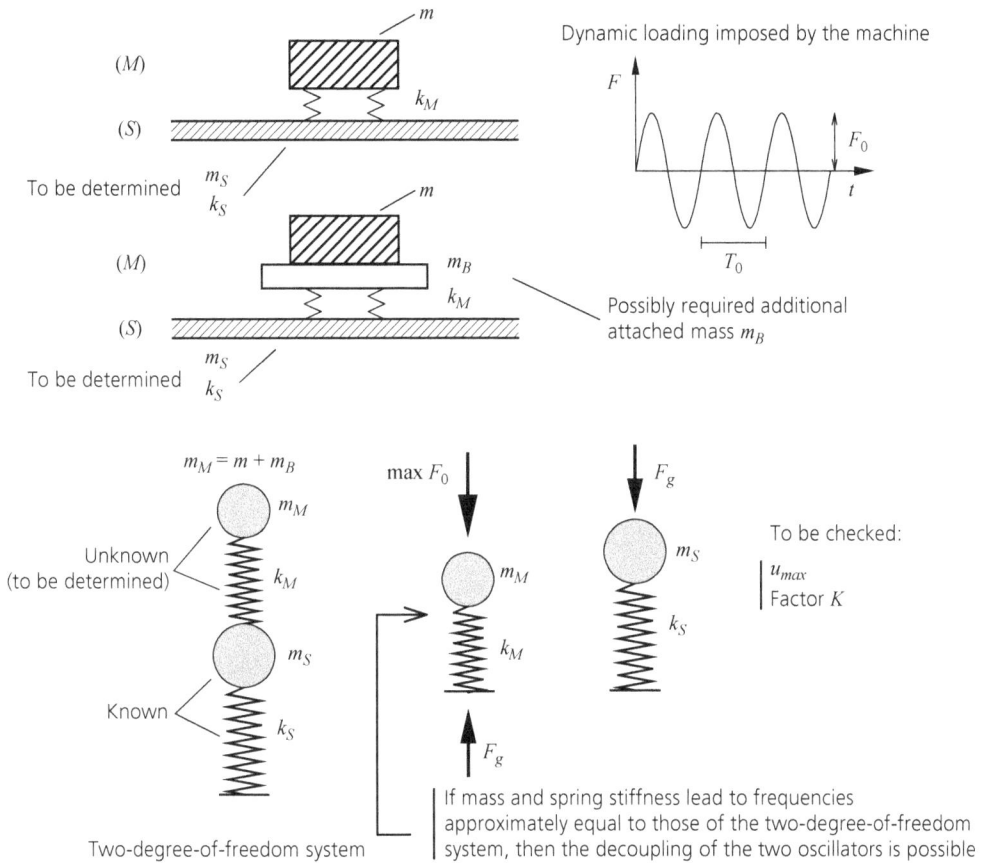

It is generally beneficial whenever possible, through appropriately selected values, for the two frequencies to be set quite clear apart – for example, by making the one a multiple of the other. This allows the two systems to be decoupled so that each mass may be considered as connected to a fixed base through its spring. Then, the frequencies f_1 and f_2 obtained from the above expression coincide – at least approximately – with the frequencies f_M and f_S of the individual oscillators, respectively.

Thus, if the two requirements:

$$f_M = \frac{1}{2\cdot\pi}\cdot\sqrt{\frac{k_M}{m_M}} = \left[\frac{1}{8\cdot\pi^2}\left(\frac{k_M}{m_M}+\frac{k_M+k_S}{m_S}-\sqrt{\left(\frac{k_M}{m_M}+\frac{k_M+k_S}{m_S}\right)^2-\frac{4\cdot k_M\cdot k_S}{m_M\cdot m_S}}\right)\right]^{1/2}$$

and

$$f_S = \frac{1}{2\cdot\pi}\cdot\sqrt{\frac{k_S}{m_S}} = \left[\frac{1}{8\cdot\pi^2}\left(\frac{k_M}{m_M}+\frac{k_M+k_S}{m_S}+\sqrt{\left(\frac{k_M}{m_M}+\frac{k_M+k_S}{m_S}\right)^2-\frac{4\cdot k_M\cdot k_S}{m_M\cdot m_S}}\right)\right]^{1/2}$$

lead – approximately – to technically plausible values for the sought-after quantities m_M and k_M, then the dynamic uncoupling of the two masses is feasible.

Then, in the system $[m_M, k_M]$, the maximum machine displacement as well as the maximum transmitted force on its base are first determined, according to Section 7.2.4 and, on the basis of this force, the system $[m_S, k_S]$ is then analysed in its relevant dynamic characteristics of interest, which concern the supporting structure itself.

More specifically, by considering for the machine part – according to the above – its total mass $(m_M = m + m_B)$, as well as stiffness, k_M, of its corresponding spring (see Figure 7.48), the corresponding eigenfrequency is obtained as:

$$f_M = \frac{1}{2\pi}\sqrt{\frac{k_M}{m_M}}$$

The mass of the equivalent oscillator with a damping coefficient, ζ, will be equal to:

$$m_S = \lambda_M \cdot M$$

The spring stiffness, k_S, is determined through the effective stiffness, k_{eff}, according to Section 7.4, as:

$$k_S = \lambda_P \cdot k_{eff}$$

The eigenfrequency, f_S, representing the fundamental eigenfrequency of the supporting structure is:

$$f_S = \frac{1}{2\pi}\sqrt{\frac{k_S}{m_S}}$$

If, now, the two systems are considered uncoupled, as previously explained, then for the mass, m_M (machine), the maximum displacement, u_{max}, as well as the maximum transmitted force, F_g, to the supporting base can be readily determined, according to Section 7.2.4, through the following relations, respectively ($\zeta = 0$):

$$u_{max} = u_{stat} \cdot \frac{1}{1-(\Omega/\omega)^2} = \frac{F_0}{k_M} \cdot \frac{1}{1-(f_0/f_M)^2} \qquad \left(\text{it must be} < \delta_{max}\right)$$

and

$$F_g = F_0 \cdot \frac{1}{1-(f_0/f_M)^2} = u_{max} \cdot k_M$$

Due to decoupling, the above force may be considered as acting directly on the supporting structural system, with a cyclic frequency, Ω, corresponding to f_M. Thus, the resulting maximum displacement may also be expressed according to Section 7.2.4 as:

$$u_{max} = \frac{F_g}{k_S} \cdot \frac{1}{\sqrt{(1-f_M/f_S)^2 + (2\cdot\zeta\cdot f_M/f_S)^2}}$$

As previously pointed out, the decisive criterion for the acceptance of the vibrational characteristics of the supporting structure is the value of the corresponding sensitivity factor, K, which may be directly calculated as:

$$K = u_{max} \frac{0.80 \cdot f_S^2}{\sqrt{1+0.032 \cdot f_S^2}}$$

REFERENCES

Bachmann H and Ammann W (1987) *Vibrations in Structures – Induced by Man and Machines.* IABSE Structural Engineering Documents, Zurich, Switzerland.

Bachmann H, Ammann W, Deischl F *et al.* (1995) *Vibration Problems in Structures.* Birkhäuser, Basel, Switzerland.

Biggs JM (1964) *Introduction to Structural Dynamics.* McGraw-Hill, New York, NY, USA.

BSI (2004) BS EN 1998-1:2004: Eurocode 8: Design of structures for earthquake resistance. BSI, London, UK.

ELOT (2003) Greek Seismic Code EAK 2003. Hellenic Organization for Standardization, Peristeri, Greece.

Müller FP and Keintzel E (1984) *Erdbebensicherung von Hochbauten.* Ernst, Berlin, Germany. (In German.)

Stavridis L (1986) Static and dynamic analysis of multistory systems. *Technika Chronika Scientific Journal of the Technical Chamber of Greece* **6(2)**: 187–219.

Stavridis L and Georgiadis K (2025) *Understanding and Designing Structures without a Computer: Plane structural systems.* Emerald Publishing, Leeds, UK.

Tedesko JW, McDougal WG and Allen RC (1999) *Structural Dynamics – Theory and Applications.* Addison Wesley Longman, Boston, MA, USA.

Leonidas Stavridis and Konstantinos Georgiadis
ISBN 978-1-83662-945-0
https://doi.org/10.1108/978-1-83662-942-920251008
Emerald Publishing Limited: All rights reserved

Chapter 8
Supporting structures on the ground

8.1.　Introduction

All structural systems may be considered as free bodies in equilibrium under the loads for which they have been designed, as well as the forces acting on them by the ground on which they are supported. These forces (i.e. the reactions) are, of course, accompanied by other equal and opposite forces that act on the soil. The forces acting on the ground should be taken by it in a safe manner regarding the induced deformations and soil stresses developed. In order to successfully transfer the forces to the ground, an adequate contact area of the structure with the soil is required. This constructional layout is called the *foundation* and consists of a concrete structure firmly connected to the superstructure. The term load-bearing structure means the whole system of superstructure plus foundation, which is actually supported on the bearing ground. In this sense, through its contact surface with the soil, the foundation must satisfy two main criteria. First, it must be designed to withstand the self-equilibrating system of soil stresses and the actions from the superstructure – these are equal and opposite to the developing reactions. Second, it must be able to introduce acceptable stresses to the soil so that they can be safely taken up by it (see Figure 8.1). In this respect, another point comes into question, namely the fact that the implied soil deformations also imposed on the structure create an additional strain in a statically indeterminate structure (see Figure 8.1). These deformations are, of course, not known in advance, but they come out as a result of the *soil–structure interaction* and, in this respect, the foundation design plays a decisive role.

Thus, in the design of a structure, the supporting soil is always directly involved through the interaction with the corresponding foundation, influencing the whole structural system.

The following sections do not intend to cover either the examination of the mechanical properties of soils or the various constructional layouts of the foundation possibilities. The aim is simply to point out particular basic structural characteristics involved in the design of some typical forms of foundations in order to facilitate the understanding of their load-bearing behaviour.

8.2.　General mechanical characteristics of soils

The great variety of soil types range between two main categories, namely non-cohesive and cohesive soils.

8.2.1　Non-cohesive soils

Non-cohesive soils (sand, gravel) in the dry condition constitute a mass of loose, unconnected grains, not smaller than 0.06 mm, mutually transmitting friction forces through their small contact areas. These friction forces are proportional to the normal pressure, σ, acting on the corresponding contact surface. They depend on the existing angle of internal friction, φ, of the soil and exhibit a maximum value of friction, τ, which is considered as the shearing strength of the soil. So, it is $\tau = \sigma \cdot \tan \varphi$. It must be pointed out that the shearing strength of soils represents the determining

Figure 8.1 The superstructure–foundation–soil system

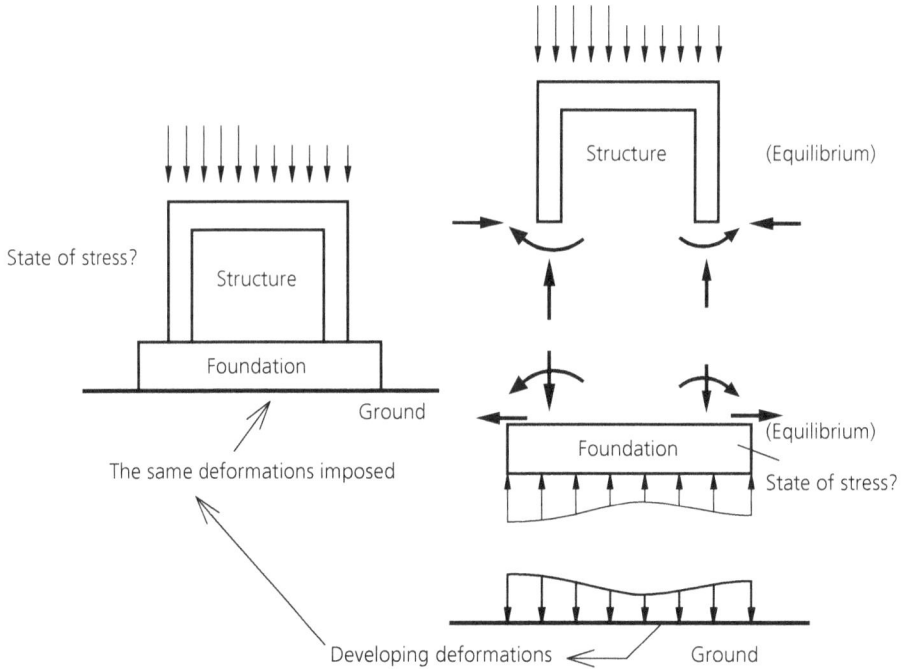

factor for their actual resistance, unlike solid materials whose strength is represented by normal tensile or compressive stresses.

Water, which can penetrate the gaps between the grains, may easily escape under pressure. This is why the settlements in these soils are developed within a short period of time, once the loads are imposed. Non-cohesive soils, in a layer of medium to high density and of adequate thickness, generally constitute a good foundation ground. However, in fine-grained, loose and water-saturated soil masses, there is a risk of losing shearing strength (i.e. loss of friction between grains) due to intense cyclic loading induced by a strong earthquake, a phenomenon known as *liquefaction*.

The specific weight, γ, of these soils in the dry condition is in the order of 20 kN/m^3. However, when the soil layer lies under water level, the soil grains are under buoyancy and the above value is reduced to $(20 - 10) = 10$ kN/m^3.

Non-cohesive soils in natural free embankments have a slope practically coinciding with the angle of internal friction, φ (see Figure 8.2). If, however, this embankment slope is prevented by a vertical front – for example, as a retaining wall against a certain soil volume – then this front all over its height, H, is under a soil pressure that increases as the angle of internal friction, φ, decreases. The horizontal component, σ_H, of these stresses is proportional to the prevailing vertical pressure, σ_V, in the considered depth, h, due either to the soil self-weight ($\sigma_V = \gamma \cdot h$) or to the vertical distributed load, p, acting on the free soil surface in addition to the soil self-weight ($\sigma_V = p + \gamma \cdot h$). Thus, it is $\sigma_H = K \cdot \sigma_V$, where the coefficient, K, is the reducing effect of the internal friction, moving between

Figure 8.2 Earth pressure in non-cohesive soils

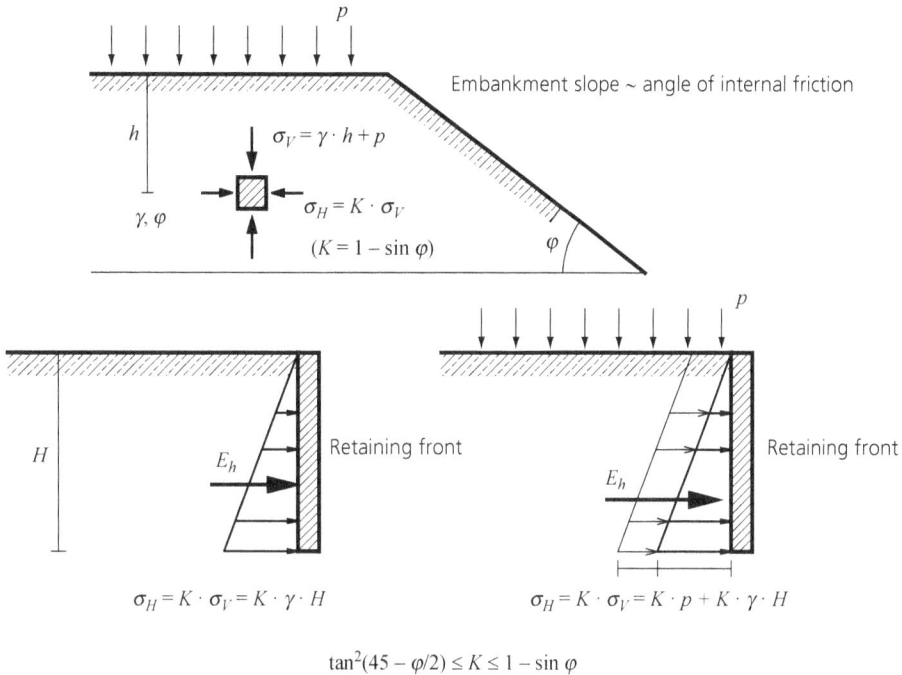

$\tan^2(45 - \varphi/2) \le K \le 1 - \sin \varphi$

a minimum value $K_a = \tan^2(45° - \varphi/2)$ and a maximum value $K_0 = 1 - \sin \varphi$ (see Figure 8.2). In the case of the presence of water in the soil mass, it is $K = 1$. The minimum value, K_a, corresponds to the case in which the possibility of a slight movement of the vertical front is allowed, referring to the so-called *active pressure*, whereas the maximum value, K_0, corresponds to an absolutely unyielding front, referring to the so-called *pressure at rest*. The resultant horizontal force, E_h, acting on the vertical front due to the soil self-weight under the triangular pressure distribution is $E_h = \gamma \cdot H^2 \cdot K/2$, whereas due to the surface load, p, it is $E_h = p \cdot K \cdot H$ (uniform pressure distribution). For the combined action of the two loadings, the above results are superposed (see Figure 8.2). In Figure 8.3(a), the influence of a differentiated value of K in the case of a layered soil mass is shown.

In the case of the presence of water in the soil mass, the reduction of the soil's specific weight, γ, to $\gamma' = \gamma - \gamma_W$ due to buoyancy, as well as the full hydrostatic action on the retaining structure, should be taken into account (see Figure 8.3(b)). It is noted that the angle of internal friction, φ, remains practically unaffected.

8.2.2   Cohesive soils
Cohesive soils (clay) constitute a consistent mass even in the dry condition. They consist of particles of oblong or slab form with diameters ranging from 0.0002 up to 0.06 mm, having many more common points of contact than non-cohesive soils. Given that the voids between the particles are much larger than the dimensions of the grains themselves, any present water cannot escape as easily as in the case of non-cohesive soils under an externally applied pressure due to the resistance

Figure 8.3 Earth pressure in a layered soil in the presence of a water table

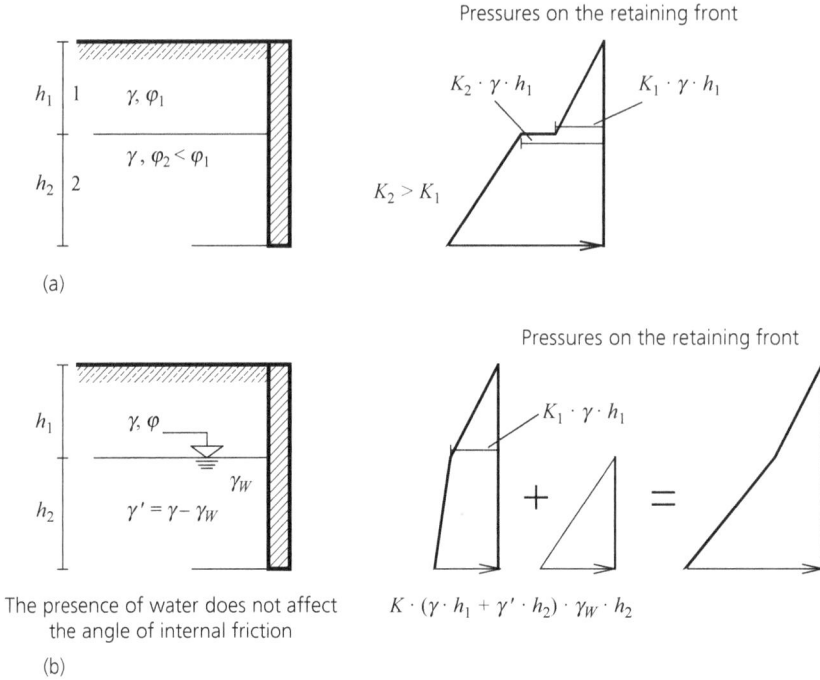

Pressures on the retaining front

(a)

Pressures on the retaining front

The presence of water does not affect the angle of internal friction

(b)

offered by the many contact points of the grains. Thus, any settlement develops very slowly and may be continued for longer periods of time (months or even years).

A basic characteristic of cohesive soils is that, apart from certain friction forces between grains, cohesion forces are developed as a result of the electrical nature of the particles. The friction forces decrease in the case of the existing water pressure, u, in the voids between the soil grains, called *pore pressure*. The pore pressure, in the case of the presence of a water table, is identical to the hydrostatic pressure at the point considered.

Thus, in the case of cohesive soils, the shearing strength, τ, basically consists of the cohesion, c (kN/m^2), as well as the friction resistance $\sigma'\cdot\tan\varphi$. The stress, σ', represents the actual normal pressure between the solid grains, obviously resulting from the externally imposed normal pressure, σ, reduced by the existing pore pressure, u. Thus, it is $\sigma' = (\sigma - u)$. The angle of internal friction, φ, generally has more moderate values than in non-cohesive soils. The cohesion, c, in water-saturated soils has a value of the order of 10 to 40 kN/m^2, whereas for partially saturated soils this value lies between 5 and 15 kN/m^2. Thus, it is:

$$\tau = c + (\sigma - u)\cdot\tan\varphi$$

The stress $\sigma' = (\sigma - u)$ is called the *effective stress*. It is clear that the shear strength of the cohesive soils depends, to a significant degree, on the presence of water in the pores, as well as on the cohesion, c, which plays an important role, sometimes being more decisive than the internal friction. It is pointed out that the presence of water decreases the angle of internal friction, φ, contrary to what may happen in partially saturated non-cohesive soils. Moreover, it is noted that in a cohesive water-saturated soil, the additionally imposed pressure on the soil mass will not imply a direct increase of the contact pressure between the grains as the extra pressure will be completely taken up by the water mass, without thereby improving the shear strength of the soil at all.

The influx of water in cohesive soils generally adversely affects their strength, while a possible drying may cause excessive deformations, accompanied by intense cracking, which will facilitate the further influx of water in the soil mass in the future. In any case, under the same loading pressure, the settlement of cohesive soils is greater than in non-cohesive soils.

Regarding the pressures on vertical retaining walls, the presence of cohesion, c, definitely has a relieving role (see Figure 8.4). Thus, the total horizontal force on a vertical front of height, H, consists of the acting force under the existing internal friction and the relieving force due to cohesion. It is:

$$E_h = \gamma \cdot H^2 \cdot K / 2 - 2 \cdot c \cdot H \sqrt{K}$$

The contribution of self-weight to the above force is depicted by a triangular diagram, whereas the influence of cohesion is represented by a constant value over the height. The treatment of coefficient, K, is the same as in the case of non-cohesive soils.

When considering the forces that develop at the interface between a deformable soil mass and a rigid surface, it is important to note the following: if the soil mass, instead of exerting a force to that surface which resists its deformation (as in the case of a retaining wall), is subjected to an external force that pushes it to deform, it must generate a reactive force to maintain equilibrium. This situation arises, for example, in soil anchoring or when a certain thrust as an externally applied load is imposed on the soil itself.

In such cases, the resistance offered by the soil against the imposed deformation through the advancing surface – known as *passive earth pressure* – is quantified using a coefficient, K_p, greater than 1. This coefficient depends on both the angle of internal friction and the cohesion of the soil.

Figure 8.4 Earth pressure in cohesive soils

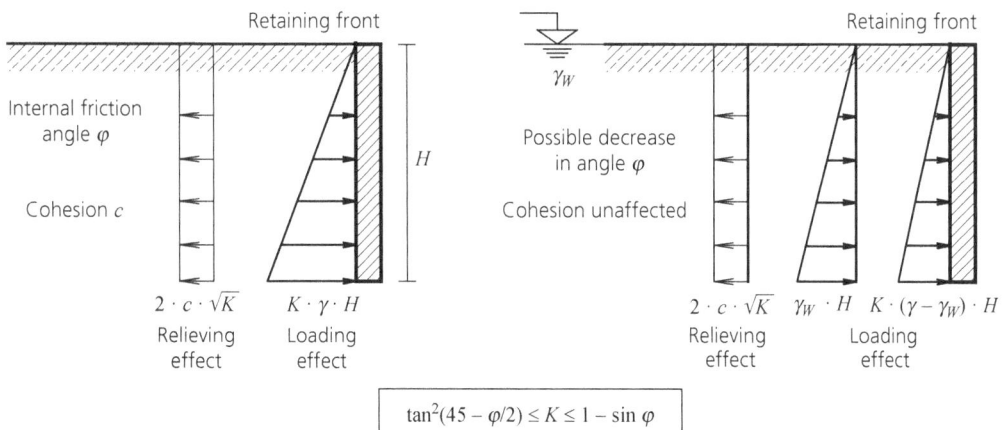

Retaining front

Internal friction angle φ

Cohesion c

H

$2 \cdot c \cdot \sqrt{K}$ — Relieving effect

$K \cdot \gamma \cdot H$ — Loading effect

γ_w

Retaining front

Possible decrease in angle φ

Cohesion unaffected

$2 \cdot c \cdot \sqrt{K}$ — Relieving effect

$\gamma_w \cdot H$ $K \cdot (\gamma - \gamma_w) \cdot H$ — Loading effect

$$\tan^2(45 - \varphi/2) \leq K \leq 1 - \sin \varphi$$

Figure 8.5 Effect of cohesion on the unbraced height of an open excavation

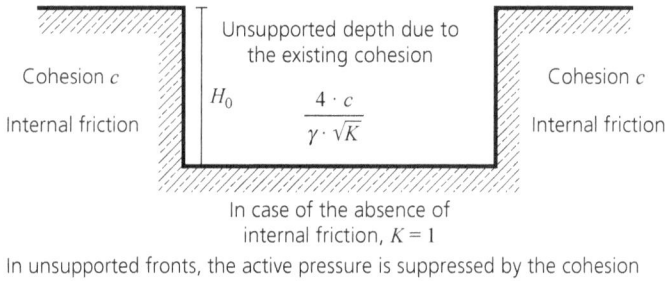

In case of the absence of
internal friction, $K = 1$

In unsupported fronts, the active pressure is suppressed by the cohesion

Moreover, in the above expression of E_h, the possible presence of a water table within the soil mass must be taken into account by the appropriate reduction of specific weight, γ, due to buoyancy. This effect is equivalent to the fact that the pore pressure, u, is equal to the acting hydrostatic pressure.

The above equation allows the estimation of the maximum height, H_0, of an unsupported vertical excavation, where it must be $E_h = 0$ (see Figure 8.5). This results in:

$$H_0 = 4 \cdot c / (\gamma \cdot \sqrt{K})$$

In the absence of internal friction ($\varphi = 0$), where, of course, $K = 1$, the maximum allowable height of an unbraced vertical excavation of a cohesive soil is:

$$H_0 = 4 \cdot c / \gamma$$

8.3.   Shallow foundations

It is reasonable, in order to provide a foundation to a structure, to always seek its safe support on the shallowest level possible since the cost of a foundation drastically increases with its depth. This, of course, presumes that the bearing capacity of the soil in the selected foundation level is adequate to carry all the anticipated loads on the structure. Such a foundation is called a *shallow foundation* and comprises the majority of foundations. However, if such a foundation is not feasible, a special structural layout should be provided so that the applied loads are transmitted to lower soil layers with satisfactorily safe bearing capacity. The foundation is then characterised as a *deep foundation*. The cost of such a foundation constitutes a significant percentage of the total cost of the structure.

What is presented below mainly concerns shallow foundations. However, some basic structural characteristics of deep foundations will be examined later in this chapter.

8.3.1   The deformational behaviour of elastic soil under vertical loads

Soil may be considered as an elastic medium – possibly consisting of intermediate layers – acted on by vertical loads on its free surface. Under these loads, soil behaves elastically. This assumption must be supported by the results of geotechnical tests performed on site. Thus, each soil layer is characterised by its elastic modulus, E_s, as well as by the corresponding Poisson's ratio, v, determined by laboratory tests on soil samples from the site. The elastic modulus, E_s, is referred to as the *compression modulus* of the soil and represents a quantity that cannot be precisely determined in contrast to construction materials such as steel. It must be stressed that soils behave as anisotropic materials without any resistance in tension and as such they have only an elastic modulus for compression. Its value corresponds to the geotechnical characteristics of the soil examined. Thus, non-cohesive soils exhibit values from 30 000 up to 300 000 kN/m^2, depending on the grain size and the degree of consolidation (sand: 30 000 up to 100 000 kN/m^2; gravel: 70 000 up to 300 000 kN/m^2), whereas in cohesive clay soils the values of E_s are clearly lower, ranging between 10 000 and 40 000 kN/m^2. The Poisson's ratio has much narrower variation margins and a smaller influence on the deformation. Its order of magnitude for sand is about 0.3 and for clay is about 0.4.

Assuming that the examined soil mass, of theoretically infinite depth, has a uniform compression modulus, E_s, and its free surface is loaded by a uniformly distributed load, p_0, over a rectangular area ($a \cdot b$), then the developed settlement, w, at a distance, x, from its central point in a parallel direction to the direction of side, a, is (see Figure 8.6):

$$w = \frac{\left(1 - v^2\right)}{E_s \cdot 2\pi} \cdot p_0 \cdot a \cdot \lambda$$

where,

$$\lambda = \left[(1 - \xi) \cdot \ln \frac{\sqrt{(1-\xi)^2 + \alpha^2} + \alpha}{\sqrt{(1-\xi)^2 + \alpha^2} - \alpha} + (1 + \xi) \cdot \ln \frac{\sqrt{(1+\xi)^2 + \alpha^2} + \alpha}{\sqrt{(1+\xi)^2 + \alpha^2} - \alpha} \right.$$

$$\left. + 2\alpha \cdot \ln \frac{\sqrt{(1-\xi)^2 + \alpha^2} + (1-\xi)}{\sqrt{(1+\xi)^2 + \alpha^2} - (1+\xi)} \right]$$

and $\alpha = b/a$, $\xi = \dfrac{2x}{a}$ (Stavridis, 1997). ■

On the basis of the above result and in view of the later examination of soil–structure interaction, a strip of length, L, and constant width, b, is considered, being divided into a number of n equal

Figure 8.6 Settlements of the elastic subspace due to a uniformly loaded orthogonal area

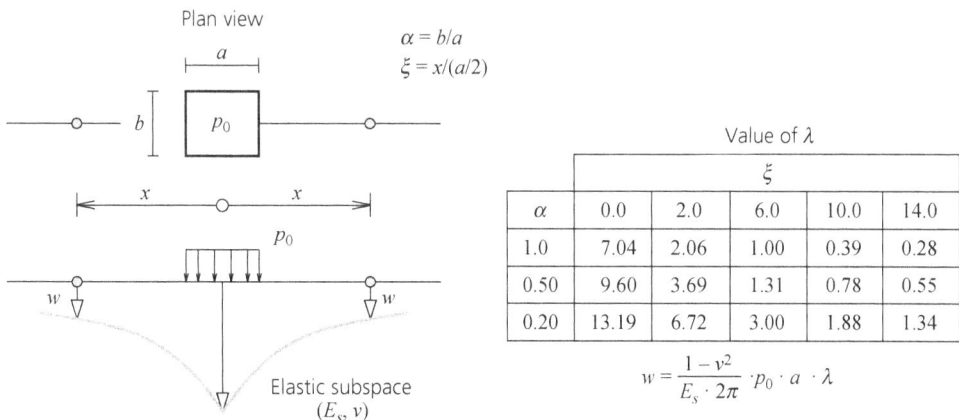

Plan view

$\alpha = b/a$
$\xi = x/(a/2)$

Value of λ

α	ξ				
	0.0	2.0	6.0	10.0	14.0
1.0	7.04	2.06	1.00	0.39	0.28
0.50	9.60	3.69	1.31	0.78	0.55
0.20	13.19	6.72	3.00	1.88	1.34

$$w = \frac{1 - v^2}{E_s \cdot 2\pi} \cdot p_0 \cdot a \cdot \lambda$$

Elastic subspace
(E_s, v)

segments of length (L/n), as shown in Figure 8.7, whereby the mutual deformational influence of each of the above segments with all others can be determined as follows.

A unit vertical load $P = 1$ kN is applied consecutively on each segment's midpoint, assumed as uniformly distributed over that segment with a value:

$$p_0 = \frac{1}{(L/n)\cdot b}$$

The settlements at the midpoint of all other segments can be readily determined on the basis of the last equation. If the segments are numbered in ascending order from 1 up to n, a number of n^2 settlements, f_{sr}, may be determined, where f_{sr} represents the settlement that develops in segment number s, when the segment number r is loaded with $P = 1$. Thus, with a total number of segments equal to $n = 3$ (see Figure 8.8), loading, for example, segment number 1 on the basis of the above relation, quantities f_{11}, f_{21} and f_{31} may be directly determined. Similarly, by loading segment number 2, quantities f_{12}, f_{22} and f_{32} may be calculated and, similarly, by loading segment number 3, settlements f_{13}, f_{23}, f_{33} are analogously obtained.

The above deformations allow the determination of the settlement at all midpoints, if each segment is under a certain load, P_r. Thus, in this example (see Figure 8.8), the settlement, w, of each segment is:

$$w_1 = f_{11}\cdot P_1 + f_{12}\cdot P_2 + f_{13}\cdot P_3$$
$$w_2 = f_{21}\cdot P_1 + f_{22}\cdot P_2 + f_{23}\cdot P_3$$
$$w_3 = f_{31}\cdot P_1 + f_{32}\cdot P_2 + f_{33}\cdot P_3$$

The extension to a greater number of segments is obvious:

$$w_1 = f_{11}\cdot P_1 + f_{12}\cdot P_2 + ... + f_{1n}\cdot P_n$$
$$w_2 = f_{21}\cdot P_1 + f_{22}\cdot P_2 + ... + f_{2n}\cdot P_n$$
$$...$$
$$w_n = f_{n1}\cdot P_1 + f_{n2}\cdot P_2 + ... + f_{nn}\cdot P_n$$

Figure 8.7 Influence of a loaded area on the settlement of a point lying on the soil surface

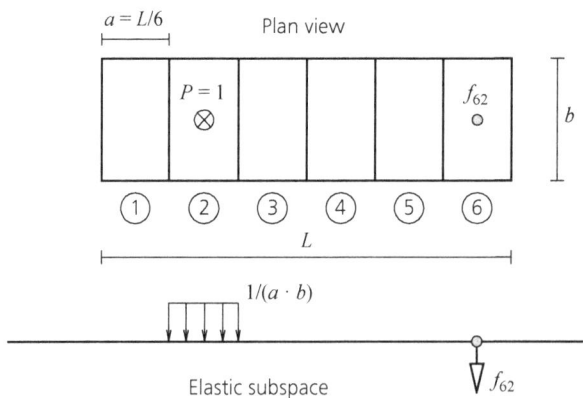

By assembling settlements 'f' in a square matrix $\mathbf{F}$ $(n \times n)$, which may be called the *flexibility soil matrix*, and further assembling settlements 'w_i' and loads 'P_i' in the column matrices $(n \times 1)$ $\mathbf{w}$ and $\mathbf{P}$, respectively, the above expressions may be written in the compact matrix form (see Chapter 7, Section 7.3.1):

$$w = \mathbf{F} \cdot \mathbf{P}$$

The assumption of a constant compression modulus, E_s, over the whole depth of a certain soil mass is not always realistic. The case of consecutive vertically arranged horizontal layers, having their own thickness with a corresponding compression modulus, is not unusual. On the other hand, even in a *homogeneous* soil mass, the compression modulus increases approximately linearly with depth. Although for preliminary design purposes the consideration of an estimated 'average' compression modulus may be sufficient, a more 'correct' determination of the settlement, w, is presented below in the case of a layered soil profile (Stavridis, 2002).

As a basic tool, the following result is used regarding the settlement, w, developed at any corner of an orthogonal area ($a \cdot b$) under a uniform load, p_0, lying over a soil layer of thickness, z, and a compression modulus, E_s. This layer is assumed to rest on an absolutely rigid base (see Figure 8.9). It is:

$$w(a,b) = u(z)/Y$$

where,

$$u(z) = \frac{a \cdot p_0}{4 \cdot \pi} \left[\zeta \cdot \arctan\left(\frac{\beta}{\zeta \cdot \gamma}\right) + \ln\left(\frac{(\gamma - \beta)}{(\gamma + \beta)} \cdot \frac{(\omega + \beta)}{(\omega - \beta)}\right) + \beta \cdot \ln\left(\frac{(\gamma - 1)}{(\gamma + 1)} \cdot \frac{(\omega + 1)}{(\omega - 1)}\right) \right]$$

Figure 8.8 Determination of settlements due to specific concentrated loads

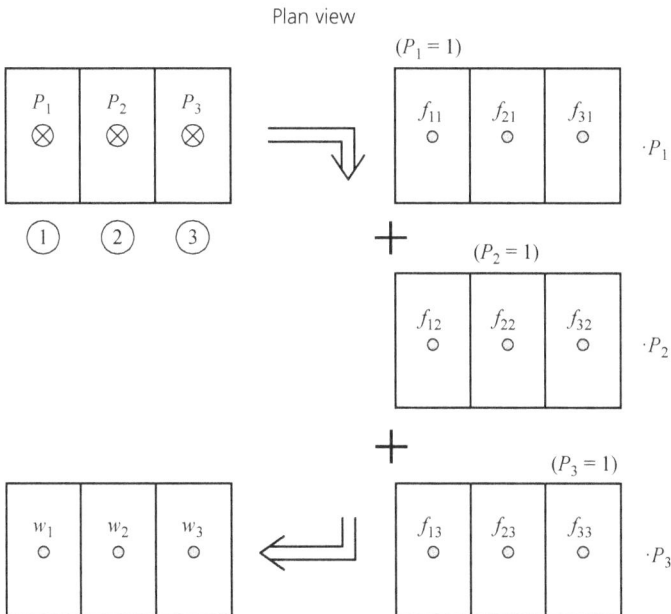

Figure 8.9 Settlement of an orthogonal area over a soil layer of finite thickness

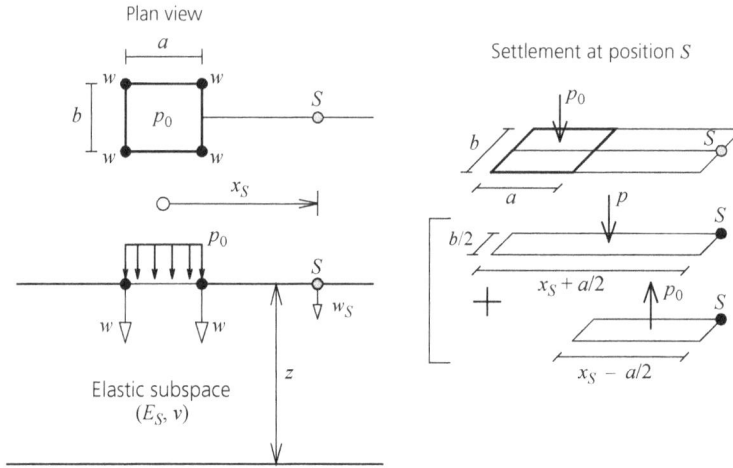

and

$$Y = \frac{E_S}{1-v^2}, \quad \beta = b/a, \quad \zeta = 2 \cdot z/a, \quad \omega = \sqrt{1+\beta^2}, \quad \gamma = \sqrt{1+\beta^2+\zeta^2}$$

It may be found now that the settlement of a point S on the soil surface at a distance, x_S, from the central point of the area ($a \cdot b$) in parallel direction to side, a, can be expressed as:

$$w_S = 2 \cdot [w(x_S + a/2, b/2) - w(x_S - a/2, b/2)]$$

where quantities $w(.,.)$ inside the brackets refer to the corner settlements of the orthogonal areas [$(x_S + a/2)$ x $(b/2)$] and [$(x_S - a/2)$ x $(b/2)$], respectively, according to the above expression of $w(a,b)$ (see Figure 8.9).

The contribution of a layer to the settlement of a particular point of the soil surface can be considered as the difference between the contribution of the entire soil mass above its bottom border and the one above its top border (see Figure 8.10). Thus, by superposing the contribution of the compressibility of each soil layer to the development of the final settlement of point S (see Figure 8.10), the following result is obtained:

$$w_S = u(z_1) \cdot \left(1/Y_1 - 1/Y_2\right) + u(z_2) \cdot \left(1/Y_2 - 1/Y_3\right) + \ldots$$
$$+ u(z_{k-1}) \cdot \left(1/Y_{k-1} - 1/Y_k\right) + u(z_k) \cdot 1/Y_k$$

It is clear that the formulation of the foregoing mutual interaction of the adjacent strip segments, according to Figure 8.7, in the case of a stratified soil remains Practically the same.

8.3.2 Rectangular spread footings

In order to provide a foundation to the base of a column element, where a specific vertical force, V_1, a horizontal force, H, and a moment, M_1, must be transmitted, an orthogonal concrete footing with a specific area and a depth is usually constructed, normally having its larger base dimension

Figure 8.10 Contribution of the existing layers to the soil surface settlement

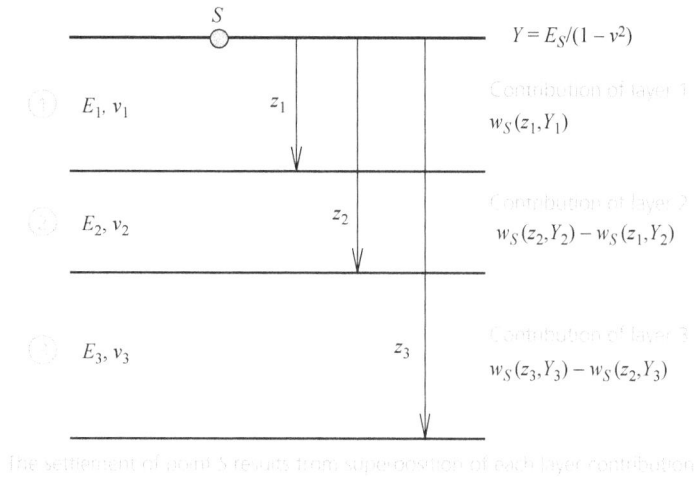

$Y = E_S/(1 - v^2)$

Contribution of layer 1
$w_S(z_1, Y_1)$

Contribution of layer 2
$w_S(z_2, Y_2) - w_S(z_1, Y_2)$

Contribution of layer 3
$w_S(z_3, Y_3) - w_S(z_2, Y_3)$

The settlement of point S results from superposition of each layer contribution

parallel to the plane of action of moment, M_1 (see Figure 8.11). As in reality, the horizontal force and the moment may be applied in two directions, the larger base dimension having to be parallel to the dominant moment.

The reason why such a footing is necessary is rather obvious. Direct support on the soil of a column without footing developing at its base the sectional forces (V_1, H, M_1), is generally impossible; the soil cannot take up the stresses required to take up the equal and opposite to the above forces within a soil surface identical to the cross-section of the column. Thus, for the proper design of a footing, the following points have to be taken into account:

(a) The maximum contact stress developed with the soil, which should not exceed the safe value determined by the geotechnical study of the site, characterised as the allowable bearing pressure.
(b) The absolute magnitude of the maximum soil settlement developed, as well as the angle of rotation of the footing.
(c) The state of stress of the footing, which equilibrates as a free body under the acting sectional forces V_1, H and M_1, its self-weight, G, as well as the acting contact pressures from the soil. The concrete footing must be reinforced and checked appropriately so that it can take up all the above forces.

Of course, on the contact area between footing and soil, a vertical force $V = V_1 + G$, a horizontal force, H, and a moment $M = M_1 + H \cdot d$ are transmitted. It is under these forces that the above points (a) and (b) should be checked.

Moreover, the horizontal force, H, must be 'covered' by the *friction resistance* ($V \cdot \mu$) between soil and foundation. A safe value for the friction coefficient, μ, is 0.5 for cohesive soils and 0.6 for non-cohesive soils. However, by adopting a safety factor, γ – usually taken equal to 1.4 – the following relation must be satisfied:

$$H < \gamma \cdot (V \cdot \mu)$$

(see Figure 8.11). Normally, the horizontal force, H, is not further involved in the design.

Figure 8.11 Footing equilibrium and soil pressures

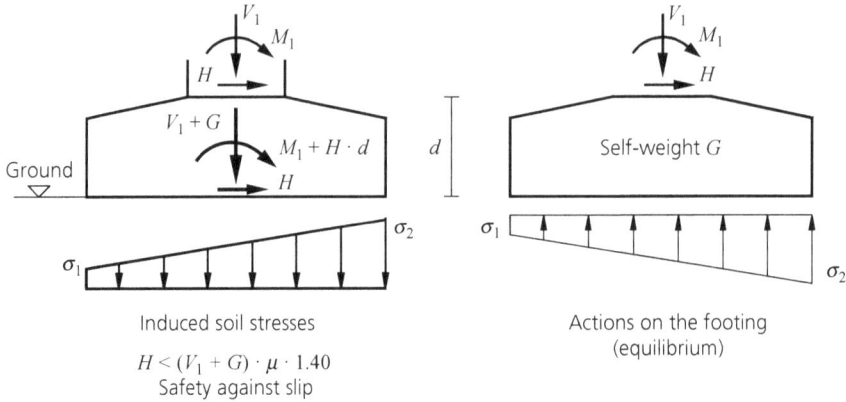

Induced soil stresses

$H < (V_1 + G) \cdot \mu \cdot 1.40$
Safety against slip

Actions on the footing
(equilibrium)

8.3.2.1 Soil pressures and settlements

In a first attempt to estimate the developed soil pressure, the contact area of the footing is considered as a normal orthogonal section under the forces V and M. It is assumed that the force, V, also includes the weight of the overlying soil mass. Force, V, is assumed to act on the centroid of the footing area (see Figure 8.12). It is, of course, clear that the moment, M, causes a shift, e, of the force, V, along its plane of action, being equal to $e = M/V$. Thus, the pair of forces (V, M) is statically equivalent to force, V, alone, shifted, however, by distance, e. As compression contact pressures make sense only over the foundation area, it is appropriate to check whether the shifted force, V, lies within the central kern of the section or not – that is, whether $e \leq a/6$. In that case, the maximum and minimum contact (compression) stresses are, according to the bending formula (Stavridis and Georgiadis, 2025, Section 2.2.1):

$$\sigma_{max} = \frac{V}{A} + \frac{M \cdot (a/2)}{I_f} = \frac{V}{a \cdot b} + \frac{6 \cdot M}{a^2 \cdot b}$$

$$\sigma_{min} = \frac{V}{a \cdot b} - \frac{6 \cdot M}{a^2 \cdot b}$$

where I_f is the second moment of area of the foundation base. Of course, if there is no moment, M, a uniform pressure $\sigma = V/(a \cdot b)$ is developed.

In the case in which the shifted force, V, overruns the central kern limits – something that should generally be avoided – the above relations do not apply since σ_{min} becomes tensile. Then, the force, V, must be equilibrated by a triangular pressure diagram, having its maximum value on the nearest side to the force, V, while its zero value to which the pressure diagram extends will obviously lie within the foundation area (see Figure 8.12). As the distance, c, of the force, V, to the near side is $c = (a/2 - e)$, it becomes clear that the triangular diagram of the contact pressures extends over a length $(3 \cdot c)$ and the above maximum pressure results from the equilibrium requirement:

$$\frac{1}{2} \cdot \sigma_{max} (3 \cdot c) \cdot b = V \qquad \blacksquare$$

Figure 8.12 Influence of the location of the resultant force on the resulting soil pressures

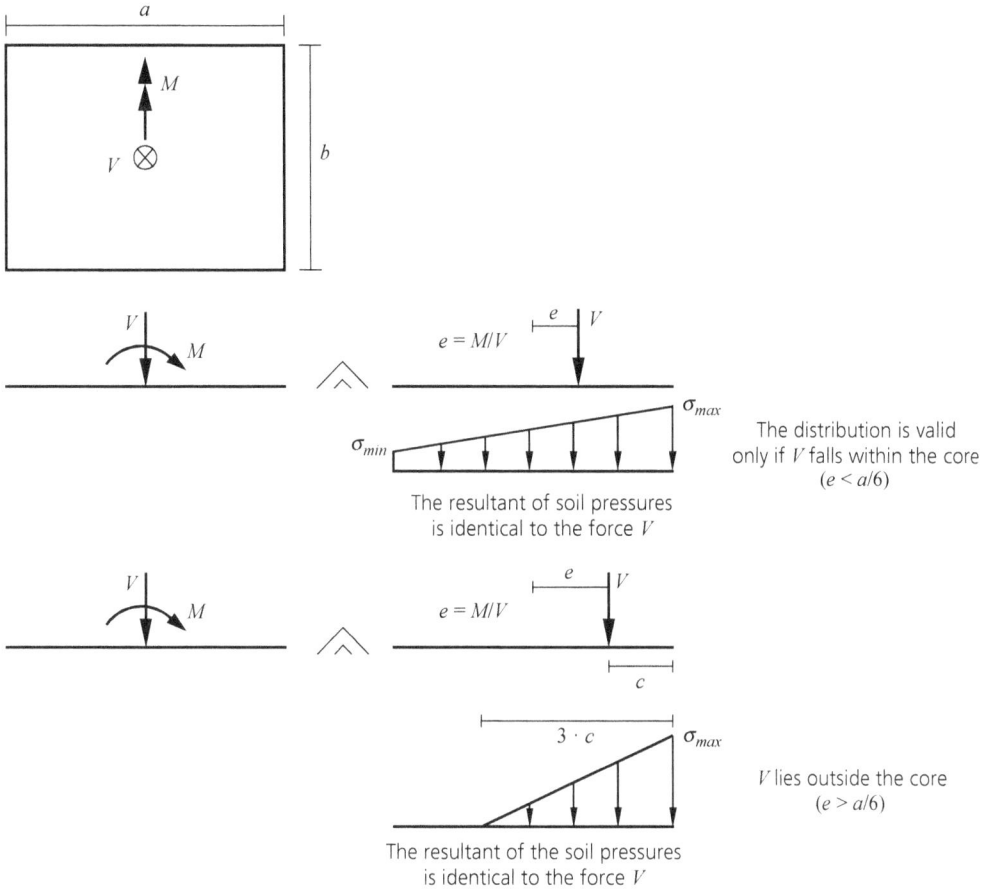

The distribution is valid only if V falls within the core $(e < a/6)$

The resultant of soil pressures is identical to the force V

V lies outside the core $(e > a/6)$

The resultant of the soil pressures is identical to the force V

Although it is clear that all the above pressure distributions satisfy the corresponding equilibrium conditions with the magnitudes V and M, it can be easily seen that they lead to deformations, which do not correspond to reality. More specifically, as the footing may be practically considered undeformable, the soil settlements, being identical to those of the foundation, should lie on a single plane exhibiting a certain rotation angle because of the presence of moment, M (Stavridis, 2009). But even in the 'simple' case of a centrally applied force, V, ($M = 0$), it is clear that the resulting uniform pressure $V/(a \cdot b)$ on the soil over a rectangular base does not lead to a uniform settlement of all the points of the soil surface.

Thus, for the case of a centrally applied force, V, after dividing the foundation surface to a certain number n of transverse strips of width (a/n), the loads, P_i, for each strip have to be determined, which sum up to the value, V, on one hand, and lead to the same settlement (w) for all strips, on the other (see Figure 8.13).

Figure 8.13 More accurate determination of soil pressures and uniform settlement

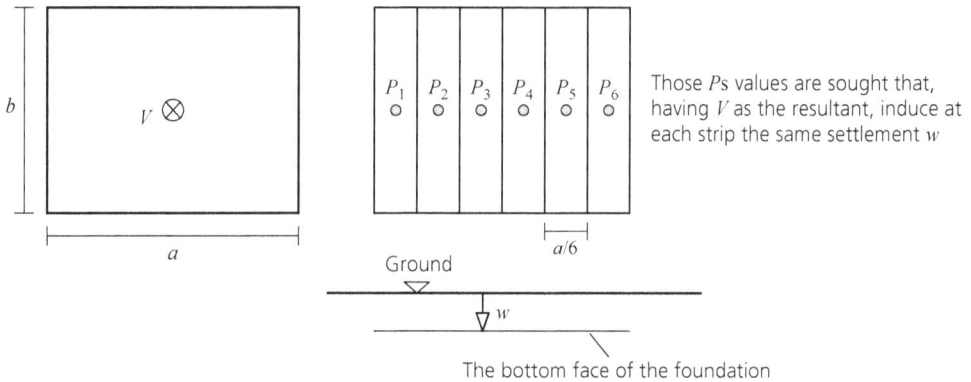

Considering, for example, three transverse strips ($n = 3$) and, on the basis of the result in Section 8.3.1, the last requirement is expressed by the following equations:

$$w = f_{11} \cdot P_1 + f_{12} \cdot P_2 + f_{13} \cdot P_3$$
$$w = f_{21} \cdot P_1 + f_{22} \cdot P_2 + f_{23} \cdot P_3$$
$$w = f_{31} \cdot P_1 + f_{32} \cdot P_2 + f_{33} \cdot P_3$$

while, in addition, the following equilibrium condition must be valid:

$$V = P_1 + P_2 + P_3$$

The (3 + 1) unknown quantities P_1, P_2, P_3 and w may be determined through the above linear system of equations. It comes out that the forces P_1 and P_3 have a greater value than P_2, so that under a central compression force, V, forces P that represent directly the soil pressures, instead of being equal, clearly increase towards the edges of the footing (see Figure 8.14).

8.3.2.2 Support on an elastic base

A less reliable but clearly more practical alternative to describe the deformational behaviour of the soil is offered by the concept of modulus of subgrade reaction represented by the coefficient, k. It is based on the assumption that the settlement, w, of the soil surface under a distributed load, p, is proportional to it – that is, it is equal to (p/k). The modulus of subgrade reaction, k (kN/m³), represents then the pressure (kN/m²) required to be applied to a soil surface in order to produce a settlement of 1 m, thus expressing the stiffness of the soil surface (see Figure 8.15).

Although the concept of subgrade modulus, k, is very convenient as being consistent with the concept of elastic support (see Stavridis and Georgiadis, 2025, Section 2.3.7), it does not constitute a measurable soil property. More specifically, according to the above, the settlement for example of a rectangular surface under a pressure, p, is equal to (p/k) – that is, it is independent of the surface itself. This is clearly not correct – as can easily be deduced from Figure 8.6 – since, under the same pressure, p, an increase of the loaded area always leads to an increase of the corresponding settlement (see Section 8.3.1 and Figure 8.6). To give an example, considering two square surfaces on a

Figure 8.14 Determination of soil pressures under a uniform settlement

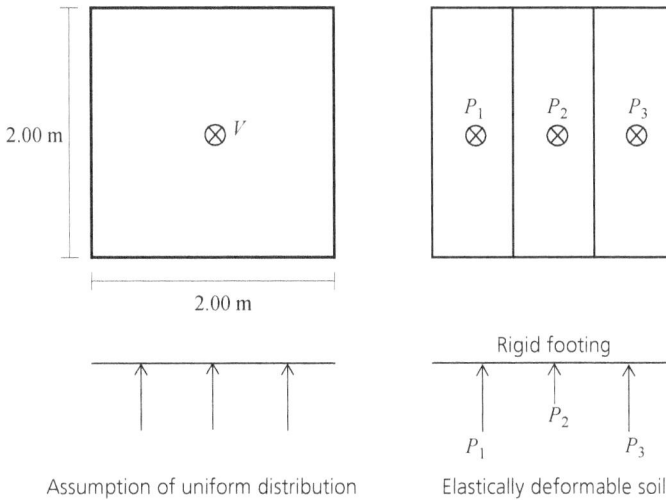

soil with compression modulus 5000 kN/m², one having dimensions 1.0 × 1.0 m and the other one dimensions 2.0 x 2.0 m, in order to produce in both cases a settlement of 1.0 m, the former surface has to be loaded with a pressure of 4900 kN/m², whereas the latter needs a pressure of 2400 kN/m² to be applied to it. Thus, the determination of a subgrade modulus, k, has no meaning as a characteristic parameter of a given soil layer as it directly depends on the specific loaded area. In addition, the assumption that all the points of the loaded area undergo the same settlement, while all the surrounding region remains undeformed, is not realistic.

However, the computational convenience together with the possibilities of the qualitative and quantitative estimation of the structural behaviour offered by the use of the parameter, k, often outweigh in practice its disadvantages due to lack of physical meaning.

Certainly, a logical value of k for a specific orthogonal area, A, may be obtained if the settlement, w, at its central point, under a pressure equal to 1 kN/m² is calculated, according to Section 8.3.1, by also using the results given in the table shown in Figure 8.6. For an acting pressure, σ, on a square soil surface the following is obtained:

$$w = \frac{(1-v^2)}{E_s \cdot 2\pi} \cdot p_0 \cdot a \cdot \lambda = \frac{\sigma \cdot \sqrt{A}}{E_s \cdot 2\pi} \cdot 7.04$$

or approximately,

$$w = \frac{\sigma \cdot \sqrt{A}}{E_s}$$

Since $k = 1/w$, for a rectangular surface, A, it can be written as:

$$k = \frac{E_s}{\sqrt{A}}$$

297

Figure 8.15 Concept of subgrade modulus

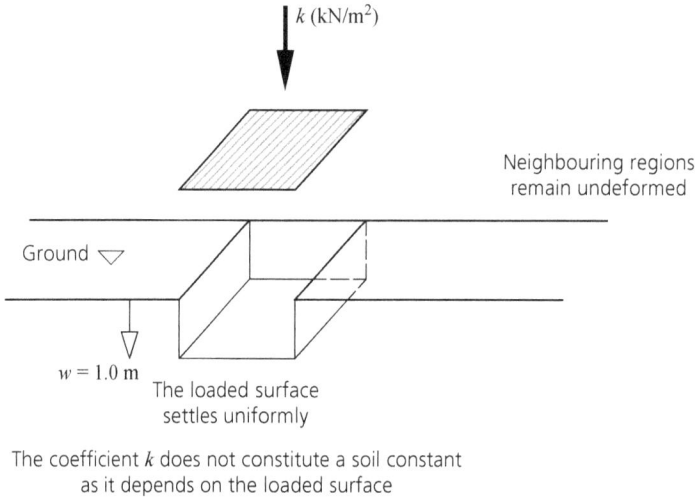

k (kN/m²)

Neighbouring regions
remain undeformed

Ground ▽

$w = 1.0$ m

The loaded surface
settles uniformly

The coefficient k does not constitute a soil constant
as it depends on the loaded surface

and for a square surface with a side equal to a, it is:

$$k \approx E_s / a \, (\text{kN/m}^3)$$

In this way, the soil under each single footing having an area, A, may be practically simulated by a spring with a stiffness $k_s = k \cdot A$ since, according to the above, a centrally acting load, P, on the footing causes a settlement $w = (P/A)/k$. Then, the whole structure may be considered as including all the corresponding springs to its footings, being itself naturally supported on a rigid base (see Figure 8.16). ∎

It is clear that the induced rotation of a footing under the action of a moment, M, having been determined – as usual – by assuming the column base is fixed, implies a reduction of this moment, according to Stavridis and Georgiadis (2025, Sections 3.1.7 and 3.2.7) regarding indeterminate structures. This footing rotation and the ensuing moment reduction should be assessed, since it will lead to more favourable soil stresses.

Thus, it is examined whether the resilience of the footing can be described by a rotational spring with a specific elastic rotation flexibility, f_φ, as defined in Stavridis and Georgiadis (2025, Section 2.3.7.2), namely as the developed footing rotation under the action of a unit moment. Obviously, such a moment should be accompanied by a vertical load, V, in order to ensure the development of contact pressures over the whole footing (see Figure 8.17).

In the case of a rigid footing, transmitting the sectional forces V and M over its bearing area, A, and, provided that over the whole foundation surface only compressive pressures are developed – meaning that the shifted resultant, V, falls within the central kern of the rectangular area – then, on the basis of a subgrade modulus, k, and, according to Section 8.3.2, the settlements s_1, s_2 at its two edges will be:

$$s_1 = \frac{\sigma_{max}}{k} = \frac{V}{A \cdot k} + \frac{M \cdot (a/2)}{I_f \cdot k} \qquad s_2 = \frac{\sigma_{min}}{k} = \frac{V}{A \cdot k} - \frac{M \cdot (a/2)}{I_f \cdot k} \qquad (I_f = b \cdot a^3 / 12)$$

Figure 8.16 Simulation of soil deformability through translational springs

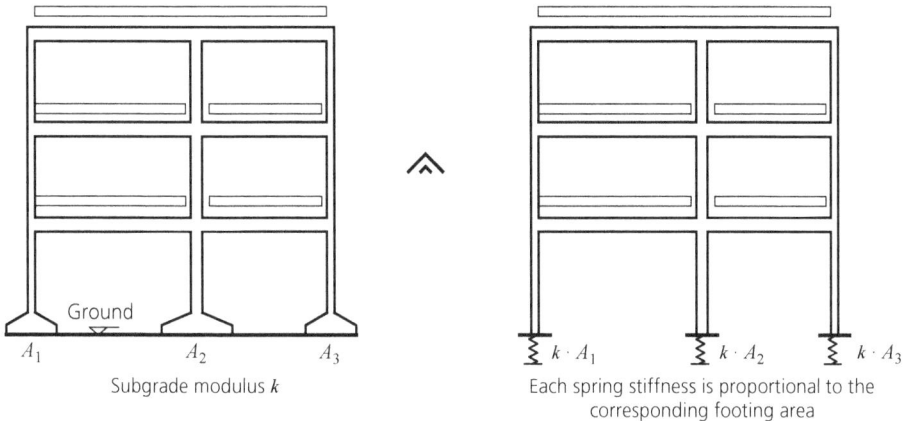

Subgrade modulus k — A_1, A_2, A_3, Ground

Each spring stiffness is proportional to the corresponding footing area — $k \cdot A_1$, $k \cdot A_2$, $k \cdot A_3$

Figure 8.17 Conditions for soil simulation through a rotational spring

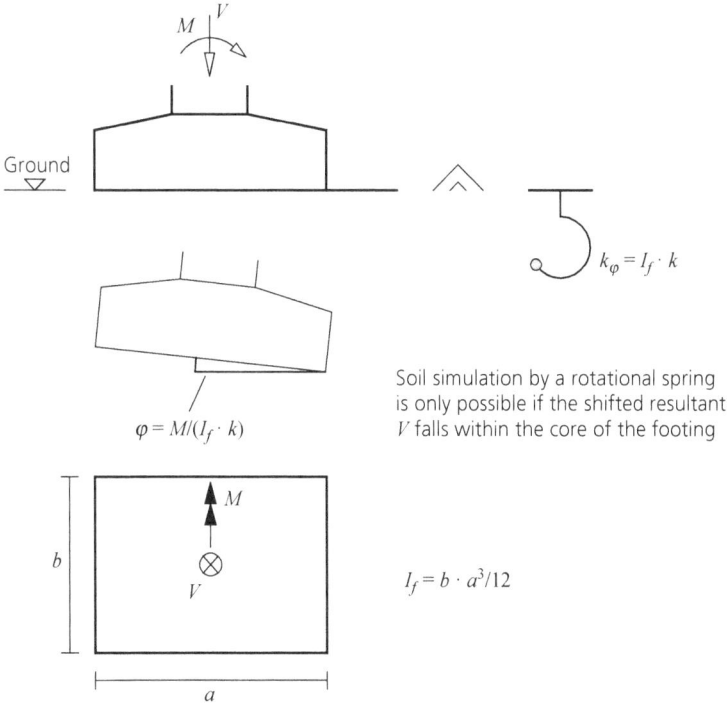

M, V, Ground

$k_\varphi = I_f \cdot k$

$\varphi = M/(I_f \cdot k)$

Soil simulation by a rotational spring is only possible if the shifted resultant V falls within the core of the footing

b, M, V, a

$I_f = b \cdot a^3/12$

Consequently, the rotation, φ, of the footing will be:

$$\varphi = \frac{s_1 - s_2}{a} = \frac{M}{I_f \cdot k}$$

This expression of φ allows the determination of the coefficient of elastic rotation, f_φ – according to Stavridis and Georgiadis (2025, Section 2.3.7.2) – or, equivalently, of the stiffness coefficient with respect to rotation $k_\varphi = 1/f_\varphi$, simulating in this way the soil through a rotational spring:

$$f_\varphi = \frac{1}{I_f \cdot k} \quad \text{and} \quad k_\varphi = I_f \cdot k \quad \text{(see Figure 8.17)}$$

However, it is pointed out that, while, on the basis of the previous more exact approach of soil deformability (see Section 8.3.1) through the compression modulus, E_s, the load eccentricity may exceed the central kern limits of the foundation surface and yet maintain the compression contact stresses everywhere, the adoption of a subgrade modulus, k, does not lead to such a consequence, as can be ascertained by the above expressions.

When the resultant force, V, acts outside of the central kern, the settlement, s_1, according to the above, is:

$$s_1 = \frac{\sigma_{max}}{k} = \frac{V}{3 \cdot (a/2 - \frac{M}{V}) \cdot b \cdot k}$$

and the corresponding rotation, φ, of the footing will be:

$$\varphi = \frac{s_1}{3 \cdot c} = \frac{V}{9 \cdot \left(\frac{a}{2} - \frac{M}{V}\right)^2 \cdot b \cdot k}$$

It is clear that the rotation, φ, is not proportional to the acting moment – depending also on the value of the vertical force, V – and, therefore, in this case the soil cannot be represented by a rotational spring. ∎

On the basis of the resistance, k_φ, offered by the rotational spring against the imposition of a unit rotation, the reduction of the moment, M, previously mentioned may be estimated as follows.

The rotational spring introduces the developed moment to the column, according to the rotation of their common joint, which obviously rotates together with the spring. If an external moment is considered to act on this node so that its rotation vanishes, then this moment, M, must clearly be equal to the support reaction of the clamped column base (see Figure 8.18). Thus, in order to make this moment disappear, since it does not actually exist, an equal and opposite moment, M, must be superposed, according to the process used in Stavridis and Georgiadis (2025, Section 3.2.1).

The above moment, M, causes a rotation, φ, both to the rotational spring and the column base, which may be determined from the equilibrium requirement of their common joint. Assuming the upper end of the column to be fixed (see Stavridis and Georgiadis, 2025, Section 3.2.4.2):

$$\frac{\varphi}{f_\varphi} + \frac{4 \cdot EI}{H} \cdot \varphi = M$$

Figure 8.18 Interaction between column–footing–ground

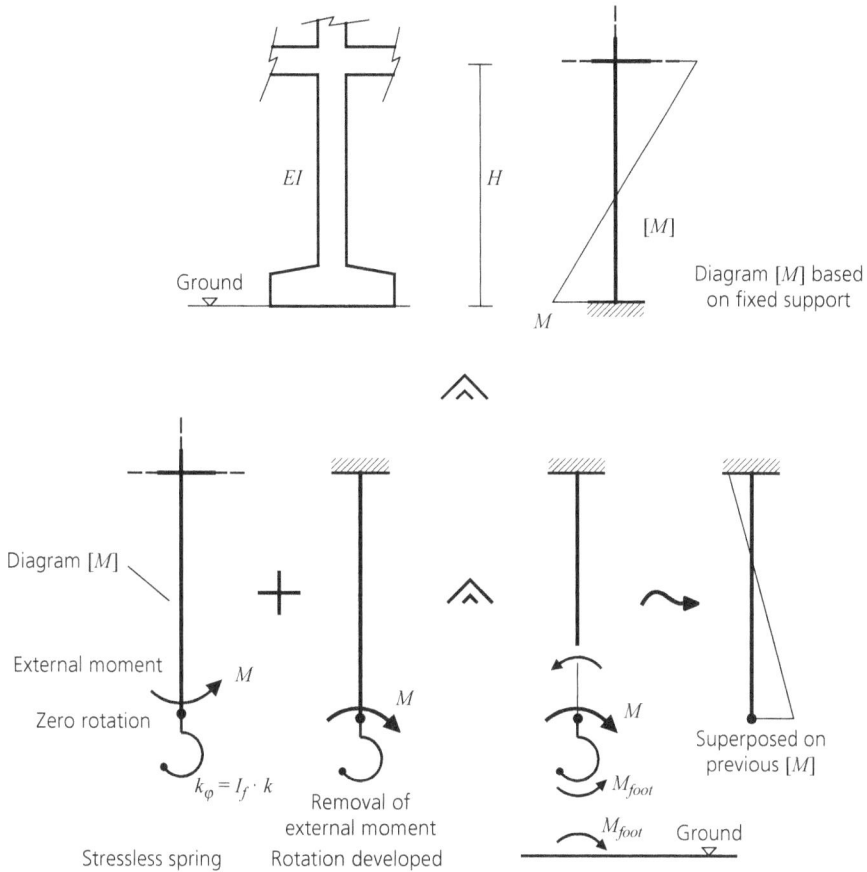

Thus, the developed moment in the spring – that is, the moment, M_{foot}, taken up by the footing and transmitted to the ground – is:

$$M_{foot} = \frac{\varphi}{f_\varphi} = \frac{M}{1 + \dfrac{4 \cdot EI}{H} \cdot f_\varphi}$$

which is clearly reduced compared with M (see Figure 8.18).

This result is identical to the one based on Stavridis and Georgiadis (2025, Section 3.2.8), namely that the moment, M, is distributed in proportion to the 'adjacent' stiffnesses:

$$M_{foot} = M \cdot \frac{1/f_\varphi}{\dfrac{1}{f_\varphi} + \dfrac{4 \cdot EI}{H}}$$

Figure 8.19 Structural action of the connecting ribs

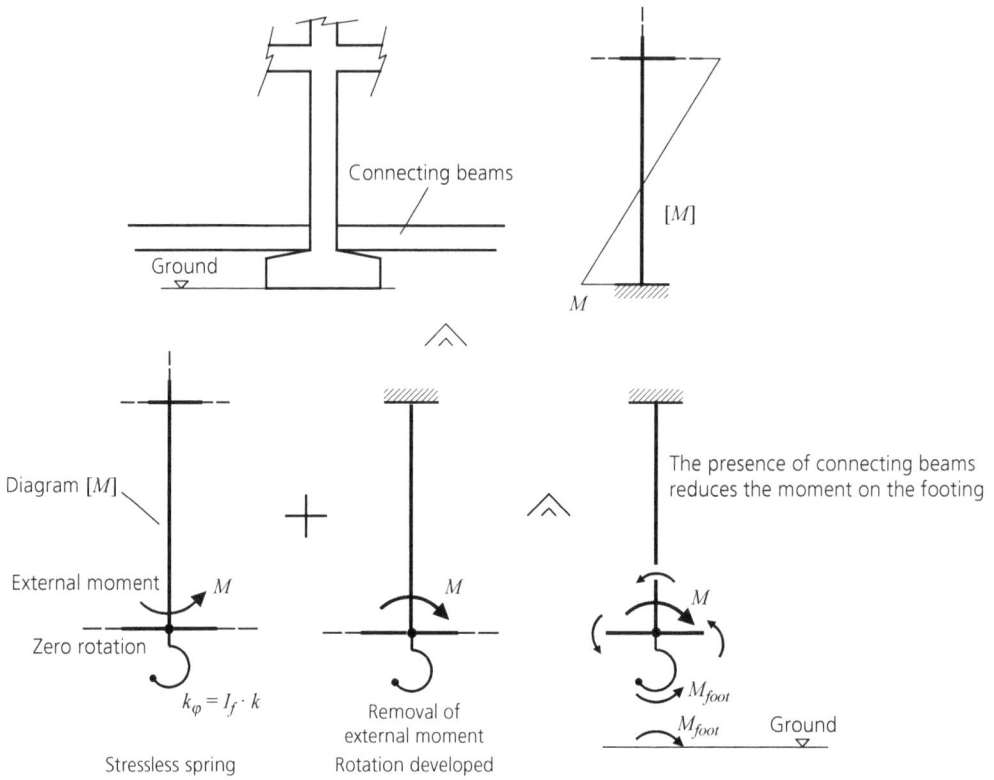

The moment, M_{foot}, may be further reduced if beams are inserted that connect the lower end of the examined column with the neighbouring ones (see Figure 8.19).

In this case, too, the equilibrium of the joint of the rotational spring that connects the column and the two connecting beams with their far ends fixed leads to the distribution of moment, M, proportionally to the stiffness of the three adjacent members (see Figure 8.19). The corresponding portion, M_{foot}, assigned to the footing surface is:

$$M_{foot} = M \cdot \frac{1/f_\varphi}{\frac{1}{f_\varphi} + \Sigma \frac{4 \cdot EI}{H}} = M \cdot \frac{1}{1 + f_\varphi \cdot \Sigma \frac{4 \cdot EI}{H}}$$

In the above expressions concerning the transmitted bending moment, M_{foot}, to the footing, the beams are assumed as fixed-fixed at both ends. However, more conservatively they could be considered as simply fixed beams, in which case in the denominator of the above expressions a factor of 3 instead of 4 should be applied.

Generally, it is advisable to provide such *connecting beams* between the lower ends of the columns. In this respect, the designer may choose the appropriate cross-section for these beams in order to

Figure 8.20 Treatment of eccentric footing

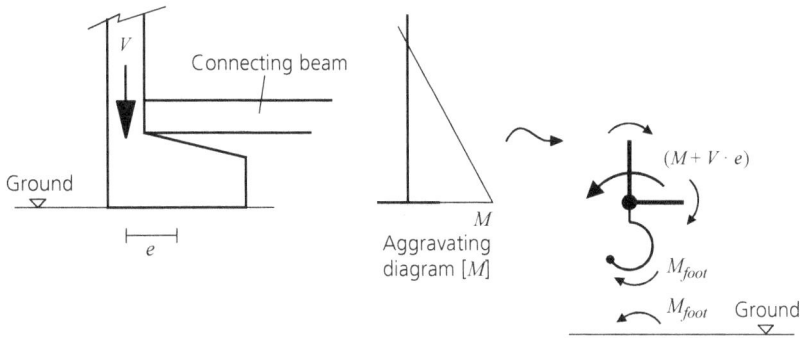

duly moderate the bending moment, M_{foot}, transmitted to the footing, so that the column load, V, falls within its core of the section. However, these connecting beams are indispensable in cases where an eccentric layout of the footing with respect to the column is constructionally imposed – for example, along property lines where each column load, V, necessarily exhibits a certain eccentricity, e (see Figure 8.20).

Of course, the isolated footings normally follow a symmetric layout with respect to the column section because of the possible shift of the column load, V, to either side due to a possible alternating moment action, M, such as during an earthquake.

Returning now to the constructional eccentricity, e, of a footing, it is clear that the column moment, M, deduced by assuming its base as fixed, must – in the most adverse case – be increased by $V{\cdot}e$ and, consequently, the last expression of the moment taken up by the foundation surface becomes:

$$M_{foot} = \frac{M + V \cdot e}{1 + f_\varphi \cdot \Sigma \dfrac{4 \cdot EI}{H}}$$

where, in the denominator the column member and the connecting rib (two members in total) must be taken into account. It is clear that in evaluating the contact pressures according to the pair (M_{foot}, V), the force, V, is assumed to act centrally (see Figure 8.20). ■

It is noted that the coefficient, f_φ, may be determined through its previous expression as $f_\varphi = 1/(I_f \cdot k)$, which in the case of a square footing with side, a, is $f_\varphi = a/(I_f \cdot E_s)$, as discussed above.

It should be pointed out here that the more yielding the foundation soil, the smaller the moment, M_{foot}, acting on the ground. This is structurally convenient, because smaller E_s generally means lower allowable soil pressure, but also a stronger reduction of the effective moment acting on the foundation surface. However, in the case of a large E_s the coefficient, f_φ, has a small value, and the reduction of the moment $(M + V{\cdot}e)$ is also small. On the other hand, the capacity of soil to safely take up higher compression stresses is increased.

303

8.3.2.3 Design of footings

The footing is subjected as a free body to the soil pressures and to the transmitted sectional forces of the column base (equal and opposite to the respective reactions). Under the above forces, the footing acts like a slab loaded 'upwards'. This loading causes bending with the bottom face of the footing being in tension, whereas the compressed region is practically limited by the corresponding width of the column. The tensile force, clearly equal to the compressive one, is offered by reinforcement distributed over the whole width of the footing (see Figure 8.21). The selection of the reinforcement refers to a section having the same width, b, as the width of the column and the same depth, d, as that of the footing itself (see Figure 8.21). It is clear that the maximum compressive force in the footing due to bending, as conditioned by the width, b, must be kept below the corresponding allowable value.

However, if the soil pressures transferred to the column through the truss mechanism shown in Figure 8.21 require a tensile force that can be offered by the concrete itself, then the footing may be constructed without reinforcement (providing just the minimum reinforcement for cracking). This is the case if the inclination angle of the corresponding strut in the above truss model is at least 60° (see Figure 8.21).

On the other hand, if the reinforced footing has been formed so that the corresponding resultant of the soil pressures outside the outline of the column may be directed to its edges under an angle greater than 45°, then the established truss mechanism, as shown in Figure 8.21, transfers the soil pressures to the column without needing any shearing reinforcement (Menn, 1990).

Figure 8.21 Design of a concrete footing

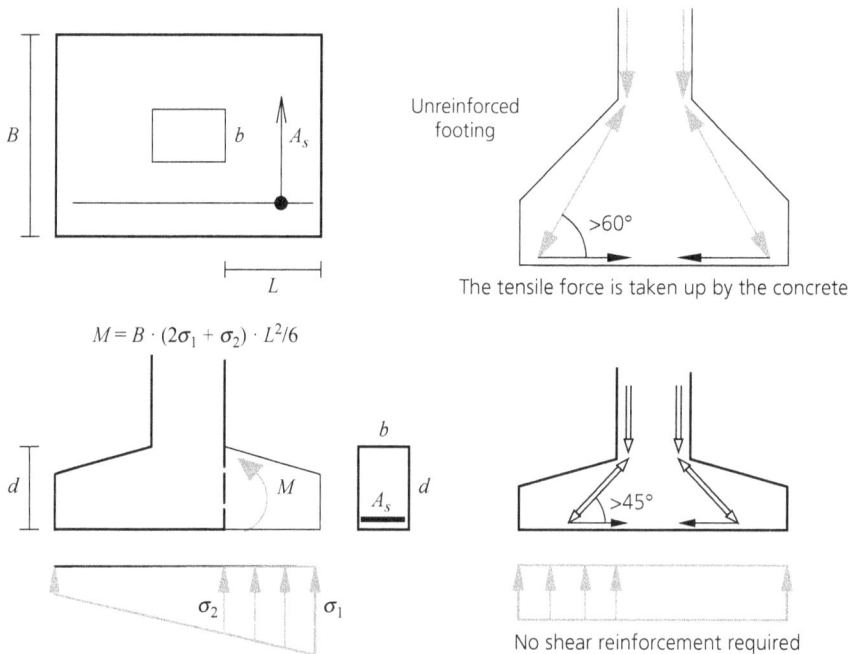

$$M = B \cdot (2\sigma_1 + \sigma_2) \cdot L^2/6$$

If the above conditions are not met, then the footing must be checked not only in bending, but also in punching shear, according to what has been examined in Chapter 2, Section 2.5.2.

Worked example

SINGLE FOOTING

The foundation of a concrete column 40 × 40 cm consists of a single square footing with dimensions 1.50 × 1.50 m, resting on a sand soil exhibiting, according to a geotechnical study, a compression modulus 50 000 kN/m². The column has a height of 3.50 m belonging to the ground floor of a frame. It transmits a vertical load of 500 kN and a computed bending moment of 350 kNm due to alternating seismic action, assuming that the base of the frame to which the column belongs is fully clamped.

The rotational spring stiffness of the soil interface is estimated on the basis of the subgrade modulus, k, and the corresponding moment of inertia, I_f. It is $k = 50000/1.50 = 33300$ kN/m² and $I_f = 1.50^3 \cdot 1.50/12 = 0.42$ m⁴ and thus $k_\varphi = 33\,300 \cdot 0.42 = 14\,000$ kNm/rad.

The bending moment, M_{foot}, transmitted to the footing is:

$$M_{foot} = 350\frac{14\,000}{14\,000 + 4 \cdot 3.0 \cdot 10^7 \cdot (0.40^4/12)/3.5} = 56.2 \text{ kNm}$$

Although the initial data where the column load falls outside the core ($M/V = 350.0/500.0 = 0.7$ m > $1.50/6 = 0.25$ m) excludes the possibility of considering an elastic rotational support, the estimated bending moment that is transmitted to the footing justifies the use of k_φ, since $56.2/350.0 < 0.20$ m.

SINGLE FOOTING WITH CONNECTING BEAMS

The same column as in the previous example carries a vertical load of 600 kN and develops a bending moment of 550 kNm at its base having assumed clamped supports for the respective frame. The considered eccentricity of $550.0/600.0 = 0.91$ m moves the column load outside the footing core and, consequently, causes excessive soil stresses. In order to reduce the induced bending moment to the footing, two concrete connecting beams on either side of the column having a length 5.50 m and a rectangular cross-section, 0.40 m width and 0.40 m depth are considered.

The bending moment, M_{foot}, transmitted to the footing is:

$$M_{foot} = 550\frac{14000}{14000 + 3 \cdot 3.0 \cdot 10^7 \cdot (0.40^4/12)/3.5 + 2 \cdot (3 \cdot 3 \cdot 10^7 \cdot 0.40^3 \cdot 0.40/(12 \cdot 5.5))} = 55.0 \text{ kNm}$$

The column load thus undergoes a lateral displacement $55.0/550.0 = 0.10$ m, which falls obviously within the core. In this case, the produced maximum soil stress:

$$\sigma = \frac{V}{A} + \frac{M}{I_f}\left(\frac{a}{2}\right) = \frac{600}{1.50 \cdot 1.50} + \frac{55.0 \cdot 1.50/2}{0.42} = 365.0 \text{ kN/m}^2$$

is not acceptable and a greater depth for the connecting beams must be chosen.

8.3.3 Foundation beams

As pointed out in Section 8.3.2, the foundation of columns on single footings should take into account not only the maximum developed soil pressures, but also the associated settlements and rotations.

The settlement of each single footing is generally also affected by its neighbouring ones (see Section 8.3.1), except when the clear distance between them is more than about one and a half times their dimensions. In that case, the footings settle independently of each other. However, it should be kept in mind that the founded structure is – almost always – a statically indeterminate one and as such it is sensitive to differential settlements of its supports. Of course, in the case of a relatively low allowable bearing pressure, the resulting footing dimensions could lead to footings lying either very near to each other or even overlapping (see Figure 8.22).

For all the above reasons, when designing the foundation of columns belonging to the same frame, it is often useful to construct a beam of an appropriate width to serve as a common foundation for the considered group of columns. This beam, called the *foundation beam*, is loaded by the upward soil pressures and supported on its columns. Strictly speaking, the corresponding – multi-storey – frame containing the columns and the foundation beam is supported as a complete 'closed' structure on the ground (see Figure 8.22).

Certainly, in such a frame, the tendency of a column to settle down as a single element is prevented by the bending stiffness of the foundation beam, tending to share this settlement between the other columns, clearly limiting the differential settlements in this way. Thus, in order to ensure the necessary bending stiffness, a foundation beam with an adequate depth is clearly superior to one consisting of single spread footings. However, while the settlement behaviour of single footings is treated simply through appropriate spring supports – translational or/and rotational – supported on a fixed base, the treatment of a foundation beam is a more complicated case, as the problem consists in the determination of the soil pressures along the foundation beam itself, which implies the examination of the interaction of the whole superstructure with the soil. This interaction affects not only the response of the foundation beam, but also that of the whole structure.

The basic concepts and procedures that allow the investigation of the interaction between the soil and the frame with its foundation beam are examined below.

8.3.3.1   Soil simulation, according to the Winkler model
The behaviour of a foundation beam resting on an elastic base may be approached through the concept of subgrade modulus, k, as defined in Section 8.3.2.2. This concept, assuming that a segment of a foundation beam undergoes a settlement, w, implies that the soil is under a pressure, p_{soil},

Figure 8.22 Incorporation of the foundation beam in the carrier system

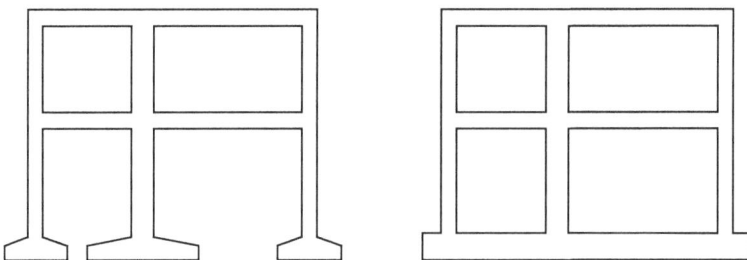

In relatively deformable soil, the foundation beam layout is preferable

equal to $p_{soil} = k \cdot w$, which, representing the resistance offered by the soil to the imposition of the above settlement, is obviously acting upwards on the beam segment itself. Expressing this pressure through the distributed load, p (kN/m), along the foundation beam of width, b, it is:

$$p_{soil} = p / b = k \cdot w$$

and, consequently,

$$p = K \cdot w$$

where $K = b \cdot k$ (expressed now in kN/m²) represents the subgrade coefficient along the beam (see Figure 8.23).

The foundation beam having rigidity (EI) is subjected to a downward distributed load $q(x)$ as well as to the upward soil reaction $(K \cdot w)$ and thus, according to Stavridis and Georgiadis (2025, Section 2.3.6), its behaviour obeys the classical beam equation (see Figure 8.23):

$$q(x) - K \cdot w = EI \frac{dw^4}{dx^4}$$

This differential equation, known as the *Winkler equation*, permits, through its closed solution, the determination of analytic expressions for the stress response of such a beam supported on an elastic soil of the type examined under some typical loading patterns (Hetényi, 1946). However, apart from this possibility, it is usually more practical to support the beam on springs with a certain stiffness (k_s) at an equidistance (s) and make use of appropriate computer software for the evaluation of the response either of the beam itself under the acting column loads or of the whole frame

Figure 8.23 Foundation beam on a Winkler foundation

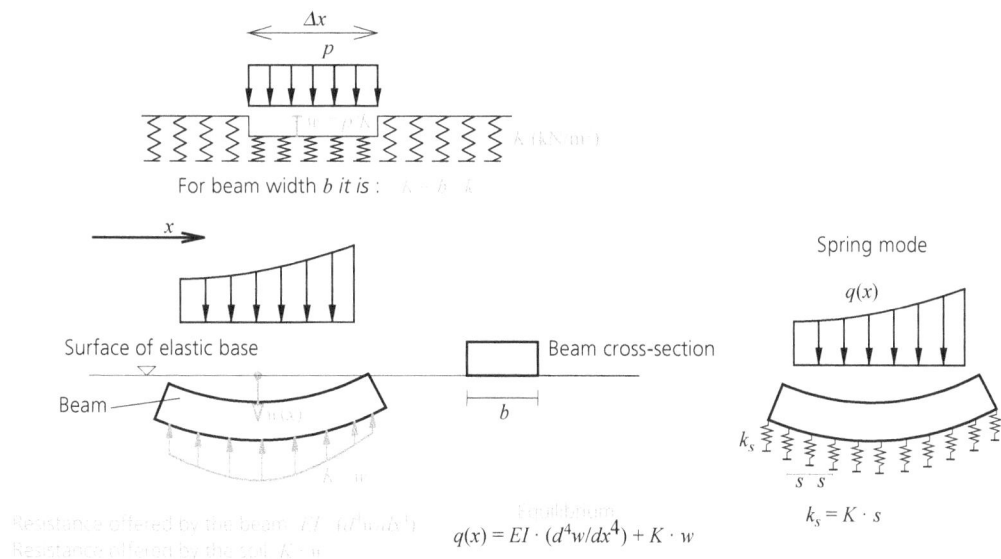

For beam width b it is :

Surface of elastic base

Beam cross-section

Beam

Spring mode

$q(x)$

k_s

s s

$k_s = K \cdot s$

$$q(x) = EI \cdot (d^4 w / dx^4) + K \cdot w$$

connected monolithically to it. By expressing the upward soil reaction on the beam as the spring forces ($k_s \cdot w$) divided by the distance, s (i.e. $p = k_s \cdot w/s$), and recognising it as the upward load ($K \cdot w$) in the above equation – that is, ($K \cdot w$) = ($k_s \cdot w/s$) – it comes out that the appropriate spring stiffness is $k_s = K \cdot s$ (see Figure 8.23).

To give an example, the bending response of the foundation beam of a three-storey frame under a uniform girder loading of 80 kN/m is shown in Figure 8.24. The outer and inner columns have cross-sectional dimensions 40/40 cm and 60/40 cm, respectively. The girders have a uniform section of 40/80 cm, whereas the section of the foundation beam is 60/120 cm. The soil is assumed to have a compression modulus $E_s = 10\,000$ kN/m², while its subgrade coefficient, K, according to the above, is considered equal to (see Section 8.3.2.2):

$$K = b \cdot k = 0.60 \cdot (10\,000 / 0.60) = 10\,000 \text{ kN/m}^2$$

For an equidistance of springs of 0.8 m, the corresponding spring stiffness is $k_s = 0.80 \cdot 10000.0 = 8000$ kN/m. The results shown in Figure 8.24 are obtained using appropriate computer software. ∎

The following points must be again emphasised. The subgrade modulus, k, does not constitute in any case a soil constant. The consistency to its own definition requires reference to a specific surface, which may be obvious for a footing of specific dimensions (see Section 8.3.2.2), but in the case of a foundation beam it is rather unclear. Moreover, loading a single foundation beam with a uniform load causes a uniform settlement over the whole length of the beam, meaning the absence of

Figure 8.24 Bending response of a foundation beam of a frame on an elastic base

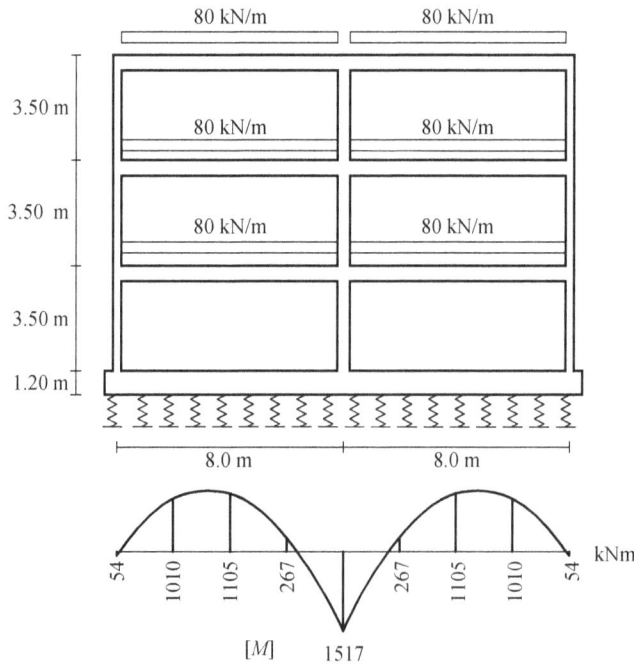

any bending along the beam, which obviously cannot be valid. Indeed, the fact that the centre of the beam settles more than its ends implies a certain curvature, which means the development of some bending ($1/r = M/EI$). However, this bending obviously does not exist. The concept of subgrade coefficient, k, although it has no physical meaning – as mentioned in Section 8.3.2.2 – nevertheless allows an approach to the determination of the soil pressures and to the response of the whole frame. In addition, it should be recognised that the development of a settlement at some location of the soil surface does indeed imply the soil deformation outside of the considered region too, due to the stiffness of the loaded beam, compensating in some way for the weakness of the soil model.

The adoption of a subgrade modulus, k, and the subsequent examination of the previously derived differential equation leads to the important concept of the so-called *characteristic* or *elastic length, L_s*, of the beam (Hetényi, 1946) (see Chapter 3, Section 3.3.1 and Stavridis and Georgiadis, 2025, Section 9.7):

$$L_s = \sqrt[4]{\frac{4 \cdot E \cdot I}{b \cdot k}}$$

The concept of the elastic length applies directly in the case of a foundation beam with infinite length loaded by a concentrated load, P, which may represent a column load causing a maximum bending moment directly under it equal to $M = P \cdot L_s/4$. The bending response of the beam extends from either side of the load, P, by the distance $\pi \cdot L_s/4$ and, consequently, the above bending moment will be not affected by the action of another load at a greater distance than ($\pi \cdot L_s/4$) (see Figure 8.25).

The physical significance of the elastic length may be further visualised if one considers a beam with a length equal to double the elastic length ($2 \cdot L_s$) resting freely on the soil and loaded at its

Figure 8.25 Physical meaning of the elastic length

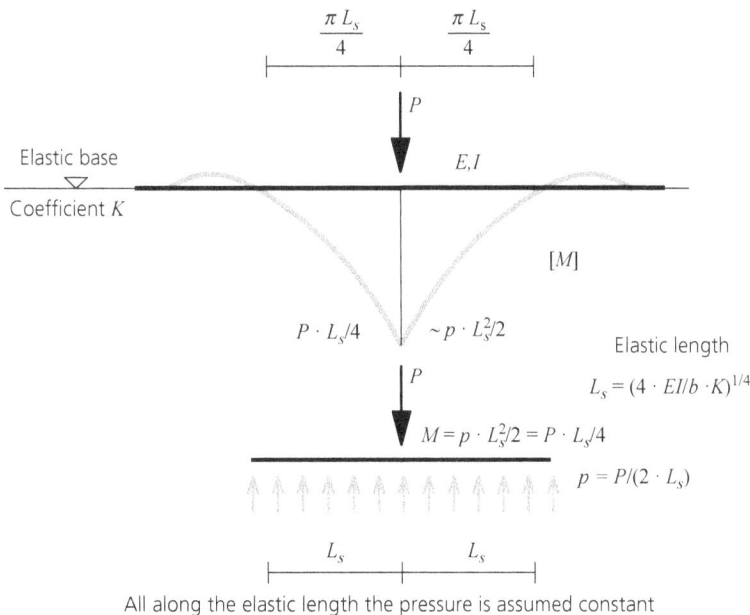

All along the elastic length the pressure is assumed constant

middle by the load, P, and assuming a uniform soil pressure, p, according to the relation $p = P/(2 \cdot L_s)$ (see Figure 8.25). It is seen that the bending moment from either of the two resulting cantilevers under the concentrated load represents exactly the above-mentioned value of the bending response under the load, P.

$$M = (P/2L_s) \cdot L_s^2 / 2 = P \cdot L_s / 4$$

Actually, the bending response of a foundation beam subjected to certain column loads can be estimated by superposing the effect of each column load separately. In this respect, for a beam of length, L, the following accurate expression of the bending moment, M_c, just under the applied column load, P, at an arbitrary position provides the possibility of superposing the effects of the column loads irrespective of their mutual distances (Hetényi, 1946) (see Figure 8.26).

$$M_c = \frac{P}{4\lambda} \cdot \frac{1}{\sinh^2 \lambda L - \sin^2 \lambda L} [(\cosh^2 \lambda a - \cos^2 \lambda a)(\sinh(2\lambda b) - \sin(2\lambda b))$$
$$+ (\cosh^2 \lambda b - \cos^2 \lambda b)(\sinh(2\lambda a) - \sin(2\lambda a)]$$

where $\lambda = 1/L_s$.

In the common case where two equal column loads, P, are applied at the two ends of the foundation beam, the bending moment at the middle section, C, is given by the following expression (see Figure 8.26):

$$M_c = -\frac{2P}{\lambda} \cdot \frac{\sinh\left(\frac{\lambda L}{2}\right) \sin\left(\frac{\lambda L}{2}\right)}{\sinh(\lambda L) + \sin(\lambda L)}$$

Referring now to the example shown in Figure 8.24, the load, P_m, applied by the central column to the midpoint of the foundation beam, is approximately:

$$P_m \approx 3 \cdot (5 \cdot 80.0 \cdot 8.0 / 8) \cdot 2 = 2400.0 \, \text{kN}$$

whereas the load, P_e, from the external column may be analogously estimated as:

$$P_e \approx 3 \cdot (3 \cdot 80.0 \cdot 8.0 / 8) = 720.0 \, \text{kN}$$

The elastic length is $L_s = \sqrt[4]{4 \cdot 3.0 \cdot 10^7 \cdot \frac{0.086}{10000}} = 5.67 \, \text{m}$.

Thus, according to the above, the bending moment at the location of the concentrated load is $M = 2400.0 \cdot 5.67/4 = 3402$ kNm. By taking into account the above more 'accurate' expression, this

Figure 8.26 Beams of finite length

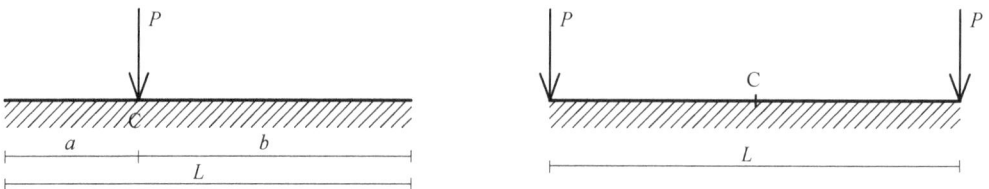

value changes only insignificantly (to 3370 kNm). However, the contribution of the two external columns produces a hogging moment, M_c, at the middle, according to the last expression:

$$\lambda = 1/5.67 = 0.176/\,\text{m},\ \lambda \cdot L = 2.82$$

$$M_c = -\frac{2 \cdot 720}{0.86} \cdot \frac{\sinh(1.41) \cdot \sin(1.41)}{\sinh(2.82) + \sin(2.82)} = -893\,\text{kNm}$$

The total value of bending moment $3370 - 893 = 1580$ kNm is very close to the computer result (1517 kNm).

8.3.3.2   Soil simulation as an elastic medium

According to the soil model examined in Section 8.3.1, the structural frame is supported with its foundation beam on an infinitely extended elastic subspace, either having a constant compression modulus, E_s, with a Poisson's ratio, v, or consisting of a certain number of horizontal layers with a corresponding E_s for each layer.

According to the precedents, a foundation beam of width, b, automatically poses the problem of determination of the distribution of soil pressures $p_{soil}(x)$ along its length. These pressures expressed as a distributed load $p(x) = p_{soil}(x) \cdot b$ (kN/m) must satisfy the following two conditions:

- It must be in equilibrium with the vertical loads of the structure including those acting on the foundation beam.
- The deformation of the foundation beam must follow exactly the surface deformation of the soil under the load $p(x)$.

Both these conditions are satisfied by the Winkler model, with the only remark that for the reasons mentioned earlier, the description of the soil surface deformation through the insertion of springs is unrealistic. Thus, in order to take into account the soil behaviour as an elastic medium in a more consistent manner, the model shown in Figure 8.27 is considered.

The oblong orthogonal contact surface of the foundation beam of length, L, is divided into a certain number n of strips of equal width (L/n) at the midpoint of which the resultant of the corresponding soil pressures is acting. The whole structure is assumed to be simply supported at the above n midpoints. These rigid supports under the loads of the superstructure develop the reactions P_i^0, which may be assembled into the column matrix P_0 ($n \times 1$).

For a certain group of imposed settlements $s_1, s_2,..., s_n$, the statically indeterminate frame will develop reactions $R_1, R_2,..., R_n$ that, together with those due to the structural loads, produce the total reactions at each fictitious support:

$$V_1 = R_1 + P_1^0,\quad V_2 = R_2 + P_2^0,..., V_n = R_n + P_n^0$$

The forces described above can also be interpreted as acting in the opposite direction on the ground. In this context, the problem reduces to determining the set of imposed settlements $s_1, s_2,..., s_n$ on a fictitiously supported frame, such that the resulting support reactions $V_1, V_2,..., V_n$, if applied in reverse

Figure 8.27 The problem of frame foundation through a foundation beam on an elastic subspace

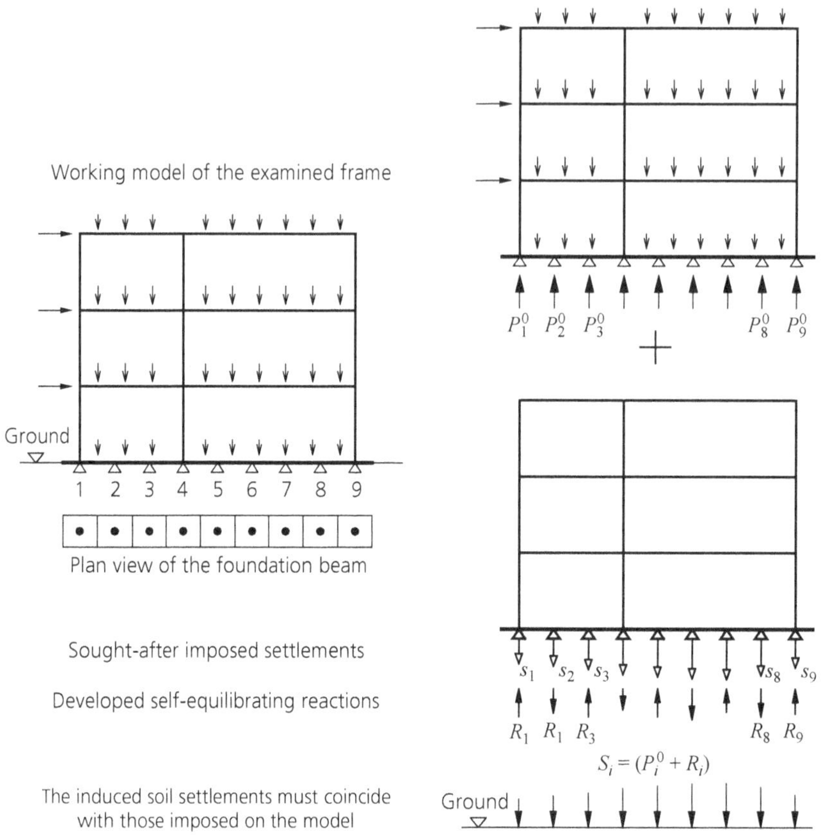

Working model of the examined frame
Ground
1 2 3 4 5 6 7 8 9
Plan view of the foundation beam
Sought-after imposed settlements
Developed self-equilibrating reactions
The induced soil settlements must coincide with those imposed on the model
P_1^0 P_2^0 P_3^0 P_8^0 P_9^0
s_1 s_2 s_3 s_8 s_9
R_1 R_1 R_3 R_8 R_9
$S_i = (P_i^0 + R_i)$
Ground

on the soil, generate deformations identical to them (see Figure 8.27). This is achieved by following the procedure described below, as shown in Figure 8.28 (Stavridis, 2002).

Imposing on the first support (1) a unit settlement $s_1 = 1$, a set of self-equilibrating reactions ($k_{11}, k_{21}, k_{31},\ldots, k_{n1}$) is developed. Next, a unit settlement $s_2 = 1$ is imposed on support (2), leading to another set of self-equilibrating reactions ($k_{12}, k_{22}, k_{32},\ldots, k_{n2}$). Proceeding consecutively in the same way up to the last support (n), a unit settlement $s_n = 1$ is imposed, causing the self-equilibrating reactions ($k_{1n}, k_{2n}, k_{3n},\ldots, k_{nn}$).

If the set of settlements $s_1, s_2,\ldots, s_n$ is now imposed on the supported frame, the developed reactions $R_1, R_2,\ldots, R_n$ may be expressed through the following relations:

$$R_1 = s_1 \cdot k_{11} + s_2 \cdot k_{12} + s_3 \cdot k_{13} + \ldots + s_n \cdot k_{1n}$$

$$R_2 = s_1 \cdot k_{21} + s_2 \cdot k_{22} + s_3 \cdot k_{23} + \ldots + s_n \cdot k_{2n}$$
$$R_3 = s_1 \cdot k_{31} + s_2 \cdot k_{32} + s_3 \cdot k_{33} + \ldots + s_n \cdot k_{3n}$$
$$R_n = s_1 \cdot k_{n1} + s_2 \cdot k_{n2} + s_3 \cdot k_{n3} + \ldots + s_n \cdot k_{nn}$$

Figure 8.28 Practical analysis of soil-frame interaction

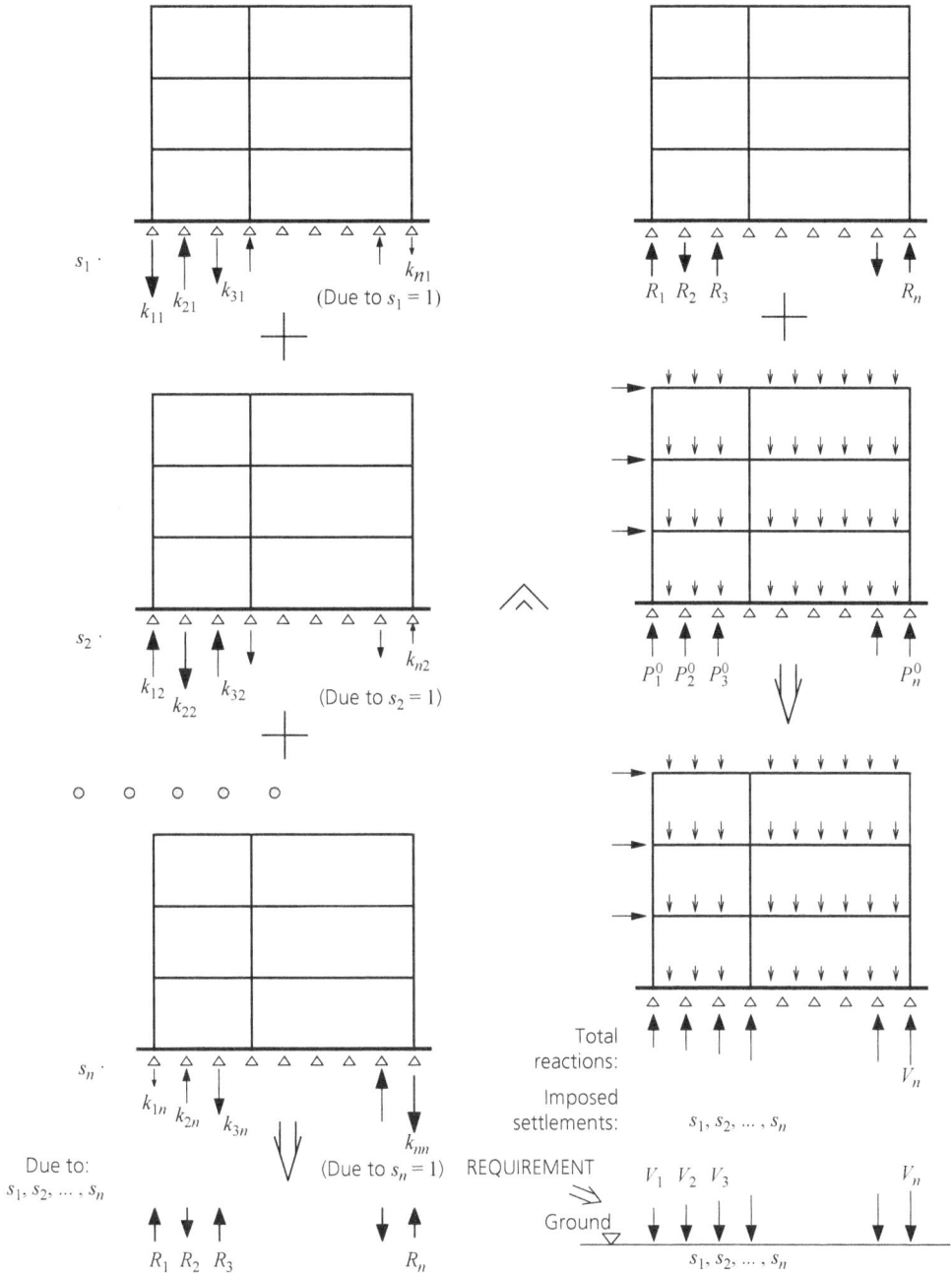

Assembling the forces 'k_{ij}' into a square matrix $\mathbf{K}_{sup}$ ($n \times n$) and further assembling settlements 's_i' and reactions 'R_i' in the column matrices ($n \times 1$) s and $\mathbf{R}$, respectively, the above expressions may be written in matrix form (see 7, Section 7.3.1):

$$R = \mathbf{K}_{sup} \cdot s$$

Taking also into account the reactions $\mathbf{P_0}$ due to structural loading, as previously mentioned, then the above expressions for the totally developed reactions 'V_i' in the fictitiously supported frame being assembled in the ($n \times 1$) column matrix V may be expressed in matrix form as:

$$V = \mathbf{P_0} + \mathbf{R} = \mathbf{P_0} + \mathbf{K}_{sup} \cdot s$$

According to the above, the forces V_1, V_2, ..., V_n, if applied in the opposite direction to the ground, must produce the soil settlements s_1, s_2, ..., s_n (see Figure 8.28). In order to apply this requirement, the relation describing the deformational behaviour of the soil surface is recalled from Section 8.3.1 as:

$$w = \mathbf{F} \cdot \mathbf{P}$$

Since $w = s$ and $\mathbf{P} = V$, this relation can be written as:

$$s = \mathbf{F} \cdot V$$

which, after multiplying both sides by $\mathbf{F}^{-1}$, results in the following expression for the forces V:

$$V = \mathbf{F}^{-1} \cdot s$$

and, finally, by substituting into the last equation the following linear system with respect to the unknown terms of s, the following is obtained:

$$(\mathbf{I} - \mathbf{F} \cdot \mathbf{K}_{sup}) \cdot s = \mathbf{F} \cdot \mathbf{P_0}$$

where $\mathbf{I}$ represents the ($n \times n$) unit matrix, according to Section 7.3.1.2.

By determining the settlements 's_i' through this linear system, the forces 'V_i' acting on the ground are obtained from the penultimate equation and, finally, the soil pressures are determined on the basis of the respective tributary areas under each force, V. It is obvious that the whole procedure is feasible only by using appropriate computer software for plane frame analysis.

Worked example

For reasons of direct comparison, the frame examined in Section 8.3.3.1 is again considered under the same loading and soil characteristics (see Figure 8.29).

The model under consideration is supported at nine intermediate equidistant points of the foundation beam, as shown in Figure 8.29. According to the procedure described above, the following settlements are obtained:

$$s_1 = 0.0571\,\text{m}, \quad s_2 = 0.0562\,\text{m}, \quad s_3 = 0.0565\,\text{m}, \quad s_4 = 0.0577\text{m}, \quad s_5 = 0.0587\text{m}$$
$$s_6 = 0.0577\,\text{m}, \quad s_7 = 0.0565\,\text{m}, \quad s_8 = 0.0562\text{m}, \quad s_9 = 0.0571\text{m}$$

This means that the state of stress and the deformation of the frame examined are identical to those of the fictitiously supported model, under the existing loading and the imposition of the above support settlements.

The bending moment diagram of the foundation beam is shown in Figure 8.29. By comparing the results with those based on the adoption of the Winkler soil model, it is observed that in the present case in which the soil is considered as an elastic half-space, the span moments have lower values, whereas the values of the bending response in the region of the central column are greater.

PRESTRESSING

As a matter of fact, the foundation beams, being part of the overall building structure, bear the total load acting on it and are subjected to sectional forces that, depending on the number of supported levels, are much greater than those appearing in the horizontal elements (i.e. beams) of the super-structure. Despite larger cross-section dimensions, they are prone to cracking as a priori concrete structures in a possibly aggressive environment, as is the soil, and so prestressing is an issue that should be examined. Of course, the introduction of a prestressing force cannot ensure the development of axial compressive stresses, mainly because of developing friction forces by the soil that eventually prevent the shortening of the foundation beam under the prestressing forces. Prestressing can nevertheless ensure the development of deviation forces in the upward sense under the columns and downwards between them. It is clear that under these deviation forces, the bending response of the foundation beam can be drastically reduced.

Actually, the tendon profile can, in principle, have the mirror layout of that applied in prestressed continuous beams (see Stavridis and Georgiadis, 2025, Section 5.4) as shown in Figure 8.30,

Figure 8.29 Example of frame analysis resting on an elastic subspace

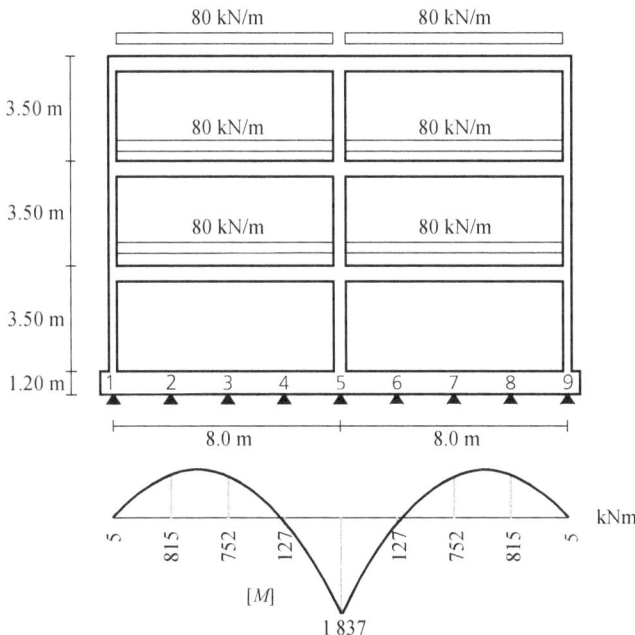

Figure 8.30 (a) Tendon layout; (b) bending response of a prestressed foundation beam under column loads; (c) bending response of a prestressed foundation beam under prestressing (Stavridis, 2010)

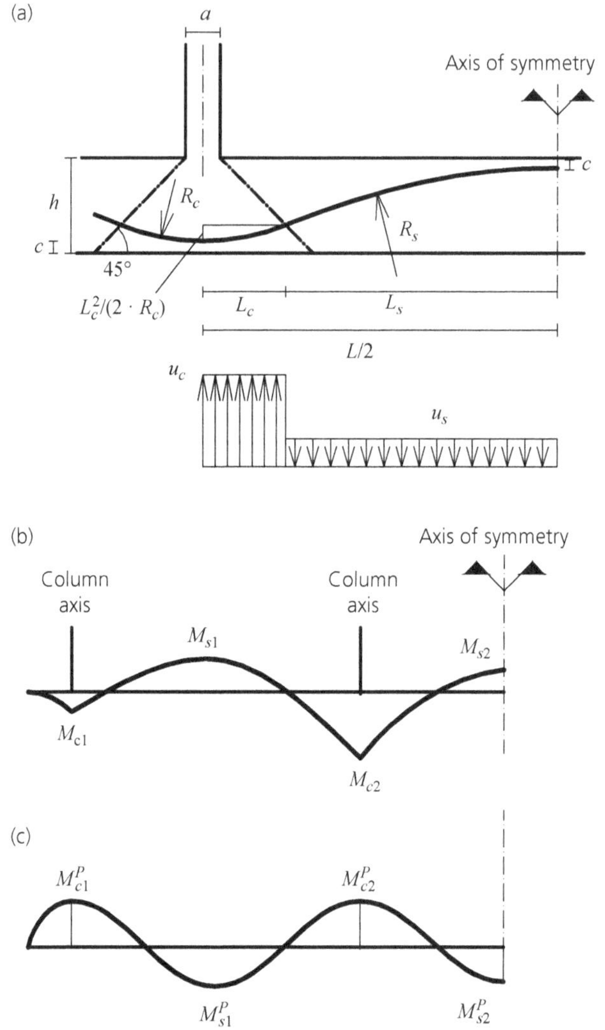

providing the counteracting bending action to that caused by the column loads through the shown deviation forces. The thus reduced bending response results in more uniform settlements as well as reduced cross-section dimensions (Stavridis, 2010).

As shown in Figure 8.30, the upwardly curved profile segment may be limited by the adopted line of load spreading under 45°. In order to achieve the best exploitation of their force, these cable segments have to assign the minimum possible radius of curvature according to the tendon diameter used. On the other hand, both this segment and the one downwardly curved have to assign a common tangent at their connection points. This is ensured by the condition $L_C/R_C = L_S/R_S$ where (L_C, R_C) and (L_S, R_S) are the horizontal length and radius of curvature of the upwardly and downwardly curved parts, respectively. Following the geometry of Figure 8.30, the length, L_C, can be determined by the relation:

$$L_C = R_C \cdot \left[\sqrt{1 + \frac{2}{R_C} \left(\frac{a}{2} + h - c \right)} - 1 \right]$$

It is clear that the final response of the whole structure results from superposition of that due to the external loads and that due to the deviation forces produced by the prestressing force introduced into the foundation beam resting on the elastically deformable soil.

8.3.4   Mat foundations

When the layout of individual footings or foundation beams – designed to distribute column loads to the ground – results in overlapping in both orthogonal directions, it is often possible to unify the foundation system into a single, continuous concrete slab, known as a *mat foundation*. This slab supports all relevant column loads from the entire superstructure and balances them with the upward-acting soil pressure.

The behaviour of a mat foundation closely resembles that of a flat slab, as described in Chapter 2, Section 2.5. All structural considerations from that section, including bending and punching shear responses, apply here in a mirrored manner due to the reversed direction of the acting loads (i.e. upward soil pressures against downward loads in typical slabs).

Given that the upward soil pressure on a mat foundation is often several times greater than the gravity loads at each structural level above, the required slab thickness must be significant to safely resist these forces. In cases where ground conditions are heterogenous or uncertain, the slab thickness should exceed the minimum derived from bending and punching shear checks. A stiffer foundation slab ensures a more uniform distribution of soil pressure beneath it.

This mirrored approach to flat slab design also extends to prestressing. As discussed in Chapter 2, Section 2.5.3, prestressing the mat foundation following similar principles to those used in foundation beams (see Section 8.3.3) can significantly reduce bending moments and practically eliminate tensile stresses. In addition, prestressing enhances uniform pressure distribution, resulting in reduced differential settlement – an effect also observed in prestressed foundation beams (see Figure 8.30).

This is precisely why particularly thick, reinforced concrete foundation slabs are adopted: to ensure structural adequacy, reduce bending and settlement, and improve overall foundation performance.

Worked example

A reinforced concrete foundation slab is to be designed with a square column raster of 8.0 m by 8.0 m and a depth of 1.10 m. The columns have a square cross-section 80 x 80 cm. The developing soil pressure on the slab is estimated as 100 kN/m². The example aims to assess the bending response with the required reinforcement, as well as to check safety against punching shear. The compressive strength of concrete is $f_c = 20$ MPa. The column load is $8.0 \cdot 8.0 \cdot 100 = 6400$ kN.

The upwards acting soil pressure causes a bending response under a linear loading $q = 6.0 \cdot 100.0 = 600$ kN/m, which is taken up in bending in both directions along a column row of 6.0 m width through a continuous beam action, in the same way as considered in Section 2.5.1.

BENDING RESPONSE

Strip load: $100.0 \cdot 8.00 = 800.0$ kN/m

Span bending moment: $M_F = (800.0 \cdot 8.0^2/24) = 2133$ kNm

Support bending moment: $M_S = (800.0 \cdot 8.0^2/12) = 4266$ kNm

SPAN REGION

Column strip (40%): $2133.0 \cdot 0.50/(0.40 \cdot 8.0) = 333$ kNm/m

Remaining width (60%): $2133 \cdot 0.50/(0.60 \cdot 8.0) = 222$ kNm/m

COLUMN REGION

Column strip (70%): $4266 \cdot 0.70/(0.40 \cdot 8.0) = 933$ kNm/m

Remaining width (30%): $4266 \cdot 0.30/(0.60 \cdot 8.0) = 266$ kNm/m

TOP REINFORCEMENT OVER THE SUPPORTS

Both directions: $A_s = 933/(0.85 \cdot 1.0 \cdot 200000) \approx 55$ cm²/m

BOTTOM REINFORCEMENT IN SPAN REGION

Both directions: $A'_s = 333/(0.85 \cdot 1.0 \cdot 200000) \approx 20$ cm²/m

PUNCHING SHEAR

Axial force in the column: $N = 6400$ kN:

$$k = \left[1 - \left(\frac{a + 2h}{L}\right)^2\right] = 1 - \left((0.8 + 2 \cdot 1.0)/8.0\right)^2 = 0.88$$

$$V_{eff} = 6400 \cdot 0.88 = 5632 \text{ kN}$$

$$\tau = \frac{5632}{4 \cdot 1.00 \cdot (0.80 + 2 \cdot 1.0)} = 502 \text{ kN} / \text{m}^2$$

The shear stress is acceptable given also that the ultimate value of the shear stress can be developed, according to Section 2.5.1.

$$\tau_u = f_c \frac{1.6 \cdot \mu \cdot (f_y / f_c)}{1 + 16 \cdot \mu \cdot (f_y / f_c)} = 20000 \cdot \frac{1.6 \cdot \dfrac{55.0}{100 \cdot 100} \cdot \dfrac{420000}{20000}}{1 + 16 \cdot \dfrac{55}{100 \cdot 100} \cdot \dfrac{420000}{2000}} = 1295 \text{ kN/m}^2$$

8.4. Pile foundations

8.4.1 Overview

As mentioned in Section 8.3, deep foundations transfer the loads of the superstructure through weak, inadequate soils to a lower stratum with sufficient bearing strength. The usual way to accomplish this transfer is through the use of piles. Obviously, piles are used when the subsurface conditions are not suitable for a shallow foundation. Piles are vertical or slightly inclined structural members made of concrete, up to 30 m long or even longer, having a circular cross-section and a diameter ranging from about 0.4 up to more than 2.5 m. Piles are arranged in a group of relatively dense layout, connected to each other at their upper end through a thick slab called the *pilecap*, which is further connected to the superstructure. Often, at their lower end, piles have a slightly enlarged base, resting on a soil layer of satisfactory bearing strength. In relatively rare cases, where this layer lies so deeply that the length required for the piles leads to a prohibitive total cost, the equilibrium of the loads transmitted to the piles from the pilecap is mainly accomplished through skin friction forces – developed between the so-called *friction pile* – and the adjoining soil through the whole length of the piles (see Figure 8.31).

Thus, the foundation system consisting of the pilecap and the piles is in equilibrium under the actions of the superstructure applied to the pilecap and the forces acting on the lower end of piles by the sound stratum. It is clear that, in the case of friction piles, the role of the acting soil force at the bottom of the pile is undertaken by the total force of the lateral friction. Usually, there is a coexistence of friction and bearing forces in a rather uncertain proportion that does not practically affect the total axial force that is required to be taken up by the pile in order to fulfill the global equilibrium conditions.

The pilecap, as shown in Figure 8.31, is connected at its top to the supported element, which transmits the actions V, M_x, M_y (i.e. a vertical force and a moment with an arbitrary horizontal vector), as well as a horizontal force, H (see Figure 8.31).

Piles are usually arranged in a symmetrical layout around the base of the founded element and are generally placed in a vertical position. Regarding the horizontal force, H, this is taken up either by the vertical piles themselves or by batter piles especially constructed for that purpose with an angle of inclination not greater than 15°.

8.4.2   Vertical loads

The symmetrical layout of N piles around the point of application O of the load, V, implies the existence of the orthogonal axes of symmetry O_x and O_y. The components M_x and M_y are referred at these two axes (see Figure 8.32).

Figure 8.31 Layout and load-bearing action of piles

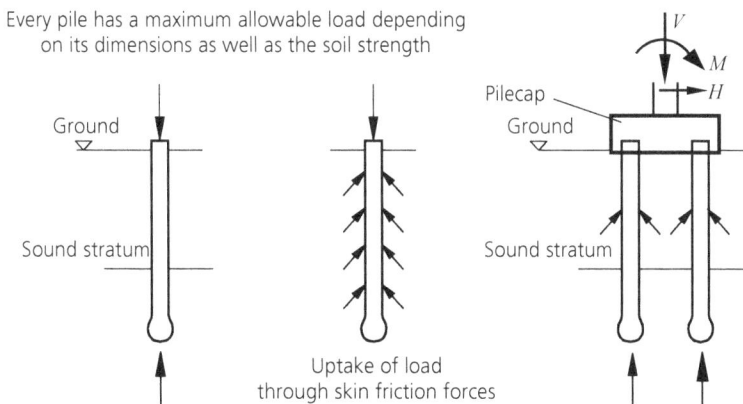

Every pile has a maximum allowable load depending on its dimensions as well as the soil strength

Ground

Sound stratum

Uptake of load through skin friction forces

Pilecap
Ground

Sound stratum

V
M
H

Figure 8.32 Loading of a pile system through a pilecap

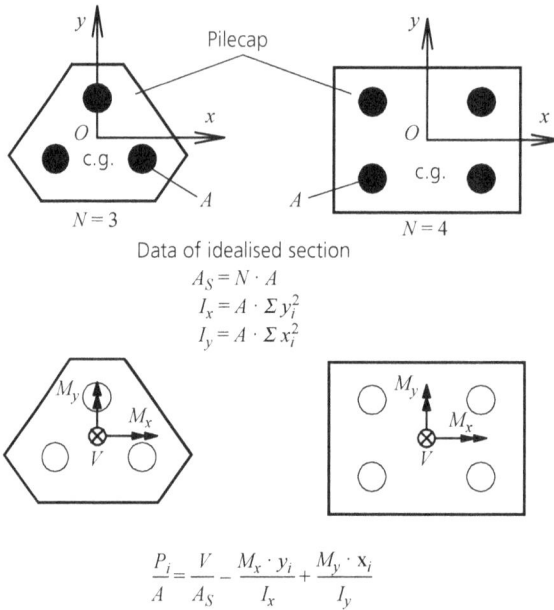

Data of idealised section

$$A_S = N \cdot A$$
$$I_x = A \cdot \Sigma y_i^2$$
$$I_y = A \cdot \Sigma x_i^2$$

$$\frac{P_i}{A} = \frac{V}{A_S} - \frac{M_x \cdot y_i}{I_x} + \frac{M_y \cdot x_i}{I_y}$$

For preliminary design purposes, it is sufficient to consider the magnitudes V, M_x, M_y as being applied to a rigid section consisting of the total layout of the circular section areas of the piles. Given the coordinates x_i, y_i of the centres of the circular pile sections, with respect to the centroid O of the section and assuming that all piles have a common section area, A, the following sectional data are derived.

Total area of section: $A_s = N \cdot A$

Moment of inertia about Ox-axis: $I_x = A \cdot \Sigma y_i^2$

Moment of inertia about Oy-axis: $I_y = A \cdot \Sigma x_i^2$

Thus, the consideration of biaxial bending to which the total section is subjected leads to the determination of the axial force, P_i, for each pile separately (see Figure 8.32):

$$\frac{P_i}{A} = \frac{V}{A_s} - \frac{M_x \cdot y_i}{I_x} + \frac{M_y \cdot x_i}{I_y}$$

The pilecap is in equilibrium under the loads V, M_x, M_y and the pile forces, P_i, and must, therefore, be designed to safely take up the above actions. For preliminary design purposes, the possible bending of the piles is ignored. Moreover, selecting large enough thickness for the pilecap – that is, equal to at least half the clear distance between the piles – the bending action of the pilecap is excluded and the load introduced from the column is transferred to the piles through a strut-and-tie model, normally a three-dimensional one (see Figure 8.33). In this model, which depicts the load

Figure 8.33 Load-bearing action of pilecap

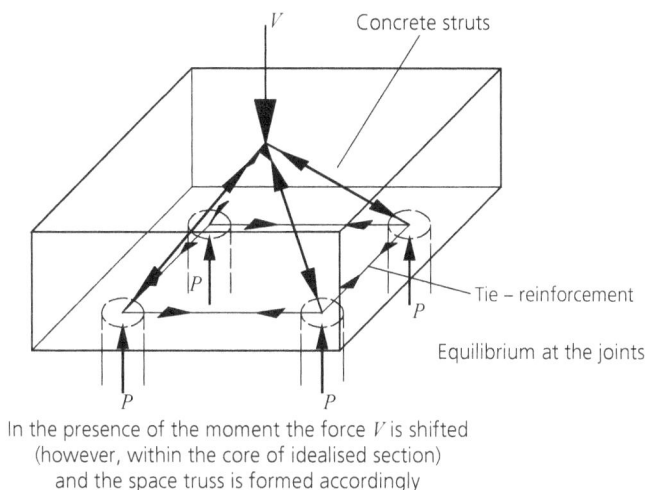

V

Concrete struts

Tie – reinforcement

P

Equilibrium at the joints

P

P   P

In the presence of the moment the force V is shifted
(however, within the core of idealised section)
and the space truss is formed accordingly

transfer through the 'stiffest' (i.e. shortest) path, the struts represent the compressed concrete between the column and the pile heads, whereas the ties are materialised through reinforcement bars, arranged within the narrow region of the pile circular section with appropriate anchorage, as shown in Figure 8.32, thus ensuring the equilibrium of the formed joints.

Designing in this way – that is, by taking up the design loads (i.e. service loads magnified by the respective safety factors) on the basis of a constructionally feasible equilibrium – ensures, according to the static theorem of plastic analysis (see Stavridis and Georgiadis, 2025, Section 6.5.2), that the collapse load is greater than that which the substitute strut-and-tie model can carry.

8.4.3   Horizontal loads

The uptake of a horizontal load by a vertical pile implies a bending response of the pile. The pile acts as laterally supported by an elastic medium, represented by a subgrade soil coefficient, k_h, and substituted accordingly through continuous distribution of lateral springs, as in the case of the foundation beam (see Figure 8.34).

In non-cohesive soils, the coefficient, k_h, may be considered as increasing linearly to a depth, t, according to the relation $k_h = n_h \cdot t/D$, where D is the pile diameter and n_h is a quantity representing the lateral stiffness of the soil, ranging from 2 up to 18 MN/m^3, depending on the layer density.

In cohesive soils, the coefficient, k_h, is considered to be constant and is expressed as $k_h = n_h/D$, where n_h is expressed in terms of cohesion, c, as $n_h = 160 \cdot c$.

Clearly, the uptake of a horizontal force, or even a moment acting on top of a pile (transmitted through the pilecap), is accomplished through the appropriate distribution of lateral soil reactions along the length of the pile, implying either tension or compression of the corresponding supporting springs (see Figure 8.33). In contrast to the foundation beam, where the development of tensile spring forces is not normally allowed, in the present case this is obviously possible, as any

Figure 8.34 Lateral response of piles

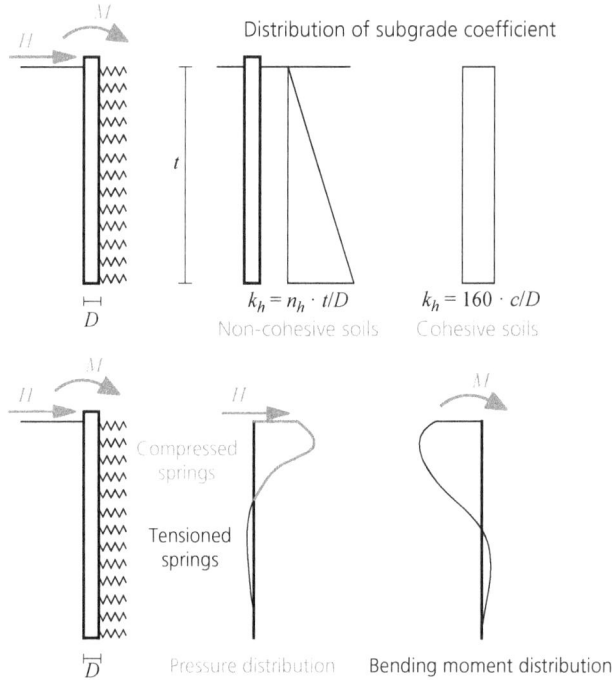

Distribution of subgrade coefficient

$$k_h = n_h \cdot t/D$$
Non-cohesive soils

$$k_h = 160 \cdot c/D$$
Cohesive soils

Compressed springs

Tensioned springs

Pressure distribution   Bending moment distribution

resulting distribution of spring forces along the length of the pile between the soil and the pile always results in compressive contact pressure. In any case, the above response can be determined through appropriate software.

REFERENCES

Hetényi M (1946) *Beams on Elastic Foundation*. University of Michigan Press, Ann Arbor, MI, USA.

Menn C (1990) *Prestressed Concrete Bridges*. Birkhäuser, Basel, Switzerland.

Stavridis L (1997) Tragwerke auf elastischem Boden. *Der Bauingenieur* **72(12)**: 565–569. (In German.)

Stavridis L (2002) Simplified analysis of layered soil-structure interaction. *Journal of Structural Engineering (ASCE)* **168(2)**: 224–230.

Stavridis L (2009) Rigid foundations resting on an elastic layered soil. *Geotechnical and Geological Engineering* **27**: 407–448.

Stavridis L (2010) Prestressed foundation beams on elastic layered soil. *Geotechnical and Geological Engineering* **29**: 431–442.

Stavridis L and Georgiadis K (2025) *Understanding and Designing Structures without a Computer: Plane structural systems*. Emerald Publishing, Leeds, UK.

Leonidas Stavridis and Konstantinos Georgiadis
ISBN 978-1-83662-945-0
https://doi.org/10.1108/978-1-83662-942-920251009
Emerald Publishing Limited: All rights reserved

Index

www.ingramcontent.com/pod-product-compliance
Lightning Source LLC
Chambersburg PA
CBHW081049220326
41598CB00038B/7040